Food Waste Valorization

Food Waste Valorization

Abebe Mekonnen
Samira Abdullahi

Kruger Brentt
Publishers
2026

Kruger Brentt Publishers UK. LTD.

Company Number 9728962

Regd. Office: 68 St Margarets Road, Edgware, Middlesex HA8 9UU

ISBN: 978-1-78715-648-7

For information on all our publications visit our website at http://krugerbrentt.com/

Preface

Food waste has emerged as a pressing global concern, with significant implications for environmental sustainability, economic efficiency, and food security. As vast quantities of edible and inedible residues are discarded each year, the need for innovative strategies to recover value from these resources has become increasingly urgent. Food waste valorization offers a transformative approach—converting waste streams into valuable products, reducing environmental impact, and supporting circular economy initiatives across the food sector.

This book, *Food Waste Valorization*, provides a comprehensive introduction to the scientific principles, technological innovations, and practical applications that enable the conversion of food waste into useful materials and energy. It is designed for students, researchers, industry professionals, and policymakers seeking a deeper understanding of sustainable waste management and resource recovery.

The initial chapters outline the types, sources, and composition of food waste generated across the supply chain—from agricultural production and processing to retail and household consumption. Subsequent sections explore a range of valorization strategies, including biochemical, thermochemical, microbial, and physicochemical processes. The book highlights the production of biofuels, bioplastics, enzymes, bioactive compounds, fertilizers, and other high-value products derived from food waste.

Special attention is given to emerging technologies such as anaerobic digestion, fermentation, pyrolysis, and novel extraction methods, along with discussions on life-cycle assessment, economic feasibility, and regulatory considerations. Real-world case studies illustrate successful valorization initiatives and demonstrate how sustainable practices can be integrated into industrial operations.

By combining scientific foundations with practical insights, *Food Waste Valorization* serves as a valuable resource for advancing circular economy practices, reducing environmental footprints, and unlocking the hidden potential of food waste.

Abebe Mekonnen

Samira Abdullahi

Contents

CHAPTER

1

Food Waste Valorization

1. FOOD WASTE VALORIZATION

Food waste is becoming a growing and important concern at both local and global levels [1,2]. According to the Food and Agriculture Organisation of the United Nations (FAO), one-third of all food production is lost or wasted globally, equivalent to 1.3 billion tons of food produced for human consumption wasted per year with an economic loss of EUR 800 billion [3]. About 44-47% is represented by fruit, vegetable, meat, and fish produced every year and wasted [4]. Because of the large availability and the composition of food waste, there is an increasing interest in their recycling and valorization.

Appropriate waste management is recognized as an essential prerequisite for sustainable development, contributing to the attainment of the global sustainability goals (SDGs 12 and 13). In this regard, Tsai et al. [5] reported an interesting study focused on the promotion policies and regulatory measures for the valorization of mandatory recyclable food waste from industrial sources in Taiwan, where the central governing agencies jointly promulgated some regulatory measures for promoting the production of bio-based products from the industrial food waste valorization such as animal feed, soil fertilizer, and bioenergy [5].

Additionally, food waste has a high potential due to its chemical composition, mainly represented by carbohydrate polymers such as starch, proteins, lipids, cellulose, and other microelements [6-9]. Due to this composition, it can be classified as a low-cost, high potency second-generation feedstock [1]. Recycling and bioconversion of food wastes represent a great opportunity in supporting sustainable development by their conversion, through a microbial fermentation process, into value-added products such as enzymes, feed and food additives, fertilizer, biofuels, animal feeds as well as other useful chemicals or products, food grade pigments, and single cell protein (SCP), enhancing food security and environmentally sustainable development [10-21].

2. FOOD WASTE VALORIZATION BY BIOFUEL AND BIOENERGY PRODUCTION

Due to its richness in moisture, carbohydrate polymers, and other constituents, food waste has been used as an excellent feedstock for the production of biofuel and bioenergy via microbial conversion [22,23]. The production of bioenergy from food waste would not only solve the environmental hazards resulting from the incineration of plants and sanitary landfill sites but would also mitigate the emissions of greenhouse gases while replacing the usage of fossil fuels with bioenergy [5].

Biomass such as rice husk, one of the most important crop residues around the world, can be converted by biochemical and thermochemical methods into useful products, as described by Tsai et al. [24]. Among the several methods applicable, pyrolysis is one of the most commonly used thermochemical conversion processes that involves the decomposition of biomass in the absence of air or oxygen at an elevated temperature [25]. The resulting biochar can be further used as solid fuel, carbon material, soil amendment, environmental adsorbent (biosorbent), functional catalyst, or feedstock for chemicals, depending on its final applications [26]. The study pointed out that rice husk-based biochar could be used as a material in environmental applications for water conservation, wastewater treatment, and soil amendment [24].

Pandit et al. [16] reviewed the bioenergy production using various types of agroindustrial wastewaters and agricultural residues utilizing the microbial fuel cell (MFC), also highlighting the techno-economics and lifecycle assessment of MFC, its commercialization, along with challenges. The use of different agricultural wastes and wastewater containing different industrial-by products for bioelectricity production in MFC seems to be a promising and alternative source of renewable energy generation. Moreover, it has been shown that different varieties of agricultural wastes and wastewater can be utilized using several different MFCs to enhance bioenergy production; thus, the conversion of agro-waste into bioenergy can be carried out by both biochemical and thermochemical MFC routes [16].

Another important issue concerning food waste biovalorization for bioethanol production is the substrate composition and the nutrients available for the microorganisms employed. It is well known that ethanol production is mainly dependent on glucose concentration (the theoretical alcohol yield is about 0.5 g of ethanol per g of glucose) and the yeast inoculums concentration, but nutrient supplementation is also an important parameter to take into consideration, since an adequate amount of specific nutrients, such as trace elements, vitamins, and nitrogen, often poor in agricultural waste, can significantly improve yeast viability and resistance to the medium, stimulating ethanol production performances [21,27]. Therefore, alcoholic fermentation is a complex biological process involving various operating factors, and the use of the classical "one factor at a time" approach for enhancing the final yield, could be time-consuming due to the large number of experiments to perform. In this regard, to implement an efficient fermentation process using industrial by-products, a predictive tool was investigated by Beigbeder et al. [17] to optimize the production of ethanol from non-treated sugar beet molasses by designing and developing a central composite design coupled with response surface methodology (CCD-RSM) statistical approach to investigate the effect of three fermentation process parameters (initial sugar, yeast, and nutrient concentrations) on ethanol productivity while considering several operating parameters such as ethanol yield and sugar utilization rate. Moreover, the second-order mathematical model obtained through the CCD-RSM was tested to evaluate its ability to make accurate predictions based on specific desired process outputs. The application of the CCDRSM statistical approach allowed to maximise the production of ethanol from non-sterilised sugar beet molasses using *Saccharomyces cerevisiae* while scaling up the experimental results up to a 100 L bioreactor scale [17].

3. FOOD WASTE VALORIZATION FOR FOOD AND FEED PRODUCTION

Feedstock and food waste, mostly represented by agricultural sources, can be used in single cell protein (SCP) production and are suitable as protein supplements in either food or feed [21]. Single cell protein technology carried out by microorganisms is designed to solve worldwide protein shortages, and it has shown a great advantage because it is independent of climate, soil characteristics and, not the least, on available land [28]. Moreover, concerns regarding the ethical and environmental implications of meat consumption have increased the demand for meat substitutes. Recently, the use of filamentous fungi as a commercial food product has gained considerable attention, due to its high protein content, the presence of essential amino acids and easy digestibility [29].

A solid-state fermentation (SSF) process carried out by the edible fungus *Neurospora intermedia* using bread waste as feedstock for the production of a protein-rich food product has been investigated by Brancoli et al. [30]. The study proposed the SSF process to be implemented as a stand-alone business, or on-site in small-scale bakeries to recover their otherwise discarded surplus bread and has integrated environmental considerations of the development of a fungal food product, showing which scenario has the best environmental performance and highlighting trade-offs and the parts of the process that are hotspots and should be in focus when optimizing the process. The research can contribute to a sustainable way to handle wasted bread, consistent with a circular economy, and it provides a broader base for the developers of the technology to make sustainable decisions during process optimization [30].

Food waste valorization is also addressed to their bioconversion in animal feed. Tropea et al. [14] reported a fermentation process using non-sterilized fish wastes, supplemented with lemon peel as a filler and prebiotic source, carried out by combined starter cultures of *Saccharomyces cerevisiae* and *Lactobacillus reuteri* for bio-transforming these byproducts into a high protein content supplement, rich in healthy microorganisms, for aquaculture feeds. The final fermented product, low in spoilage microorganisms and rich in healthy microorganisms, showed a content of protein and lipids suitable for aquafeed, reducing the problem of a lack of protein sources for aquaculture by encouraging the conversion of fish waste and lemon peel into feed [14].

An interesting review on the utilization of pomaces, waste generated from the pressing of fruits and olives to obtain juices and olive oil, was reported by Munekata et al. [31], where the valorization of this waste as a feed supplement for animal production was deeply investigated. The advances in incorporating and optimizing the use of pomaces in animal feed by generating silages and feeds that improve animal health represent a relevant alternative to using fermented pomaces. Growth performance can be affected, whereas animal health status can be improved. The absence of negative effects and the improvement in the nutritional quality of the foods obtained from animals fed with fermented pomaces is another favorable characteristic to support this strategy [31].

The evaluation of the effects of the feed obtained via fermentation on final consumers was investigated by Panyawoot et al. [32]. Their study has been evaluated the effect of fermented discarded durian peel, a seasonal fruit growing widely in tropical countries, with *Lactobacillus casei,* cellulase, and molasses separately or in combination in total mixed rations on feed utilization, digestibility, ruminal fermentation, and nitrogen utilization in growing crossbreed Thai Native-Anglo-Nubian goats. The study showed that the discarded durian peel fermented with a combination of molasses and *L. casei* had significantly greater nutrient digestibility and propionate concentration, while estimated methane production, the acetate-to-propionate ratio and urinary nitrogen decreased when compared with untreated discarded durian peel. Therefore, a combination-treated discarded durian peel with molasses and *L. casei* could add 25% of dry matter to the diet of growing goats without a negative impact [32].

4. CRUDE ENZYMES, NUTRIENT SUPPLEMENTS AND BIOPOLYMERS PRODUCTION FROM FOOD WASTE

Agricultural or animal food wastes, thanks to their natural composition, can represent an important substrate to be used as a source of enzymes, food-grade pigments, nutrient supplements, or biopolymers. Munekata et al. [31] reported an interesting review on the use of pomace from food processing for the production of high-addedvalue products via fermentation processes as a strategy applied to obtain carotenoids, fatty acids, linolenic acid, and polyphenols. The authors reviewed, in terms of industrial processes, the production of high-added value products, in particular from grape, apple, and olive, such as enzymes and organic acids for application in food processing as well as in other areas of relevant application such as the development of functional foods or the production of volatile compounds for improving the aroma of food products. The review also highlights the limitations in terms of industrial application and the additional studies that are required to define strategies for using the high-added value compounds obtained from the fermentation/biotransformation of pomaces in the development of food products [31].

The ability of "generally recognized as safe" (GRAS) microorganisms to secrete enzymes extracellularly along with featuring properties, such as high catalytic activity and reaction rate, has been demonstrated in the study of Lappa et al. [33]. The study indicates the successful development of a novel cheese whey valorization approach within the concept of circular bio-economy. A two-stage operation was established to generate crude enzymatic consortia via fungal solid-state fermentations with *Aspergillus awamori.* Fermentation conditions were optimized, and a novel biocatalyst was effectively secreted, and subsequently implemented to hydrolyze whey lactose, formulating a nutrient substrate for fermentative bioconversions. Bacterial cellulose production was also conceptualized as a transitional compound for subsequent functional food formulations, along with the protein fraction, to complement the sustainability and circularity of the process [33].

Another interesting study aimed to promote an integrated bio-refinery approach fully exploiting discarded whey from buffalo milk has been carried out by Alfano et al. [34]. In their work, they evaluated the permeate and retentate of ultra-filtered whey, both provided by a local dairy factory in the Campania region, where cheese manufacturing is one of the main industrial activities in the food sector. The permeate was further processed to investigate a potential downstream approach for obtaining reusable water with a low organic load. The retentate was evaluated to identify further potential biotechnological applications of buffalo milk whey. In particular, it was investigated as the main substrate for the growth of a probiotic strain showing several potential biomedical usages, *Lactobacillus fermentum.* Furthermore, it was investigated for the identification of active molecules for tissue repair induction by using wound healing assays on mammalian cells. The study pointed out that the concentrated ultra-filtered retentate could represent suitable support for the growth of probiotic strain, *Lactobacillus fermentum,* having an adequate sugars and proteins content; moreover, it was demonstrated to stimulate epidermis (keratinocyte) regeneration and therefore meaning potential applicability as an ingredient in skincare products [34].

The production of microbial pigments as bio-pigments for the food industry has been gradually increasing, and the evaluation of whey as an alternative low-cost sustainable fermentative substrate has been investigated by Mehri et al. [35]. The study refers to the production of red colour pigment by *Monascus purpureus* suitable for the food industry, using raw, demineralized and deproteinized whey as substrates by simultaneous hydrolysis and fermentation. The authors carried out interesting research on the evaluation of several factors affecting pigment production, such as fermentation pH, initial lactose concentration, monosodium glutamate (MSG) concentration as the nitrogen source, inoculation ratio, mycelial development, and pigment synthesis kinetics of the microorganism employed. This study pointed out that demineralized whey is a sustainable substrate in the fermentation process of the *M. purpureus* red pigment [35].

The use of a biosurfactant produced by *Bacillus cereus* as an additive in a cookie formulation, evaluating the nutritional benefits of its addition, the non-toxicity, the antioxidant potential and the effects on the physicochemical properties as well as the texture of the product has been reported by Durval et al. [36]. The study demonstrated that the biosurfactant produced by *B. cereus* grown in a medium containing waste frying oil has the potential to be used as a bioemulsifier in food systems. The addition of the biosurfactant in the formulation of cookies showed no drastic changing in the final product as the biosurfactant-containing formulations showed energetic and physical characteristics similar to those of the standard formulation. The biosurfactant was non-toxic and showed considerable antioxidant activity. Moreover, it demonstrated promising results as an ingredient for a flour-based product in terms of the physical, physicochemical, and textural properties of the cookies formulated, also ensuring good preservation [36].

Asimakopoulou et al. [18] carried out a study by assessing wheat straw from Greek agricultural residues as a feedstock for the growth of the heterotrophic microalga *Cryptheco- dinium cohnii* and the accumulation of polyunsaturated omega-3 fatty acids (PUFAs), more specifically docosahexaenoic acid (22:6n-3,DHA). The work reports an efficient, holistic approach for the integrated valorization of all sugar-containing fractions of biomass towards the production of this valuable product through fermentation, representing the first report demonstrating, as a proof of concept, the valorization of all sugar streams towards the production of omega-3 fatty acids from non-edible sources [18].

Food waste valorization through fermentation processes represents an interesting way of obtaining new value-added products in the cosmetic and pharmaceutical fields also. Ferracane et al. [37] carried out a study aimed to produce and evaluate the different ripening stages of soaps produced with non-edible fermented olive oil (NEFOO soap), evaluating the pH, color, and solubility. The results obtained were compared with those obtained from soaps produced with extra virgin olive oil (EVOO soap). The study pointed out an innovative method to produce "alternative" olive oils on a large scale, exploiting non-edible drupes currently used to produce fodder, natural fertilizer, and energy biomass [37].

The glucan and pectin contents detected in the green husks of walnuts grown in two different soil and climate areas of Southern Italy (Montalto Uffugo e Zumpano) were investigated for potential use in food, cosmetics, and pharmaceutical fields by La Torre et al. [38]. The authors reporteda biovalorization of this waste material in their study and also investigated the spectroscopic, morphological and thermal characterizations of the extracted high-value compounds in order to evaluate if the different pedoclimatic conditions of the two areas could affect both the content of glucans and pectins and their functional uses [38]. Finally, a new perspective on the bioremediation of industrial effluents was demonstrated by Costa et al. [39]. In the study, the authors reported the implementation of *Aspergillus oryzae,* a fungal strain widely exploited as an amylase producer, for the bioremediation of starch in industrial paper mill wastewater by carrying out submerged fermentation technologies (SmF) and solid-state fermentation (SSF). *A. oryzae* was found to grow on non-conventional media such as paper mill wastewater. The SSF of *A. oryzae* was performed on rice hulls. In the bioremediation of paper mill wastewater, for removing starch, the fungus maintains its amylase activity and uses reducing sugars as metabolic substrates [39].

REFERENCES

1. Papargyropoulou, E.; Lozano, R.; Steinberger, J.K.; Wright, N.; bin Ujang, Z. The food waste hierarchy as a framework for the management of food surplus and food waste. *J. Clean. Prod.* **2014**, 76,106-115. [CrossRef]

2. Elijah, A.I.; Edem, V.E. Value addition to Food and Agricultural wastes: A Biotechnological approach. *Nig. J. Agric. Food Environ.* **2017**,13,139-154.

3. United Nations Environment Programme. *Food Waste Index Report;* United Nations Environment Programme: Nairobi, Kenya, 2021; ISBN 978-92-807-3868-1.

4. Aureli, V.; Scalvedi, M.L.; Rossi, L. Food Waste of Italian Families: Proportion in Quantity and Monetary Value of Food Purchases. *Foods* **2021**,*10*,1920. [CrossRef]

5. Tsai, W.-T.; Lin, Y.-Q. Analysis of Promotion Policies for the Valorization of Food Waste from Industrial Sources in Tai wan. *Fermentation* **2021**, 7, 51. [CrossRef]

6. Lo Turco, V.; Potorti, A.G.; Tropea, A.; Dugo, G.; Di Bella, G. Element analysis of dried figs *(Ficuscarica* L.) from the Mediterranean areas. *J. Food Compos. Anal.* **2020**, 90,103503. [CrossRef]

7. Potortf, A.G.; Lo Turco, V.; Saitta, M.; Bua, G.D.; Tropea, A.; Dugo, G.; Di Bella, G. Chemometric analysis of minerals and trace elements in Sicilian wines from two different grape cultivars. *Nat. Prod. Res.* **2017**, 31,1000-1005. [CrossRef] [PubMed]

8. Tuttolomondo, T.; Dugo, G.; Leto, C.; Cicero, N.; Tropea, A.; Virga, G.; Leone, R.; Licata, M.; La Bella, S. Agronomical and chemical characterisation of *Thymbra capitata* (L.)Cav. biotypes from Sicily, Italy. *Nat. Prod. Res.* **2015**, 29,1289-1299. [CrossRef]

9. La Torre, G.L.; Potorti, A.G.; Saitta, M.; Tropea, A.; Dugo, G. Phenolic profile in selected Sicilian wines produced by different techniques of breeding and cropping methods. *Ital. J. Food Sci.* **2014**, 26, 41-55.

10. Kieliszek, M.; Piwowarek, K.; Kot, A.M.; Pobiega, K. The aspects of microbial biomass use in the utilization of selected waste from the agro-food industry. *Open Life Sci.* **2020**,15, 787-796. [CrossRef]

11. Tropea, A.; Gervasi, T.; Melito, M.R.; Curto, A.L.; Curto, R.L. Does the light influence astaxanthin production in *Xanthophyl- lomycesdendrorhous? Nat. Prod. Res.* **2013**,27, 648-654. [CrossRef]

12. Dufosse, L. Microbial Production of Food Grade Pigments. *Food Technol. Biotech.* **2006**, 44, 313-321.

13. Benavente-Valdesa, J.R.; Aguilara, C.; Contreras-Esquivela, J.C.; Mendez-Zavalab, A.; Montanez,J. Strategies to enhance the production of photosynthetic pigments and lipids in *Chlorophycae* species. *Biotechnol. Rep.* **2016**,10,117-125. [CrossRef] [PubMed]

14. Tropea, A.; Potorti, A.G.; Lo Turco, V.; Russo, E.; Vadala, R.; Rand, R.; Di Bella, G. Aquafeed Production from Fermented Fish Waste and Lemon Peel. *Fermentation* **2021**, 7, 272. [CrossRef]

15. Jarunglumlert, T.; Bampenrat, A.; Sukkathanyawat, H.; Prommuak, C. Enhanced Energy Recovery from Food Waste by CoProduction of Bioethanol and Biomethane Process. *Fermentation* **2021**, 7, 265. [CrossRef]

16. Pandit, S.; Savla, N.; Sonawane, J.M.; Sani, A.M.; Gupta, P.K.; Mathuriya, A.S.; Rai, A.K.; Jadhav, D.A.; Jung, S.P.; Prasad, R. Agricultural Waste and Wastewater as Feedstock for Bioelectricity Generation Using Microbial Fuel Cells: Recent Advances. *Fermentation* **2021**, 7,169. [CrossRef]

17. Beigbeder, J.-B.; de Medeiros Dantas, J.M.; Lavoie, J.-M. Optimization of Yeast, Sugar and Nutrient Concentrations for High Ethanol Production Rate Using Industrial Sugar Beet Molasses and Response Surface methodology. *Fermentation* **2021**, 7, 86. [CrossRef]

18. Asimakopoulou, G.; Karnaouri, A.; Staikos, S.; Stefanidis, S.D.; Kalogiannis, K.G.; Lappas, A.A.; Topakas, E. Production of Omega-3 Fatty Acids from the Microalga *Crypthecodinium cohnii* by Utilizing Both Pentose and Hexose Sugars from Agricultural Residues. *Fermentation* **2021**, 7, 219. [CrossRef]

19. Tropea, A.; Wilson, D.; Lo Curto, R.B.; Dugo, G.; Saugman, P.; Troy-Davies, P.; Waldron, K.W. Simultaneous saccharification and fermentation of lignocellulosic waste material for second generation ethanol production. *J. Biol. Res.* **2015**, 88,142-143.

20. Tropea, A.; Ferracane, A.; Albergamo, A.; Potorti, A.G.; Lo Turco, V.; Di Bella, G. Single Cell Protein Production through Multi Food-Waste Substrate Fermentation. *Fermentation* **2022**, 8, 91. [CrossRef]

21. Salafia, F.; Ferracane, A.; Tropea, A. Pineapple Waste Cell Wall Sugar Fermentation by *Saccharomyces cerevisiae* for Second Generation Bioethanol Production. *Fermentation* **2022**, 8,100. [CrossRef]

22. Kiran, E.U.; Trzcinski, A.P.; Ng, W.J.; Liu, Y. Bioconversion of food waste to energy: A review. *Fuel* **2014**,134, 389-399. [CrossRef]

23. Sen, B.; Aravind, J.; Kanmani, P.; Lay, C.H. State of the art and future concept of food waste fermentation to bioenergy. *Renew. Sustain. Energy Rev.* **2016**, 53, 547-557. [CrossRef]

24. Tsai, W.-T.; Lin, Y.-Q.; Huang, H.-J. Valorization of Rice Husk for the Production of Porous Biochar Materials. *Fermentation* **2021**, 7, 70. [CrossRef]

25. Lehmann,J.;Joseph, S. Biochar for environmental management: An introduction. In *Biochar for Environmental Management,* 2nd ed.; Lehmann, J., Joseph, S., Eds.; Routledge: New York, NY, USA, 2015; pp. 1-13.

26. Dai, Y.J.; Zhang, N.X.; Xing, C.M.; Cui, Q.X.; Sun, Q.Y. The adsorption, regeneration and engineering applications of biochar for removal organic pollutants: A review. *Chemosphere* **2019**,223,12-27. [CrossRef] [PubMed]

27. Tropea, A.; Wilson, D.; Cicero, N.; Potorti, A.G.; La Torre, G.L.; Dugo, G.; Richardson, D.; Waldron, K.W. Development of minimal fermentation media

supplementation for ethanol production using two *Saccharomyces cerevisiae* strains. *Nat. Prod. Res.* **2016**, 30, 1009-1016. [CrossRef]

28. Hulsen, T.; Hsieh, K.; Lu, Y.; Tait, S.; Batstone, D.J. Simultaneous treatment and single cell protein production from agri-industrial wastewaters using purple phototrophic bacteria or microalgae—A comparison. *Bioresour. Technol.* **2018**,254, 214-223. [CrossRef]
29. Filho, P.F.S.; Nair, R.B.; Andersson, D.; Lennartsson, P.R.; Taherzadeh, M.J. Vegan-mycoprotein concentrate from pea-processing industry by product using edible filamentous fungi. *Fungal Biol. Biotechnol.* **2018**, 5, 5. [CrossRef]
30. Brancoli, P.; Gmoser, R.; Taherzadeh, M.J.; Bolton, K. The Use of Life Cycle Assessment in the Support of the Development of Fungal Food Products from Surplus Bread. *Fermentation* **2021**, 7,173. [CrossRef]
31. Munekata, P.E.S.; Dominguez, R.; Pateiro, M.; Nawaz, A.; Hano, C.; Walayat, N.; Lorenzo, J.M. Strategies to Increase the Value of Pomaces with Fermentation. *Fermentation* **2021**, 7, 299. [CrossRef]
32. Panyawoot, N.; So, S.; Cherdthong, A.; Chanjula, P Effect ofFeeding Discarded Durian Peel Ensiled with *Lactobacillus casei* TH14 and Additives in Total Mixed Rationson Digestibility, Ruminal Fermentation, Methane Mitigation, and Nitrogen Balance of Thai Native-Anglo-Nubian Goats. *Fermentation* **2022**, 8, 43. [CrossRef]
33. Lappa, I.K.; Kachrimanidou, V.; Papadaki, A.; Stamatiou, A.; Ladakis, D.; Eriotou, E.; Kopsahelis, N. A Comprehensive Bioprocessing approach to Foster Cheese Whey Valorization: On-Site-Galactosidase Secretion for Lactose Hydrolysis and Sequential Bacterial Cellulose Production. *Fermentation* **2021**, 7,184. [CrossRef]
34. Alfano, A.; D'ambrosio, S.; D'Agostino, A.; Finamore, R.; Schiraldi, C.; Cimini, D. ConcentratedBuffalo Whey as Substrate for Probiotic Cultures and as Source of Bioactive Ingredients: A Local Circular Economy Approach towards Reuse of Wastewaters. *Fermentation* **2021**, 7, 281. [CrossRef]
35. Mehri, D.; Perendeci, N.A.; Goksungur, Y. Utilization of Whey for Red Pigment Production by *Monascus purpureus* in Submerged Fermentation. *Fermentation* **2021**, 7, 75. [CrossRef]
36. Durval, I.J.B.; Ribeiro, B.G.; Aguiar, J.S.; Rufino, R.D.; Converti, A.; Sarubbo, L.A. Application of a Biosurfactant Produced by *Bacillus cereus* UCP 1615 from Waste Frying Oil as an Emulsifier in a CookieFormulation. *Fermentation* **2021**, 7,189. [CrossRef]
37. Ferracane, A.; Tropea, A.; Salafia, F. Production and Maturation of Soaps with Non-Edible FermentedOlive Oil and Comparison with Classic Olive Oil Soaps. *Fermentation* **2021**, 7, 245. [CrossRef]

38. La Torre, C.; Caputo, P.; Plastina, P.; Cione, E.; Fazio, A. Green Husk of Walnuts (*Juglans regia* L.) from Southern Italy as a Valuable Source for the Recovery of Glucansand Pectins. *Fermentation* **2021**, 7, 305. [CrossRef]

39. Costa, S.; Summa, D.; Zappaterra, F.; Blo, R.; Tamburini, E. *Aspergillus oryzae* Grown on Rice Hulls Used as an Additive for Pretreatment of Starch-Containing Wastewater from the Pulp and Paper Industry. *Fermentation* **2021**, 7, 317. [CrossRef]

CHAPTER 2

Analysis of Promotion Policies for the Valorization of Food

1. INTRODUCTION

Growing concern about circular bioeconomy, hunger, resource conservation, and sustainable development associated with food loss and waste (FLW) has raised public awareness in recent years [1,2]. With t he changes in diet habits and the improvement of living standards, many food waste has been generated from residential, commercial, and institutional sources, such as retails, wholesales, restaurants, hospitals, schools, and hotels, as well as from industrial sources like food processing plants, animal-breeding farms, crop/vegetable/fruit farms, and employee lunchrooms [3,4]. The discarded food often contained vegetable leaves, leftover meals and grains, fruit peelings, dairy, oils/grease, salts, and water. Due to its constituents like lignocelluloses, protein, and oils and grease, such wastes without valorization for the production of value-added materials and/or bioenergy would imply the resource depletion and wastage. Moreover, the environmental concern could be derived from their compositions, which may cause negative effects on the environment (e.g., odors, vectors, emission gas emissions, or climate change) if they are illegally disposed of in dumping sites or fields [5]. In addition, these food discards may be rich in the moisture and nutrient compositions, thus causing wastewater discharge with high biochemical oxygen demand (BOD) and/or chemical oxygen demand (COD) values in the landfill leachate [6]. For these reasons, the valorization of food waste has become increasingly important in recent years due to the United Nations (UN) Sustainable Development Goals (SDGs) and national regulatory requirements [7]. Therefore, the reuse of food waste as a valuable resource for the production of materials, fertilizer, and biofuels has been reviewed recently [7,8].

As mentioned above, FLW has become one of vital issues raised by great public concern. In order to provide a target-oriented blueprint for peace and prosperity to all countries in the near future (2030), the United Nations (UN) announced 17 Sustainable SDGs on 25 December 2015 [9], reflecting the increased global awareness for the environmental issues. In this regards, the Target 12.3 of the SDGs, thus, calls for halving per capita global food waste at retail and consumer levels by 2030, as well as reducing food losses along the production and supply chains. In line with the international trends, the Taiwan government announced the Taiwan's Sustainable Development Goals in July 2019 [10], including the goals for 2030 and targets for 2020. Regarding FLW, the 12th Goal is "Responsible Consumption and Production", which involves the Goal 12.3 by reducing food lose in the supply-chain side and also reducing food wastage in the consumer side, and the Goal 12.4 by reducing (food) waste generation through green production and also promoting (food) waste valorization and its technological capacity. In addition, the 7th and 13th Goals in the Taiwan's SDGs aims at taking actions for providing renewable energy and combating climate change, respectively, which were also relevant to the food waste issue.

According to the official definition in Taiwan, waste can be categorized into general (urban) waste and industrial waste. Under the circular economy principle, the central governing agency (i.e., Environmental Protection Administration, EPA) in Taiwan has implemented the Four-in-One Resource Recycling Plan over the past two decades [11,12], which combined the efforts by communities, recycling enterprises, local governments, and the Recycling Fund. It showed that the urban waste recycling rate has increased from about 10% in 2000 to over 56% in 2019 [13]. Regarding the kitchen waste (including waste cooking oil) management, the Taiwan EPA promulgated the regulations of governing the valorization of food waste from non-industrial (urban) and industrial sources by designating it as a mandatory recyclable waste under the authorization of the Waste Management Act (WMA) [14,15]. Furthermore, over ten items of food waste valorization (or reuse) from industrial sources have defined by the central responsible agencies under the authorization of the WMA, including the Council of Agriculture (COA), Ministry of Finance (MOF), and Ministry of Economic Affairs (MOEA) [16]. Currently, most of the industrial food waste items were reused as valuable feedstocks for animal feed, organic fertilizer, or biomass energy. Meanwhile, the EPA also provided the subsidies for local governments to establish their valorization (animal feed) programs for the prevention of African swine fever (ASF) spread because the virus can persist in the kitchen waste without high- temperature (>90°C) cooking [15]. Moreover, the burden on municipal solid waste (MSW) incineration plants and sanitary landfills can be reduced in recent years [15].

Regarding the regulatory and promotional measures for mandatory valorization of food waste from the industrial sources, few studies were discussed previously [4,17]. Mirabella et al. [4] reviewed the valuable compounds and fuels derived from the solid and liquid waste in the food processing industry but lacked the promotion policies or regulatory measures for the food waste valorization. Naziri et al. [17] reported the valorization of the major agrifood (i.e., olive oil, wine, and rice) industrial by-products and waste from Central Macedonia in Greece for the recovery of value-added compounds (e.g., antioxidants) for food applications. As compared to other countries [18,19], the food waste valorization with high recycling rate in Taiwan may be a learnable case due to the adaptation of "zero waste and resource recycling" policy. In the previous studies [15,20], the author focused on the regulatory and promotion measures for the valorization of food waste from the non-industrial sources like residential and service sectors. Therefore, the present study will put emphasis on the promotion policies and regulatory measures for the valorization of food waste from industrial sources (hereinafter industrial food waste) in Taiwan. Therefore, the aim of this study was twofold. First, the updated data on the statistics and status of industrial food waste generation and treatment in Taiwan will be addressed in Section 3.1 to analyze the trends. Second, the promotion policies and regulatory measures for industrial food waste valorization were studied subsequently based on the joint-efforts by the central governing authorities under the authorization of the WMA. In addition, a case study was addressed to highlight the environmental and economic benefits regarding the valorization of urban food waste for the production of bioenergy by anaerobic digestion in Taiwan.

2. DATA MINING

In this work, the statistical database, promotion policies, and regulatory measures, and case study relevant to industrial food waste valorization were accessed on the official yearbooks and relevant websites, which were briefly summarized below.

- Activity (statistics and status) of industrial food waste generation

According to the annual yearbook of environmental protection statistics [13], the updated data on the statistics and status of industrial food waste generation and treatment in Taiwan were analyzed in the present study.

- Promotion policies and regulatory measures for industrial food waste valorization

The information about the promotion policies and regulatory measures for industrial food waste valorization was accessed on the relevant website [16], which was built by the Ministry of Justice (MOJ).

- Case study: Production of bioenergy from the anaerobic digestion of urban food waste

An official plan for establishing biogas-to-power plants from the urban food waste was addressed to highlight the environmental and economic benefits of food waste valorization by anaerobic digestion in Taiwan [21].

3. RESULTS AND DISCUSSION

3.1 Status of Industrial Food Waste Generation in Taiwan

According to the project ("Food Use for Social Innovation by Optimizing Waste Prevention Strategies") funded by the European Commission [22], the definition of food waste refers to "any food, and inedible parts of food, removed from the food supply chain to be recovered or disposed". In this regards, food waste will be often generated from the retailers, consumers, and food service providers or food processing manufacturers due to the expiration, discard during the sorting operation, leftover, and other wastage reasons [23]. Figure 1 shows the categories of food waste based on its generation sources. In the present study, the industrial sources referred to the agricultural farms (e.g., s laughterhouse, hatchery, oyster farm, aquafarm, and truck farm), and animal-based or plant-based food processing plants or sites. By contrast, the non-industrial sources included the residential, commercial (e.g., office, retail, wholesale, warehouse and distribution, hotel, and restaurant), and service (e.g., hospital) and institutional (school) facilities [3].

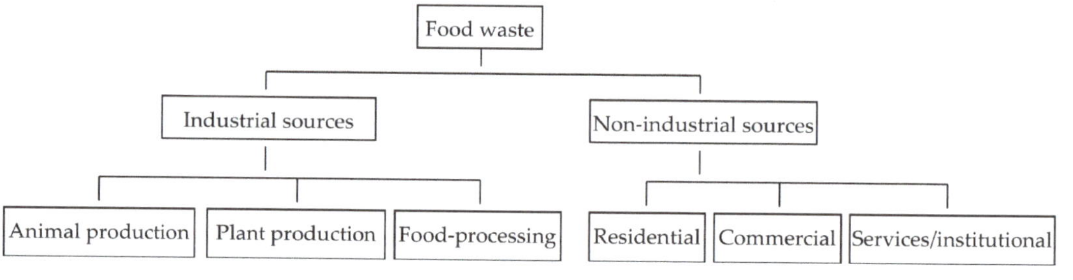

Figure 1. Food waste category and its generation sources.

In order to manage food waste from the industrial sources efficiently, the Taiwan EPA announced that the se sources with over specified amount of food waste generated annually were required to submit their industrial waste management plans by on-line reporting system based on the requirements of the WMA. According to the on-line reports during the period of 2015-2019 [24], Table 1 listed the reported amounts of industrial food waste, showing that the organic food waste from alcoholic beverage manufacturers (i.e., lees, dregs, or alcohol mash) and inorganic food waste from oyster farms (i.e., waste oyster shell) accounted for about half (about 250,000 metric ton) of industrial food waste generation in Taiwan. Table 2 showed the reported amounts of kitchen waste and waste cooking oil during the period of 2015-2019, which were categorized into the industrial sources and non-industrial sources. Obviously, the amount of kitchen waste generation was mostly from the non-industrial sources, indicating a stable amount of about 600,000 metric tons. It should be noted that the significant increases in kitchen waste generation from industrial sources and waste cooking oil from non-industrial sources in 2017 was attributed to the regulatory requirements due to the "food safety scandal" event and circular bioeconomy promotion [14,15]. It can be seen that total reported amounts of waste cooking oil significantly increased from 12,877 metric tons in 2015 to 29,507 metric tons in 2019.

Table 1. Current status of industrial food waste generation during the period of 2015-2019 [24].

Food Waste Type	Generation Amount (Metric Ton)				
	2015	2016	2017	2018	2019
Animal-based residues	19,399	26,306	30,262	36,264	46,400
Dead livestock and poultry	41,809	43,904	44,759	45,427	45,617
Fishery residues	2235	2454	2135	2049	1921
Food processing waste	31,200	31,300	31,952	32,515	14,610
Fruit and vegetable residues	26,382	25,599	26,554	28,848	22,593
Lees, dregs, or alcohol mash	149,207	152,061	139,973	142,029	127,453
Livestock and poultry slaughtering scraps	31,518	48,308	52,522	61,271	64,410
Pig hair	737	717	697	817	788
Plant-based residues	37,018	38,162	42,525	42,607	51,039
Waste oyster shell	131,196	123,966	139,068	128,574	116,352

Table 2. Current status of kitchen waste and waste cooking oil generation during the period of 2015-2019 [24].

Food Waste Type	Generation Amount (Metric Ton)				
	2015	2016	2017	2018	2019
Kitchen waste					
Industrial sources	155	1850	42,040	64,755	70,195
Non-industrial sources	609,706	575,932	551,332	594,992	498,045
Sum	609,861	577,782	593,372	659,747	568,240
Waste cooking oil					
Industrial sources	11,278	15,523	16,085	15,853	15,772
Non-industrial sources	1599	3978	12,591	12,315	13,735
Sum	12,877	19,501	28,676	28,168	29,507

3.2 Promotion Policies for the Valorization of Food Waste from Industrial Sources

Regarding the on-line reporting amount and the regulatory measures for the reuse (or valorization) status of food waste from the non-industrial (residential and commercial) sources in Taiwan, the data during the period of 2010-2017 have been addressed in the previous study [15]. Table 3 further updated the data [13], indicating that the increasing trend in composting valorization during the period of 2015-2019 could be due to the driving force by the "food safety scandal" event in September 2014 and the prevention of African swine fever (ASF) spread since 2019 [25]. Under the authorization of the WMA, the Taiwan EPA subsequently promulgated and/or revised some promotion policies for the valorization of mandatory recyclable food waste (including kitchen waste and waste cooking oil) from all sources to avoid entering the food chain and the 24 larger-scale MSW incineration plants [14]. According to the regulatory definition, the valorization (or reuse) of food waste refers to the production of value-added resources like organic fertilizer, animal feed, and biomass energy. From the regulatory joint-efforts by the central governing authorities [16], including EPA, COA, MOF, and MOEA, Tables 4 and 5 summarized the regulatory measures for the valorization (or reuse) options of urban food waste (kitchen waste and waste cooking oil) and industrial food waste, respectively.

Table 3. Amounts of urban kitchen waste in terms of recycling method over the past decade in Taiwan[1].

Year	Composting	Pig Feed	Others	Sum
2015	197,107	408,524	4076	609,706
2016	197,307	372,280	6346	575,932
2017	204,598	343,906	2828	551,332
2018	231,676	358,229	5087	594,992
2019	246,367	237,849	13,828	498,045

[1] *Source [13]; Unit: metric ton.*

3.2.1. Promotion Policies for the Production of Value-Added Materials from Food Waste

Over the past two decades, industrial ecology concepts, including Cradle to Cradle (C2C) and circular economy, played a prevailing role in the waste management [4]. The approach for the "zero waste" aimed at reusing the food waste as raw material (or feedstock) for new products and applications. In Taiwan, the promotion policies for the production of value-added materials from food waste were based on the legislations by the governing authorities, including the Council of Agriculture

(COA), the Ministry of Economic Affairs (MOEA), and the EPA [16]. Therefore, the central governing agency (i.e., COA) jointly promulgated some regulatory measures for ensuring the bio-products from the food waste valorization under the acts, including the Fertilizer Management Act, the Feed Management Act, and the Animal Industry Act. Based on the Fertilizer Management Act, the organic fertilizer produced from food waste must meet the specifications for the soil-based nutrients/compositions like organic matter, nitrogen and phosphorus, and the limits of toxic metals, including arsenic (As), cadmium (Cd), chromium (Cr), copper (Cu), mercury (Hg), nickel (Ni), lead (Pb), and zinc (Zn).

Table 4. Regulatory measures and generation sources for urban food waste (kitchen waste and waste cooking oil) valorization under the authorization of the Waste Management Act (WMA).

Central Governing Agency (Taiwan)	Food Waste Type	Valorization Option	Generation Source
Environmental Protection Administration (EPA)	Kitchen waste	Feed Feedstock for feed Feedstock for organic fertilizer Feedstock for cultivation medium (soil) Feedstock or fuel for renewable (biomass) energy	Residential, commercial and industrial sources
	Waste cooking oil	Feed for soap Feedstock for stearic acid Feedstock for biodiesel Feedstock for fatty acid methyl ester (blending with fuel oil) Feedstock or fuel for renewable (biomass) energy	Residential, commercial and industrial sources

Table 5. Regulatory measures and generation sources for industrial Food waste valorization under the authorization of the WMA.

Central Governing Agency (Taiwan)	Food Waste Type	Valorization Option	Generation Source
Council of Agriculture (COA)	Waste oyster shell	Feedstock for (mineral supplement) feed (after crushing) Feedstock for fertilizer (after crushing)	Oyster farm
	Fruit and vegetable residues	Feedstock for organic fertilizer Edible directly for livestock and poultry	Fruit and vegetable wholesale market
	Pig hair	Feedstock for feed (Not to be used as feed ingredients for ruminants) Feedstock for organic fertilizer Pig hair product	Slaughterhouse
	Livestock and poultry slaughtering scraps	Feedstock for feed (Not to be used as feed ingredients for ruminants) Feedstock for organic fertilizer Feedstock or fuel for biomass energy	Slaughterhouse
	Dead livestock and poultry	Feedstock for feed (Not to be used as feed ingredients for ruminants) Feedstock for organic fertilizer Feedstock or fuel for biomass energy	Animal farm, breeder, meat wholesale market, or slaughterhouse
	Broken eggs or hatching waste	Feedstock for organic fertilizer	Animal farm, egg washing field, egg (washing) warehouse, hatchery, breeding poultry production site
Ministry of Finance (MOF)	Lees, dregs, or alcohol mash	Feed Feedstock for feed Feedstock for organic fertilizer Feedstock for organic cultivation medium	Alcoholic beverage manufacturer

Central Governing Agency (Taiwan)	Food Waste Type	Valorization Option	Generation Source
Ministry of Economic Affairs (MOEA)	Plant-based residues	Feed Feedstock for feed Feedstock for organic fertilizer Feedstock for organic cultivation medium Feedstock for biomass energy	Food & drinking manufacturer
	Animal-based residues	Feed (Not to be used as feed ingredients for ruminants) Feedstock for feed Feedstock for organic fertilizer Feedstock for organic cultivation medium Feedstock for biomass energy	Food manufacturer

3.2.2. Promotion Policies for the Production of Bioenergy from Food Waste

Due to its richness in moisture, carbohydrate polymers and other constituents (e.g., protein, and lipids), food waste has been used as an excellent feedstock for the production of various kinds of value-added biobased materials and/or biobased products via microbial conversion, including methane, hydrogen, ethanol, organic acids, and bio-fertilizers [26,27]. More importantly, the production of bioenergy from food waste would not only solve the environmental hazards from the MSW incineration plants and sanitary landfill sites but will also mitigate the emissions of greenhouse gases while replacing the usage of fossil fuels by bioenergy. In Taiwan, the promotion policy for the production of bioenergy from food waste was based on the legislation of the Renewable Energy Development Act (REDA) in 2009, which was recently revised in 2019. Among the promotion measures in the REDA, the central governing agency (i.e., MOEA) shall announce the so-called feed-in tariff (FIT) rates for all types of renewable energy annually, which were reviewed or amended in considering related factors like technical progress in power generation, changes in cost and renewable electricity goal achievement. Table 6 listed the FIT rates for promoting electricity generation from biomass-to-power (via anaerobic digestion process or not) and waste-to-power in Taiwan since 2010. It showed an increasing trend from 2.0615 NT$/kW-h in 2010 to 5.1176 NT$/kW-h in 2021 for the biomass-to-power by anaerobic digestion. The MOEA will open up the business model for the direct supply and transfer of green power generation, and also guarantee the fixed 20-year rate which renewable electricity can be converted at the feed-in tariff (FIT) officially announced.

Table 6. Variations of feed-in tariff (FIT) for biomass-to-power and waste-to-power in Taiwan since 2010.

Year	FIT (NT$/kW-h)[a]		
	Biomass-to-Power by AD	Biomass-to-Power by Non-AD	Waste-to-Power
2010	2.0615	2.0615	2.00879
2011	2.1821	2.1821	2.6875
2012	2.6995	2.3302	2.8240
2013	2.8014	2.4652	2.8240
2014	3.2511	2.5053	2.8240
2015	3.3803	2.6338	2.8240
2016	3.9211	2.7174	2.9439
2017	5.0087	2.6000	3.9839
2018	5.0161	2.5765	3.8945
2019	5.0874	2.5765	3.8945
2020	5.1176	2.6871	3.9482
2021	5.1176	2.6884	3.9482

[a] *1 NT$ « 0.035 US$ (2020).*

In order to diversify food waste treatment options and also reduce the burdens on the MSW incineration plants and sanitary landfills in Taiwan, the EPA revised the relevant regulation ("Regulations for Collection, Clean-up and Treatment of General Waste") on 3 November 2017 and announced the new regulation ("Management Regulations for Reuse of Common Industrial Waste") on 8 January 2018. The former regulation added the energy recovery of food waste and other organic residues to one of the specified treatment methods. The food waste and waste cooking oil were listed for the industrial waste reuse items in the latter regulation. Their reuse options have been listed in Table 4. Afterwards, the EPA further launched the Resource Recycling and Reuse Plan in 2018, which was based on the cross-departmental action strategies and promotion measures under the Taiwan's SDGs [10]. In this regards, the EPA has assisted five municipal governments to build food waste bioenergy plants since 2018, which will be further addressed as a case study to discuss their environmental and economic benefits in the subsequent section.

3.3. Official Plan for the Production of Bioenergy from the Anaerobic Digestion of Urban Food Waste

In Taiwan, industrial waste refers to waste that is generated from industrial activities (or sources), but excludes waste generated by the employees themselves. In this regards, food waste (e.g., kitchen waste and waste cooking oil) derived from industrial sources can be listed as general (urban) waste. In order to reduce the negative impacts of food waste on MSW incineration plants and/or sanitary landfills, the Taiwan EPA has been funding local governments in developing diverse treatment options to increase the food waste processing capacity. For instance, the EPA has subsidized the installation of over 50 sets of food waste pretreatment (i.e., shredding and drying) composting plants since 2003. In recent years, the EPA is subsidizing the establishments of food waste-to-bioenergy plant in the five special municipalities, including Taichung City, Taoyuan City, Taipei City, New Taipei City, and Kaohsiung City [21]. Table 7 listed the comparisons of five municipalities in Taiwan, showing that the variations of food waste generated per capita in 2019 were very large. Among these biogas-to-power plants, the first case in Taichung City has been completed and was in operation because this site was to revitalize the old composting plant. Taoyuan City planned to finish the construction of its plant by the end of July 2021, and all plants were estimated to be completed by 2024. Upon completion, these plants can process 230,000 metric tons of food waste per year for about 14.5 million citizens and also generate 41,970,000 kW-h electricity per year. The expected power generation not only gains an annual revenue of NT$214.79 million from selling electricity based on the FIT rate of 5.1176 NT$/kW-h, but also reduces the emissions of carbon dioxide by 21,363 metric tons according to the official carbon emission factor of 0.509 kg CO2/kW-h [28].

Table 7. Comparisons of five municipalities in Taiwan.

Item	Taipei City	Kaohsiung City	New Taipei City	Taichung City	Taoyuan City
Area (km^2)	271.8	2951.9	2053.6	2214.9	1221.0
Population (Million)	2.61	2.77	4.03	2.82	2.26
Year of establishment	1967	1979	2010	2010	2014
District	12	28	29	29	13
Food waste generation [1] (metric ton)	61,849	30,319	124,178	41,147	34,308
Amount of food waste generated per capita (kg/capita)	23.7	10.9	30.8	14.6	15.2

[1] *Source [13].*

4. CONCLUSIONS

In Taiwan, industrial food waste was listed as one of "mandatory" recyclables under the authorization of the WMA for the production of biobased products like organic fertilizer, animal feed, and bioenergy (or biogas). This circular bioeconomy option would not only solve the environmental hazards from the traditional treatment options but will also mitigate the emissions of greenhouse gases while replacing the usage of fossil fuels by biobased materials and bioenergy. In this work, the findings showed that the organic food waste from alcoholic beverage manufacturers (i.e., lees, dregs, or alcohol mash) and inorganic food waste from oyster farms (i.e., waste oyster shell) accounted for about half (about 250,000 metric ton) of industrial food waste generation in Taiwan during the period of 20152019. Under the jointly efforts and legislations by the central governing agencies (i.e., EPA, COA, and MOEA), current regulatory and technological measures for converting industrial food waste into organic fertilizers, animal feed, biofuels, or electricity have resulted in the significant benefits from the environmental and economic viewpoints. In order to achieve the Taiwan's SDGs, one of official plans was to build five large-scale bioenergy plants at local governments via anaerobic digestion of food waste, which were estimated to process 230,000 metric tons annually and also generate 41,970,000 kW-h electricity per year. On completion by 2024, the expected power generation not only gains an annual revenue of NT$214.79 million from selling electricity based on the FIT rate of 5.1176 NT$/kW-h, but also reduces the emissions of carbon dioxide by 21,363 metric tons according to the official carbon emission factor of 0.509 kg CO_2/kW-h. Obviously, the regulatory measures for industrial food waste valorization in Taiwan play a critical role in providing environmental, economic and energy benefits, which will be more adopted in industries.

REFERENCES

1. Ishangulyyev, R.; Kim, S.; Lee, S.H. Understanding food loss and waste—Why are we losing and wasting food? *Foods* **2019**, *8, 297.* [CrossRef] [PubMed]
2. Mak, T.M.W.; Xiong, X.; Tsang, D.C.W.; Yu, I.K.M.; Poon, C.S. Sustainable food waste management towards circular bioeconomy: Policy review, limitations and opportunities. *Bioresour. Technol.* **2020**, 297,122-497. [CrossRef] [PubMed]
3. Rhyner, C.R.; Schwartz, L.J.; Wenger, R.B.; Kohrell, M.G. *Waste Management and Resource Recovery;* CRC Press: Boca Raton, FL, USA, 1995; pp. 26-39.
4. Mirabella, N.; Castellani, V.; Sala, S. Current options for the valorization of food manufacturing waste: A review. *J. Clean. Prod.* **2014**, 65, 28-41. [CrossRef]
5. Capone, R.; Berjan, S.; El Bilali, H.; Debs, P.; Allahyari, M.S. Environmental implications of global food loss and waste with a glimpse on the Mediterranean region. *Int. Food Res. J.* **2000**, 27, 988-1000.

6. Carmona-Cabello, M.; Garcia, I.L.; Leiva-Candia, D.; Dorado, M.P. Valorization of food waste based on its composition through the concept of biorefinery. *Curr. Opin. Green Sustain. Chem.* **2018**,14,67-79. [CrossRef]

7. Nayak, A.; Bhushan, B. An overview of the recent trends on the waste valorization techniques for food wastes. *J. Environ. Manag.* **2019**, 233, 352-370. [CrossRef] [PubMed]

8. Caldeira, C.; Vlysidis, A.; Fiore, G.; De Laurentiis, V.; Vignali, G.; Sala, S. Sustainability of food waste biorefinery: A review on valorisation pathways, techno-economic constraints, and environmental assessment. *Bioresour. Technol.* **2020**, 312,123-575. [CrossRef] [PubMed]

9. United Nations (UN). Take Action for The Sustainable Development Goals. Available online: https://www.un.org/ sustainabledevelopment/sustainable-development-goals/ (accessed on 24 February 2021).

10. Council for Sustainable Development. *Annual Review Report on the Taiwan's Sustainable Development Goals;* Environmental Protection Administration: Taipei, Taiwan, 2020. (In Chinese)

11. Tsai, W.T.; Chou, Y.H.; Lin, C.M.; Hsu, H.C.; Lin, K.Y.; Chiu, C.S. Perspectives on resource recycling from municipal solid waste in Taiwan. *Resour. Policy* **2007**, 32, 69-79. [CrossRef]

12. Chang, Y.M.; Liu, C.C.; Hung, C.Y.; Hu, A.; Chen, S.S. Change in MSW characteristics under recent strategies in Taiwan. *Waste Manag.* **2008**, 28, 2443-2455. [CrossRef] [PubMed]

13. Environmental Protection Administration (EPA, Taiwan). *Yearbook of Environmental Protection Statistics 2019;* EPA: Taipei, Taiwan, 2020.

14. Tsai, W.T. Mandatory recycling of waste cooking oil from residential and commercial sectors in Taiwan. *Resources* **2019**, 8, 38. [CrossRef]

15. Tsai, W.T. Turning food waste into value-added resources: Current status and regulatory promotion in Taiwan. *Resources* **2020**, 9,53. [CrossRef]

16. Laws and Regulation Retrieving System (Taiwan). Available online: https://law.moj.gov.tw/Eng/index.aspx (accessed on 23 February 2021).

17. Naziri, E.; Nenadis, N.; Mantzouridou, F.T.; Tsimidou, M.Z. Valorization of the major agrifood industrial by-products and waste from Central Macedonia (Greece) for the recovery of compounds for food applications. *Food Res. Int.* **2014**, 65,350-358. [CrossRef]

18. Ju, M.; Bae, S.J.; Kim, J.Y.; Lee, D.H. Solid recovery rate of food waste recycling in South Korea. *J. Mater. Cycles Waste Manag.* **2016**, 18, 419-426. [CrossRef]

19. Lim, W.J.; Chin, N.L.; Yusof, A.Y.; Yahya, A.; Tee, T.P. Food waste handling in Malaysia and comparison with other Asian countries. *Int. Food Res. J.* **2016**,23, S1-S6.

20. Tsai, W.T. Management considerations and environmental benefit analysis for turning food garbage into agricultural resources. *Bioresour. Technol.* **2008**, 99, 5309-5316. [CrossRef] [PubMed]

21. Taiwan's Renewable Energy News (Ministry of Economic Affairs, Taiwan). Available online: https://www.re.org.tw/news/ more.aspx?cid=198&id=3126 (accessed on 27 February 2021).

22. Food Use for Social Innovation by Optimising Waste Prevention Strategies (FUSIONS) Website (EU). Available online: https: //www.eu-fusions.org/index.php (accessed on 26 February 2021).

23. Food Lose and Food Waste (FAO). Available online: http://www.fao.org/food-loss-and-food-waste/flw-data (accessed on 26 February 2021).

24. Industrial Waste Reporting and Management Information System (Taiwan EPA). Available online: https://waste.epa.gov.tw/ RWD/Statistics/?page=Month1 (accessed on 28 February 2021).

25. Wang, W.H.; Lin, C.Y.; Ishcol, M.R.C.; Urbina, A.N.; Assavalapsakul, W.; Thitithanyanont, A.; Lu, P.L.; Chen, Y.H.; Wang, S.F Detection of African swine fever virus in pork products brought to Taiwan by travelers. *Emerg. Microbes Infect.* **2019**, 8,100-102. [CrossRef] [PubMed]

26. Kiran, E.U.; Trzcinski, A.P.; Ng, W.J.; Liu, Y. Bioconversion of food waste to energy: A review. *Fuel* **2014**,134, 389-399. [CrossRef]

27. Sen, B.; Aravind, J.; Kanmani, P.; Lay, C.H. State of the art and future concept of food waste fermentation to bioenergy. *Renew. Sustain. Energy Rev.* **2016**, 53, 547-557. [CrossRef]

28. Ministry of Economic Affairs (MOEA). *Energy Statistics Handbook-2019;* MOEA: Taipei, Taiwan, 2020.

CHAPTER 3

Valorization of Rice Husk for the Production of Porous Biochar Materials

1. INTRODUCTION

The most common biomass feedstocks for the production of energy and carbon materials are plant, wood, agricultural waste or crop residues, which are mainly composed of water, lignocellulosic components (lignin, hemicellulose and cellulose), extractives and ash [1]. Notably, the energy-containing biomass is derived from the sun by converting atmospheric carbon dioxide (CO2) and water into carbohydrates (or lignocelluloses) through photosynthesis, thus mitigating greenhouse gas (GHG) emissions by displacing fossil fuel use. In this regard, the valorization of biomass for fuels and chemicals was motivated mainly by the benefits of renewable resources, global warming (environmental protection) and social economy [1]. For example, biomass has been considered as a carbon-neutral feedstock or fuel from the viewpoint of the carbon cycle principle regarding the environment. Furthermore, biomass can be converted by biochemical and thermochemical methods into useful products [2]. Pyrolysis, one of the commonly used thermochemical conversion processes, involves decomposition of biomass in the absence of air or oxygen at an elevated temperature [3]. The resulting biochar can be further used as solid fuel, carbon material, soil amendment, environmental adsorbent (biosorbent), functional catalyst or feedstock for chemicals, depending on its final applications [4-9].

In Asia, rice is the most important crop, suggesting that rice husk is an important crop residue because it accounts for around 20% of grain weight. Approximately 150 million metric tons of rice husk are produced annually based on the world production of paddy rice (i.e., 750 million metric tons) [10]. According to the agricultural statistics [11], it was thus estimated that around 0.4 million metric tons of rice husk are generated annually in Taiwan. Due to its richness in silica and lignocellulosic constituents [12], the biomass is currently used for bioenergy (or solid fuel) in rice milling plants, as a paving/bedding material in poultry farms, animal feed and as a soil amendment in different forms in agricultural lands to increase soil fertility and crop productivity [13-15]. Rice husk is directly reused without converting it into useful materials by thermochemical or biochemical processes. As compared to uncontrolled burning on fields, these direct reuse approaches do not valorize the energy content of the material and may generate toxic emissions without leading to valuable applications such as porous carbon materials.

In order to increase the pore properties, mediate environmental pollution and also mitigate the carbon dioxide release as GHG forms by valorizing rice husk, the potential to enhance the porous structure of resulting biochar products at limited pyrolysis conditions has been widely investigated [16-26]. Vassileva et al. pyrolyzed rice husk at 250,350,480 and 700°C at a heating rate of 4°C/min and subsequently maintained this temperature for 4 h [16]. Jindo et al. charred rice husk for 10 h at different temperatures (400-800°C) at a heating rate of 10°C/min [17]. Phuong et al. investigated the effects of pyrolysis temperature (350,450 and 550°C) and heating rate (10 and 50°C/min) on the yield and properties of the resulting biochar derived from rice husk [18]. Ahiduzzaman and Sadrul Islam produced rice husk biochar at 650°C for 60 min, which was further activated to produce activated carbon [19]. Wei et al. prepared rice husk biochars at 300,500 and 750°C, which were used as adsorbents for comparing the adsorption performance of herbicide metolachlor with their physicochemical characteristics [20]. Zhang et al. reported the physicochemical properties of rice husk biochars prepared under different temperatures (200-800°C) at a fixed heating rate of 10°C/min and then kept for 60 min [21]. Dissanayake et al. conducted pyrolysis experiments on rice husk at 350, 500 and 650°C at a heating rate of 10°C/min [22]. In this case, the experiment at 350°C took around 2 h to complete pyrolysis, while the experiments at 500°C and 650°C completed the process in around 25 min after reaching the pyrolysis temperature. Jia et al. produced rice husk biochar under 300,400,500, 600 and 700°C (heating rate of 15°C/min) for 3 h [23]. Shi et al. investigated adsorption interactions between lead ion and biochars produced at 300,500 and 700°C [24]. Singh et al. used rice husk biochars, prepared at 300, 450 and 600°C, as adsorbents of nutrient nitrogen (i.e., urea), showing the huge sorption potential of the biochar due to high functionality and porosity [25]. Saeed et al. performed a pyrolysis experiment at a constant temperature of 500°C for 60 min [26]. It is clear that the pore properties of rice husk biochar will be more developed at higher pyrolysis temperatures because of the greater formation of turbostratic crystallites [27]. Regarding the applications of rice husk biochar, it has been used as an effective adsorbent for the removal of trichloroethylene [28], a good medium for the growth of soursop *(Annona muricata L.)* seedlings [29] and a soil amendment for increasing nutrient retention [30,31].

In Taiwan, rice husk biochar has been extensively applied to agricultural soils for enhancing soil fertility and crop yields in recent years due to the promotion of organic farming [30]. However, these biochars showed poor pore properties [32]. For example, the values of the Brunauer-Emmett-Teller (BET) surface area were lower than 5 m^2/g. In addition, few studies on the production of rice husk biochar at higher temperatures (e.g., 900°C) for different holding times have been reported in the literature, as mentioned above. Therefore, this work focused on investigating the variations in the yields and pore properties of rice husk biochar in the pyrolysis process as a function of temperature (400, 500, 600, 700, 800 and 900°C) and residence time (0, 30, 60 and 90 min) at a commonly used heating rate (10°C/min). The instrumental analyses, including nitrogen adsorption-desorption isotherms, true density (gas pycnometry using helium displacement principle), scanning electron microscopy-energy-dispersive X-ray spectroscopy (SEM-EDS) and Fourier-transform infrared spectroscopy (FTIR), were performed to determine the physicochemical properties of the resulting rice husk biochar.

2. MATERIALS AND METHODS

2.1 Material

Rice husk (RH), used as a precursor for producing biochar, was obtained from an agricultural research and extension station (Pingtung County, Taiwan). The as-received biomass was cut by a rotary knife-type shredder. The shredded RH was further sieved to a broad range of mesh screen sizes, ranging from mesh no. 20 (opening size of 0.841 mm) to mesh no. 40 (opening size of 0.420 mm). The sieved RH was first dried in a forced air-circulation oven and then used for the pre-pyrolysis test involving thermogravimetric analysis (TGA) and the pyrolysis experiments. In line with the previous study [33], the dried RH contained high contents of carbon (45.28 wt%), hydrogen (5.51 wt%), oxygen (~44 wt%, estimated) and silicon (3.90 wt%), indicating that this biomass has a moderate calorific value (16.4 MJ/kg) [12].

2.2 Pre-Pyrolysis Test via Thermogravimetric Analysis (TGA)

In order to obtain the proper pyrolysis temperature range of the dried RH, a thermal analyzer (TGA-51; Shimadzu Co., Japan) was used to obtain the TGA data. Around 0.2 g of the RH sample was placed on a quartz crucible and then heated externally to a maximum temperature of 1000°C at a fixed rate of 10°C/min under a nitrogen (N2) flow rate of 50 cm^3/min. During the TGA operation, ongoing data on the sample weight and temperature were collected to determine the thermal stability and reaction mechanism by the curves of TGA and its derivative thermogravimetry (DTG). The TGA curve was normalized based on the weight of RH sample fed.

2.3 Pyrolysis Experiments

Based on the TGA/DTG data (discussed in Section 3.1), the pyrolysis temperature had to be above 400°C because the peak temperature of DTG was found to be around 380°C. This result was consistent with those reported by Johar et al. [34], indicating the complete devolatilization of hemicellulose and cellulose at around 400°C. Therefore, the pyrolysis experiments were performed by the matrix combinations of temperature (400-900°C, increased by the interval of 100°C) and holding time (0-90 min, increased by the interval of 30 min) in the present study. Herein, a null holding time means that the heating was stopped (powered off) when the specified temperature was reached. Based on previous studies for producing biochar [35-40], a vertical fixed-bed electric furnace is suitable equipment for the production of biochar from various biomass feedstocks. For each pyrolysis experiment, around 5 g of the RH sample was pyrolyzed at a heating rate of around 10°C/min under an inert atmosphere by passing N2 flow rate of 500 cm^3/min. The biochar yield was obtained from the difference between the initial RH sample weight and the resulting biochar weight based on the weight of RH sample fed. In order to describe the RH-based biochar (i.e., BRH), the resulting biochar was denoted as BRH-pyrolysis temperature-holding time. For example, the product BRH-400-30 refers to the pyrolysis conditions at temperature of 400°C for holding time of 30 min.

2.4 Physicochemical Properties of Resulting Biochar

In this work, an accelerated surface area and porosimetry system (ASAP 2020; Mi- cromeritics Co., USA) was used to determine the pore properties of the BRH products, including the Brunauer-Emmett-Teller (BET) surface area, pore volume and pore size distribution. Herein, the calculation of specific surface area was performed using the BET equation, using a range of relative pressure from 0.05 to 0.30. The total pore volume was taken as the nitrogen liquid volume adsorbed at a relative pressure of ca. 0.995. Using the *t*-plot method, the data on micropore area and micropore volume were obtained by the Halsey equation [41]. According to the definition by the International Union of Pure and Applied Chemistry (IUPAC) [41], micropores refer to pores with an internal diameter or width of less than 2 nm. Mesopores are defined as pores with an internal diameter or width between 2 and 50 nm. Regarding the pore size distributions of the BRH products, the Barrett-Joyner-Halenda (BJH) method was employed to calculate them in the range of mesopores and small macropores from experimental N2 isotherms (desorption branch) using the Kelvin model of pore filling [41].

The porosity of a material is defined as the ratio of the total pore volume to the apparent volume of the particle or powder (excluding inter-particle voids). Therefore, this property can be estimated by subtracting the ratio of particle (apparent or skeletal) density to true density from 1 [42,43]. In this work, a gas (helium) pycnometer (AccuPyc 1340; Micromeritics Co., USA) was used to determine the true density. Although mercury (Hg) porosimetry is often used to measure particle density due to the high surface tension of Hg, this property was estimated by using the measured data on the true density and total pore volume [43].

The textural morphology on the surface of the BRH product was observed by the SEM system (S-3000N; Hitachi Co., Japan). Prior to the SEM analysis, the BRH sample was coated by a conductive gold film using an ion sputter (E1010; Hitachi Co., Japan). An accelerating potential of 15.0 kV (electron beam) in a vacuum chamber was applied to the specimen surface during the SEM analysis. In addition, EDS analysis was used to quantify the elemental compositions rapidly and simply when scanning the BRH sample during the SEM analysis.

The oxygen-containing functional groups on the surfaces of the BRH products were analyzed by FTIR (FT/IR-4600; JASCO Co., Japan). Prior to the FTIR analysis, the dried BRH sample was mixed with spectroscopic-grade potassium bromide (KBr) powder, and then ground in an agate mortar. The finely uniform mixture (around 1 wt% BRH) was pressed in a hydraulic machine to form a disc with a diameter of ca. 1.2 cm and thickness of ca. 1 mm. The FTIR spectra were obtained by scanning with a wave number range of 4000-400 cm^{-1} and a resolution of 4 cm^{-1}.

3. RESULTS AND DISCUSSION

3.1. Thermogravimetric Analysis (TGA) of Rice Husk (RH)

As mentioned above, a pre-pyrolysis test of the RH sample was carried out by a thermal analyzer under an inert atmosphere at a fixed heating rate of 10 °C/min. The test conditions were very similar to the pyrolysis experiments for producing RH biochar. Figure 1 presents the TGA/TGA curves for the RH sample, showing that the significant degradation temperature was less than 400 °C. As shown, there were two apparent peaks and one shoulder as the pyrolysis temperature increased from room temperature to 400 ° C. In the first peak, mass loss should occur in the form of water vapor (moisture) between 60 and 200 °C. For a lignocellulosic biomass, hemicellulose and a smaller amount of cellulose may be the most labile polymeric components as compared to lignin [1]. This implies that the TGA shoulder appeared at a temperature of around 300 °C, which is lower than that of lignin. During the stage of pyrolysis, the complex reactions (e.g., cracking, condensation) involve depolymerization and scission, thus causing a continual mass loss in the form of vapors such as moisture, CO2 and volatile organics [44]. The final peak around 400 °C can be attributed to the thermal degradation of most cellulose and a smaller amount of lignin. In order to produce porous biochar from RH, the pyrolysis experiments thus started from 400 °C, where more lignin was pyrolyzed to form the products of char and tar. At higher temperatures (400-1000 °C), the mass loss can be attributed to the thermal degradation of most lignin and inorganic minerals (e.g., carbonates, chlorides, oxides) [26].

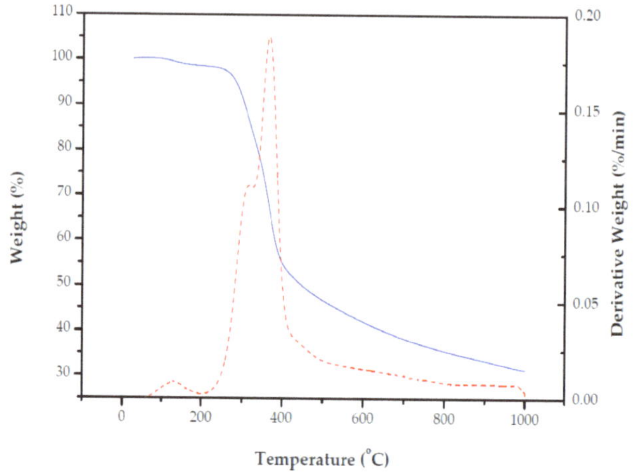

Figure 1. Thermogravimetric analysis (TGA) curve (blue color) and derivative thermogravimetry (DTG) curve (red dash line) of rice husk (RH) at a heating rate of 10°C/min.

3.2. Yieldo and Pore Properties ofResulting Biochar

3.2.1. Yie lds of Resulting Biochar

The biochar yield was calculated by the ratio of biochar mass to RH mass loaded into each pyrolysis experiment (around 5 g). As the pyrolysis temperature increased, more condensable products were formed as water and o rganic components, but there rema ined less carbonized solids as biochar due to the complex degradation reactions in progress. Figure 2 illustrates tXe variation in ihe yield of the resulting RH-based biochar as a function of temperature (400-900°C) for a holding time of 30 min, indicating a decreasing trend. Although not shown here, tire yields of the resulting RH-based biochat prepared ait 900°C for four holding times (0, 30, 60 and 90 min) also andicated this trend.

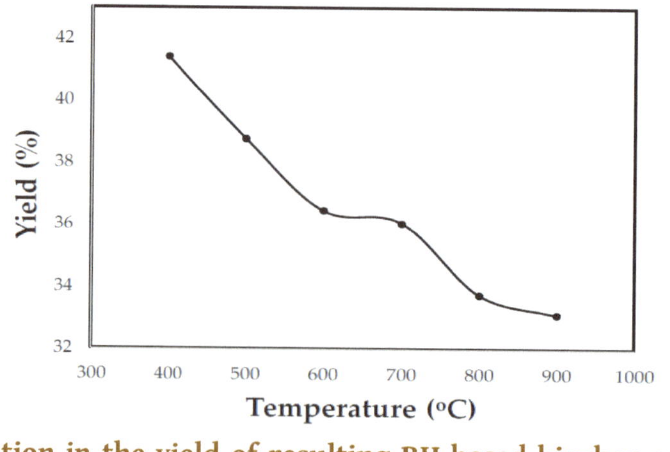

Figure 2. Variation in the yield of resulting RH-based biochar as a function of temperature.

3.2.2. Pore Properties of Resulting Biochar

The potential applications of biochar for water retention, sorption of contaminants and as a habitat for microorganisms are strongly dependent on its pore properties, which include specific surface area, density, porosity and morphology [45]. In the present study, all pore properties of the biochar samples were based on N_2 gas adsorption-desorption at -196°C and helium gas displacement. Table 1 lists the pore properties of the resulting biochar (BRH series), which included the BET surface area (S_{BET}), total pore volume (V_t), true density (p_s), particle density (p_p) and porosity (εp). Although particle density can be determined by a mercury (Hg) pycnometer due to the high surface tension of Hg and its inability to enter any pore of the porous sample [41], it was calculated by the values of total pore volume and true density in the present study [42].

$$\rho_p = \rho_s / (V_t \times \rho_s + 1) \quad (1)$$

Table 1. Pore properties of resulting RH-based biochar.

Biochar [a]	BET Surface Area [b] (m^2/g)	Total Pore Volume [c] (cm^3/g)	True Density[d] (g/cm^3)	Particle Density[e] (g/cm^3)	Porosity[f] (-)
BRH-400-30	35.4	0.049	1.662	1.537	0.075
BRH-500-30	210.9	0.161	1.643	1.299	0.209
BRH-600-30	225.6	0.145	1.717	1.375	0.199
BRH-700-30	219.5	0.157	1.852	1.435	0.225
BRH-800-30	244.3	0.154	1.987	1.521	0.234
BRH-900-30	258.6	0.196	2.071	1.473	0.289
BRH-900-0	242.8	0.175	2.075	1.522	0.266
BRH-900-30	258.6	0.196	2.071	1.473	0.289
BRH-900-60	274.6	0.222	2.074	1.420	0.315
BRH-900-90	278.9	0.223	2.076	1.419	0.316

[a] Sample notation indicated the resulting biochar (BRH-temperature-time) produced at the temperature of 400-900°C for a holding time of 0-90 min using 5 g dried rice husk (RH).[b] BET surface area (S_{bet}) was based on relative pressure range of 0.05-0.30.[c] Total pore volume (V_t) was obtained at relative pressure of around 0.99.[d] True density (p_s) was measured by the helium displacement method. [e] Particle density (pp) was calculated from the total pore volume (V_t) and true density (p_s).[f] Particle porosity (tp) was calculated from the particle density (pp) and true density (p_s).

Furthermore, the porosity (tp) can be obtained by subtracting the ratio of particle density to true density from 1 [43]:

$$\varepsilon_p = 1 - (\rho_p / \rho_s) \quad (2)$$

As listed in Table 1, there were obvious variations in the pore properties of the biochar as a function of pyrolysis temperature (400-900°C) and holding time (0-90 min). Similar to numerous studies [45,46], there was a positive correlation between the BET surface area (or porosity) and pyrolysis temperature. When the pyrolysis temperature increased from 400 to 900°C, there was greater formation of pyrogenic amorphous biochar [27], thus causing a greater pore space or more pores with the pyrolysis temperature as charring intensity increased. Noticeably, the pore properties of the biochar continuously increased as the pyrolysis temperature increased up to 900°C (for a fixed holding time of 30 min), or as the holding time was extended up to 90 min (at a fixed pyrolysis temperature of 900°C). This implies that the structural breakdown of the resulting biochar produced at higher temperatures, probably due to sintering or fusion [45], was not observed in this work [47]. In the present study, the process parameter of pyrolysis temperature had a more significant effect on the pore properties of biochar as compared with the holding time (or reaction residence time). This is because the extent of physical changes (e.g., mass loss) for the biochar was highly dependent on the temperature during pyrolysis processing (seen in Figure 1). From the viewpoint of efficient energy use, the pyrolysis conditions at 500°C for 30 min may be suitable for the production of RH-based biochar with a high BET surface area of 211 m^2/g when compared with the resulting biochar (BET surface area of 279 m^2/g) prepared at 900°C for 90 min.

The BET surface area of the resulting biochar was also linked to its N2 adsorption desorption isotherms, which can be further transformed into its pore size distribution based on the BJH method [41]. Figure 3 depicts the N2 adsorption-desorption isotherms and pore size distributions of the resulting biochar prepared at various pyrolysis temperatures for a holding time of 30 min. Furthermore, Figure 4 shows the similar characteristics of the resulting biochars preparnd at 900 °C for different holding times. As shown in Figures 3 and 4, the corresponding Na adsorption-desorption isotherms and pore size distributions od the resulting biochars were consistent with those in Table 1. Furthermore, the biochar pyrolyzed at higher temperatures showed more pronounced pore development, including micropores and mesopores. Therefore, these biochars possess the so-called Type I and Type-IV shapes (isotherms) [41]. This mesoporous feature will be beneficial in providing water retention, adsorption capacity and microbial growth space.

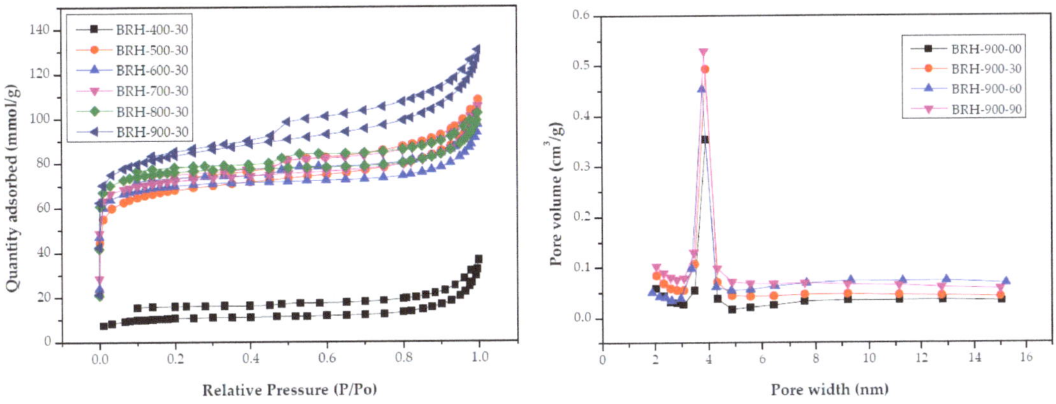

Figure 3. N_2 **adsorption-desorption isotherms (upper) and pore size distributions (lower) of resulting biochar prepared at vtrious temperatures for a dixed holding time of 30 min.**

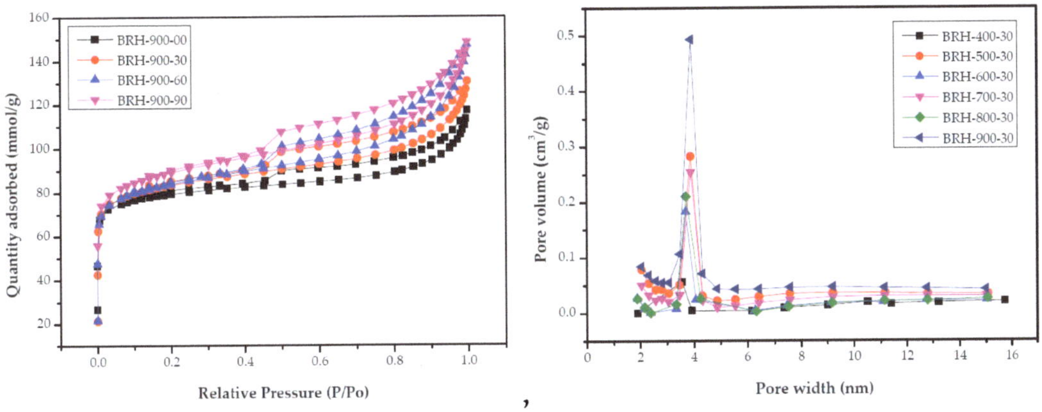

Figure 4. N_2 **adsorption-desorption isotherms (upper) and pore size distributions (lower) of resulting biochar prepared at 900°C for various holding; times.**

In order to observe the porous texture of the resulting biochar, SEM analyses at 15 kV were performed on its gold-coated surface with different magnifications (i.e., 200 and 1000). As shown in the SEM images (Figure 5), thee representative biochar (i.e., BRH-900-30) has fine pores on the surface. These pores will be more formed from gas vesicles at higher pyrolysis temperatures, resulting in better pore properties such as specific surface area and porosity (Table 1). In addition, the resulting biochar also exhibits a rigid, irregular and rough surface due to its rigid silica composition and the rigorous carbonization at a higher temperature (i.e., 900°C) for a longer holding time (i.c., 30 min). In this regard, the RHibased biochar may be a good medium ior aossible applications in water retention and wastewater treatment in the soil environment due to its highly porous structure.

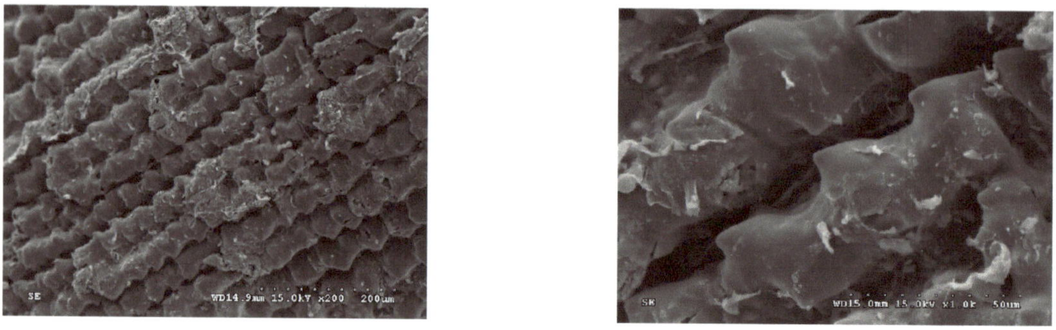

Figure 5. SEM images (x200, left; xl000, right) of the resulting biochar (BRJI-900-3()).

3.3. Chemical Characterization of Resulting Biochar

As described in Section *2.4,* the EDS sysSem, which is commonly used alongside SEEM, was used to enable semi-quantitative elemental analysis on the surSace of the resulting biochar. Figure 6 shows an EDS spectrum from the resulting biochar (BRH-900-30), observing the presence of carbon (C), oxygen (O) and silicon (Si). As a preliminary quantification, the corresponding contriOutions to the elemental composition are 18.15, 4.66 and 40.1 9 wt%, respectively. The presence of C and O in the RH-based btochar should arise from its lignocellulosic precursor (i.e., RH). The most significant peak is assigned to the presence ot Si due ta the high content of silica (SKZt) in the RH. The rich presence of silica in the RH-based biorhar can be furthe r tdnntified by X-ray di fftac tio n (XRD) [48] or X- ray photoelecfron spectroscopy (XPS) analysis [49].

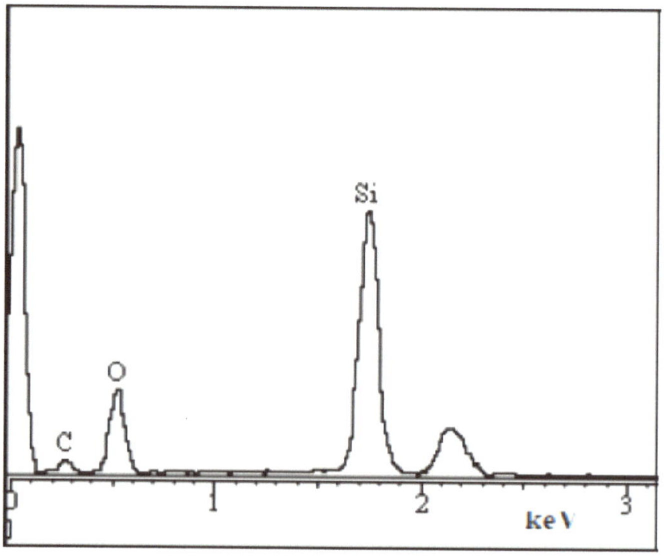

Figure 6. EDS spectrum of the resulting biochar (BRH-900-30).

In general, the chemical characteristics of biochar mainly comprise aromatic C and inorganic minerals, which ara dependent on the starting feedstock and pyrolysis conditions applied. The presence of functional groups on the surface of biocher plays a vital role in the soil and water environments. For example, the addition of biochar to soil has been proven to enhance the sorption capacities for heavy metal ions by the electrostatic attraction [45], which can be attributed to the high contents of oxygen-containing functional groups on tho surface due to the negatively charged surface. The FTIR spectrum of the resulting biochar iBRH-900-30) shown in 9iguro 7 has the following features [23,50,51]. The peak at 3460 cm^{-1} can be assigned to the presence of adsorbed water and hydrogen-bonded biochar O-H groups. The band between t500 and 1700 cm^{-1} could be attributed to C=O stretching or C=C groups in aromatic rings. The sharp peak around 1385 cm^{-1} could be assigned to the symmetric deformation band of $-CH_3$ groups. The peak at around 1115 cm^{-1} could be the stretching vibration of C-O and the vibration of silica bonds. Si-O framework bands at around 1115 cm^{-1} and below 800 cm^{-1} were evident [17]. These FTIR results were in accordance with those in the EDS? analysis (Figure 6), enhancing the slight hydrophilicity on the surface of the RH-based biochac. These functional groups will imp act the soil pH and interaction with ionic contaminants in soil, causing higher cation exchange capacity (CEC) and adsorption capacities for cations (e.g., metal ions).

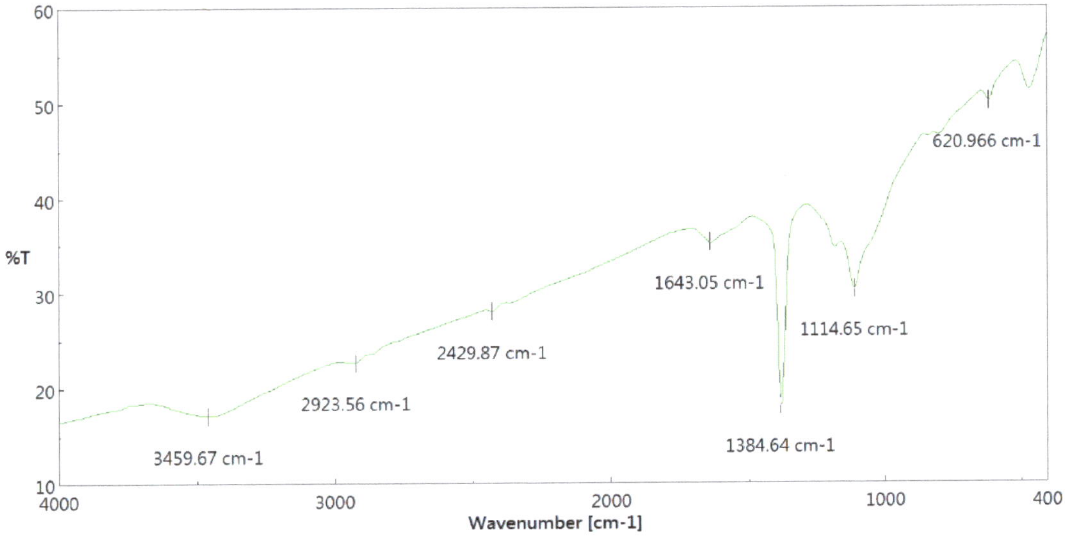

Figure 7. FTIR spectrum of the resulting biochar (BRH-900-30).

4. CONCLUSIONS

In this work, a series of biochars were prepared from RH at various pyrolysis conditions , which were designed by the matrix of temperature (400-900°C) and holding times (i.e., 0-90 min) based on the results of a pre-pyrolysis test by TGA. The pore properties (i.e., BET[1] surface area and porosity)) of the resulting RH-based biochar were clearly positively linked to the studied ranges of pyrolysis temperature and holding time. The maximal pore properties with the BET surface area of around 280 m^2/g could be obtained from tine conalitions at 900°C for a holding time of 90 min. The porosity of the optimal biochar was as high as 0.316. In addition, the resulting biochair showed characteristic of microporoos and mesoporous structures based on the N_2 adsorption-desorptlon isotherms and pore size distributions. The results of the EtDS and FTIR analyses also supported the slighf hydrophilicity on the surface of the RH-based biochar due to the richness in oxygen/silioa-containing functional groups. Based on the physicochemical properties determined, the RH-based biochar could be an excellent material for poscible applications in water conservation, wastewater treatment and soil amendment.

REFERENCES

1. Basu, P *Biomass Gasification, Pyrolysis and Torrefaction,* 2nd ed.; Academic Press: San Diego, CA, USA, 2013.
2. Demirbas, A. Biomass resource facilities and biomass conversion processing for fuels and chemicals. *Energy Convers. Manag.* **2001**, 42,1357-1378. [CrossRef]
3. Lehmann,J.;Joseph,S.Biochar for environmental management:An introduction. In *Biochar for Environmental Management,* 2nd ed.; Lehmann, J., Joseph, S., Eds.; Routledge: New York, NY, USA, 2015; pp. 1-13.
4. Dai,Y.J.;Zhang,N.X.;Xing,C.M.;Cui,Q.X.;Sun,Q.Y.The adsorption,regeneration and engineering applications of biochar for removal organic pollutants: A review. *Chemosphere* **2019**,223,12-27. [CrossRef] [PubMed]
5. Zhang, Z.K.; Zhu, Z.Y.; Shen, B.X.; Liu, L.N. Insights into biochar and hydrochar production and applications: A review. *Energy* **2019**,171, 581-598. [CrossRef]
6. Wang, J.; Wang, S. Preparation, modification and environmental application of biochar: A review. *J. Clean. Prod.* **2019**, 227, 1002-1022. [CrossRef]
7. Fdez-Sanroman, A.; Pazos, M.; Rosales, E.; Sanroman, M.A. Unravelling the environmental application of biochar as low-cost biosorbent: A review. *Appl. Sci.* **2020**,10, 7810. [CrossRef]
8. Shan, R.; Han, J.; Gu, J.; Yuan, H.R.; Luo, B.; Chen, Y. A review of recent developments in catalytic applications of biochar-based materials. *Resour. Conserv. Recycl.* **2020**,162,105036. [CrossRef]
9. Wang, D.;Jiang, P.K.; Zhang, H.B.; Yuan, W. Biochar production and applications in agro and forestry systems: A review. *Sci. Total Environ.* **2020**, 723,137775. [CrossRef]

10. Food and Agriculture Organization FAOSTAT. Available online: http://www.fao.org/faostat/en/#data/QC/visualize (accessed on 18 March 2021).

11. Council of Agriculture (COA). *Agriculture Statistics Yearbook*; COA: Taipei, Taiwan, 2020.

12. Jenkins, B.M.; Baxter, L.L.; Miles, T.R., Jr.; Miles, T.R. Combustion properties of biomass. *Fuel Process. Technol.* **1998**, 54,17-46. [CrossRef]

13. Moraes, C.A.M.; Fernandes, I.J.; Calheiro, D.; Kieling, A.G.; Brehm, F.A.; Rigon, M.R.; Fiho, J.A.B.; Schneider, I.A.H.; Osorio, E. Review of the rice production cycle: By-products and the main applications focusing on rice husk combustion and ash recycling. *Waste Manag. Res.* **2014**, 32,1034-1048. [CrossRef]

14. Soltani, N.; Bahrami, A.; Pech-Canul, M.I.; Gonzalez, L.A. Review on the physicochemical treatments of rice husk for production of advanced materials. *Chem. Eng. J.* **2015**, 264, 899-935. [CrossRef]

15. Quispe, I.; Navia, R.; Kahhat, R. Energy potential from rice husk through direct combustion and fast pyrolysis: A review. *Waste Manag.* **2017**, *59*, 200-210. [CrossRef] [PubMed]

16. Vassileva, P.; Detcheva, A.; Uzunov, I.; Uzunova, S. Removal of metal ions from aqueous solutions using pyrolyzed rice husks: Adsorption kinetics and equilibria. *Chem. Eng. Comm.* **2013**, 200,1578-1599. [CrossRef]

17. Jindo, K.; Mizumoto, H.; Sanchez-Monedero, M.A.; Sonoki, T. Physical and chemical characterization of biochars derived from different agricultural residues. *Biogeosciences* **2014**,11, 6613-6621. [CrossRef]

18. Phuong, H.T.; Uddin, M.A.; Kato, Y. Characterization of biochar from pyrolysis of rice husk and rice straw. *J. Biobased Mater. Bioenergy* **2015**, 9, 439-446. [CrossRef]

19. Ahiduzzaman, M.; Sadrul Islam, A. Preparation of porous bio-char and activated carbon from rice husk by leaching ash and chemical activation. *SpringerPlus* **2016**, 5,1248. [CrossRef] [PubMed]

20. Wei, L.; Huang, Y.; Li, Y.; Huang, L.; Mar, N.N.; Huang, Q.; Liu, Z. Biochar characteristics produced from rice husks and their sorption properties for the acetanilide herbicide metolachlor. *Environ. Sci. Pollut. Res.* **2017**,24, 4552-4561. [CrossRef] [PubMed]

21. Zhang, Y.; Ma, Z.; Zhang, Q.; Wang, J.; Ma, Q.; Yang, Y.; Luo, X.; Zhang, W. Comparison of the physicochemical characteristics of bio-char pyrolyzed from moso bamboo and rice husk with different pyrolysis temperatures. *BioResources* **2017**, 12, 4652-4669. [CrossRef]

22. Dissanayake, D.K.R.P.L.; Dharmakeerthi, R.S.; Karunarathna, A.K.; Dandeniya, W.S. Changes in structural and chemical properties of rice husk biochar co-pyrolysed with Eppawala rock phosphate under different temperatures. *Trop. Agric. Res.* **2018**, 30,19-31. [CrossRef]

23. Jia, Y.; Shi, S.; Liu, J.; Su, S.; Liang, Q.; Zeng, X.; Li, T. Study of the effect of pyrolysis temperature on the Cd'[2] adsorption characteristics of biochar. *Appl. Sci.* **2018**, 8,1019. [CrossRef]

24. Shi, J.; Fan, X.; Tsang, D.C.W.; Wang, F.; Shen, Z.; Hou, D.; Alessi, D.S. Removal of lead by rice husk biochars produced at different temperatures and implications for their environmental utilizations. *Chemosphere* **2019**, 235, 825-831. [CrossRef]

25. Singh, S.V.; Chaturvedi, S.; Dhyani, V.C.; Kasivelu, G. Pyrolysis temperature influences the characteristics of rice straw and husk biochar and sorption/desorption behaviour of their biourea composite. *Bioresour. Technol.* **2020**, 314,123674. [CrossRef] [PubMed]

26. Saeed, A.A.H.; Harun, N.Y.; Sufian, S.; Afolabi, H.K.; Al-Qadami, E.H.H.; Roslan, F.A.S.; Rahim, S.A.; Ghaleb, A.A.S. Production and characterization of rice husk biochar and Kenaf biochar for value-added biochar replacement for potential materials adsorption. *Ecol. Eng. Environ. Technol.* **2021**, 22,1-8. [CrossRef]

27. Keiluweit, M.; Nico, P.S.; Johnson, M.G.; Kleber, M. Dynamic molecular structure of plant biomass-derived black carbon (biochar). *Environ. Sci. Technol.* **2010**, *44*,1247-1253. [CrossRef]

28. Rossi, M.M.; Silvani, L.; Amanat, N.; Papini, M.P. Biochar from pine wood, rice husks and *iron-Eupatorium* shrubs for remediation applications: Surface characterization and experimental tests for trichloroethylene removal. *Materials* **2021**,14,1776. [CrossRef]

29. Harun, N.S.N.; Jaafar, N.M.; Sakimin, S.Z. The effects of rice husk biochar rate on arbuscular mycorrhizal fungi and growth of soursop (*Annona muricata* L.) seedlings. *Sustainability* **2021**,13,1817. [CrossRef]

30. Tsai, C.C.; Chang, Y.F. Effects of rice husk biochar on carbon release and nutrient availability in three cultivation age of greenhouse soils. *Agronomy* **2020**,10, 990. [CrossRef]

31. Selvarajh, G.; Ch'ng, H.Y.; Md Zain, N.; Sannasi, P.; Mohammad Azmin, S.N.H. Improving soil nitrogen availability and rice growth performance on a tropical acid soil via mixture of rice husk and rice straw biochars. *Appl. Sci.* **2021**,11,108. [CrossRef]

32. Tsai, C.C.; Chang, Y.F Carbon dynamics and fertility in biochar-amended soils with excessive compost application. *Agronomy* **2019**, 9, 511. [CrossRef]

33. Tsai, W.T.; Lee, M.K.; Chang, Y.M. Fast pyrolysis of rice husk: Product yields and compositions. *Bioresour. Technol.* **2007**, 98, 22-28. [CrossRef]

34. Johar, N.; Ahmad, I.; Dufresne, A. Extraction, preparation and characterization of cellulose fibres and nanocrystals from rice husk. *Ind. Crops Prod.* **2012**, 37, 93-99. [CrossRef]

35. Touray, N.; Tsai, W.T.; Chen, H.L.; Liu, S.C. Thermochemical and pore properties of goat-manure-derived biochars prepared from different pyrolysis temperatures. *J. Anal. Appl. Pyrolysis* **2014**,109,116-122. [CrossRef]

36. Tsai, W.T.; Huang, C.N.; Chen, H.R.; Cheng, H.Y. Pyrolytic conversion of horse manure into biochar and its thermochemical and physical properties. *Waste Biomass Valori.* **2015**, 6, 975-981. [CrossRef]

37. Liu, S.C.; Tsai, W.T. Thermochemical characteristics of dairy manure and its derived biochars from a fixed-bed pyrolysis. *Int. J. Green Energy* **2016**,13, 963-968. [CrossRef]

38. Hung, C.Y.; Tsai, W.T.; Chen, J.W.; Lin, Y.Q.; Chang, Y.M. Characterization of biochar prepared from biogas digestate. *Waste Manag.* **2017**, 66, 53-60. [CrossRef]

39. Tsai, C.H.; Tsai, W.T.; Liu, S.C.; Lin, Y.Q. Thermochemical characterization of biochar from cocoa pod husk prepared at low pyrolysis temperature. *Biomass Convers. Biorefin.* **2018**, 8, 237-243. [CrossRef]

40. Tsai, W.T.; Huang, C.P.; Lin, Y.Q. Characterization of biochars produced from dairy manure at high pyrolysis temperatures. *Agronomy* **2019**, 9, 634. [CrossRef]

41. Lowell, S.; Shields, J.E.; Thomas, M.A.; Thommes, M. *Characterization of Porous Solids and Powders: Surface Area, Pore Size and Density;* Springer: Dordrecht, The Netherlands, 2006.

42. Smith, J.M. *Chemical Engineering Kinetics,* 3rd ed.; McGraw-Hill: New York, NY, USA, 1981.

43. Suzuki, M. *Adsorption Engineering*; Elsevier: Amsterdam, The Netherlands, 1990.

44. Masek, O.; Johnston, C.T. Thermal analysis for biochar characterisation. In *Biochar: A Guide to Analytical Methods;* Singh, B., Camps-Arbestain, M., Lehmann, J., Eds.; CRC Press: Boca Raton, FL, USA, 2017; pp. 283-293.

45. Mukome, F.N.D.; Parikh, S.J. Chemical, physical, and surface characterization of biochar. In *Biochar: Production, Characterization, and Applications;* Ok, Y.S., Uchimiya, S.M., Chang, S.X., Bolan, N., Eds.; CRC Press: Boca Raton, FL, USA, 2016; pp. 67-96.

46. Chia, C.H.; Downie, A.; Munroe, P. Characteristics of biochar; Physical and structural properties. In *Biochar for Environmental Management,* 2nd ed.; Lehmann, J., Joseph, S., Eds.; Routledge: New York, NY, USA, 2015; pp. 89-137.

47. Brown, R.A.; Kerche, A.K.; Nguyen, T.H.; Nagle, D.C.; Ball, W.P. Production and characterization of synthetic wood chars for use as surrogates for natural sorbents. *Org. Geochem.* **2006**, 37, 321-333. [CrossRef]

48. Singh, B.; Raven, M.D. X-ray diffraction analysis of biochar. In *Biochar: A Guide to Analytical Methods;* Singh, B., Camps-Arbestain, M., Lehmann, J., Eds.; CRC Press: Boca Raton, FL, USA, 2017; pp. 245-252.

49. Smith, G.C. X-ray photoelectron spectroscopy analysis of biochar. In *Biochar: A Guide to Analytical Methods;* Singh, B., Camps- Arbestain, M., Lehmann, J., Eds.; CRC Press: Boca Raton, FL, USA, 2017; pp. 229-244.

50. Johnston, C.T. Biochar analysis by Fourier-transform infra-red spectroscopy. In *Biochar: A Guide to Analytical Methods;* Singh, B., Camps-Arbestain, M., Lehmann, J., Eds.; CRC Press: Boca Raton, FL, USA, 2017; pp. 199-228.

51. Islam, M.S.; Ang, B.C.; Gharehkhani, S.; Afifi, A.B.M. Adsorption capability of activated carbon synthesized from coconut shell. *Carbon Lett.* **2016**, 20,1-9. [CrossRef]

CHAPTER

4

Agricultural Waste and Wastewater as Feedstock for Bioelectricity Generation Using Microbial Fuel Cells: Recent Advances

1. INTRODUCTION

Today, several challenges are besieging the environment, and as such, an equal mea¬sure to address such challenges must be in place to counter environmental and ecological degradation [1]. For example, in maintaining a clean and safe environment, waste reduction and recovery of valuable products [2] and/or their repurposing is a must [3]. Nutrient-rich agro-waste is usually produced from agro-processing industries [4]. Similarly, one of the major wastes is agro-based wastewater containing many carbon-based compounds [5], which in turn affects the receiving water bodies when released untreated [5]. Agricul¬tural residues are also considered one of the most prominent substrates in energy and carbon source content. Their sugars are obtained either via treatment with dilute acids or enzymes [6].

An example of an important agricultural residue is wheat straw containing about 34-40% cellulose organic carbon content, hemicellulose containing about 21-26% of organic carbon, and lignin-containing about 11-23% of organic carbon. All of these can undergo hydrolysis yielding wheat straw hydrolysate, generating a substantial amount of electricity in a microbial fuel cell (MFC) [7]. Similarly, raw corn stover is another important agricultural waste that contributes immensely to the production of electricity when a single-chamber MFC is used. However, the treatment thereof was effective, but the power output was much lower compared to MFC in which glucose is used as a substrate [8].

Agricultural waste, even in agro-industrial wastewater, is produced during agriculture produce pre-harvesting, harvesting, and processing activities. Agricultural processing activities and industrial food processing operations contribute to agro/food-waste and wastewater generation. Agricultural waste can easily undergo biodegradation as it contains a high level of organic matter and many other different macro-and micro-nutrients suitable for microbial growth. Many agro-industrial wastewaters also contain a high concentration of organic pollutants, including a large amount of waste effluent produced from livestock and agro-products processing [9]. However, these agricultural residues and wastewater can be considered new alternative sources of renewable energy that can be converted into biofuels, biogas, bioelectricity, bio-bricks, fertilizer, and biochar [9] suitable technology such as MFCs, among others.

1.1 The Availability of Various Agricultural Waste

Various types of agro-waste can be found in the environment, which depends upon the source and availability. They can be derived from many different sources such as municipal solid waste works, livestock excrements, lignocellulosic and agro-wastes, food crops, etc. Thus, such waste can be classified into four main generations based on their ability to produce different types of products [6]:

First-generation: This comprises various classes of food crops such as wheat, corn, rice, and sorghum. The direct utilization of these crops as a primary feedstock of interest is often associated with energy generation and the production of various products. However, one of the major challenges associated with this generation is the competition between its utilization in fuel and food production. Fuel production is viewed to be of a higher return on investment than food production.

Second-generation: This generation generally consists of lignocellulosic wastes like sugarcane bagasse, wood chips, crop residues, and organic waste that can be employed to generate bioenergy using different waste beneficiation techniques. This type of waste is associated with the overcoming of major limitations identified with the first-generation biomass.

Third generation: Microalgal biomass, which is used in engineered energy source production systems as a feedstock. Hence, its cultivation can easily be achieved in lagoons and open ponds using a high nitrogenous compound containing agro-waste containing wastewater.

Fourth generation: This type of biomass is from metabolically engineered species such as bacteria, including algae generated from cleaner disposal, or emissions control processes such as CO2 capture systems. This increases the value of this generation as it can be used in high-value product production associated with higher polymeric hydrocarbon content requirements or any other bioenergy products.

1.2 Current Status of Agricultural Wastes

A majority of agro-wastes are usually derived from oilseed crops, wheat, rice, fruits, and vegetable processing. This usually causes several health-related problems to the human population, animals, and the environment, particularly whereby their means of disposal is through landfilling. Even though most of the chosen techniques are cheap and easy to implement, there is a harmful impact on downstream agro-systems and environmental outcomes due to toxic leachate production. This further contributes to adverse climatic conditions as other unwanted gaseous by-products are also produced. Additionally, combustion also generates different types of disturbances to the environment, including smoke that simultaneously causes air pollution by releasing greenhouse gases contributing to global warming. Furthermore, smog is also produced and does mobilize particulate matter into the atmosphere, with the residual as haltering the soil's physical, chemical, and biological structure, including microbial community. Therefore, a necessary major intervention is needed for the sustainable disposal of agro-waste. This can take the form of sustainable energy technology development and generation in renewable energy technology.

1.3 Characteristics of Agricultural Wastes and Wastewaters

Today, agro-wastes are known as the best alternative source for renewable energy production. This depends upon their classification and physicochemical properties, as they consist of different proportions of lignocellulosic hemicellulosic, including lignin in some complex and hardy agro-waste materials. These play an important role in converting the agro-waste to hydrolysate used in bioenergy production processes. The properties conferred to agro-waste depends upon its sources' location, climatic conditions, characteristics, etc. Thus, the physical and chemical properties of such agro-waste are discussed in the following sub-sections. Carbohydrates (2300-3500 mg/L), sugars (0.65-1.18 percent), proteins (0.12-0.15 percent), and starch (65-75 percent) are discharged in starch processing wastewater (SPW), which is an important energy-rich feedstock that may be converted to a wide variety of useful products [10]. Worldwide, the varied structure of lipids, proteins, fibers, excessive organic matter, parasites, meat processing effluents, and veterinary medicines is recognized as hazardous [11]. Due to the broad range of slaughterhouse wastewater (SWW) and pollutant concentrations, SWW is often evaluated using bulk criteria. SWW contains substantial amounts of biological oxygen demand (BOD), chemical oxygen demand (COD), total organic carbon (TOC), total nitrogen (TN), total phosphorus (TP), and total suspended solids (TSS) [12]. Substrates such as cellulose and chitin are readily available and cheap biopolymeric resources that may be used to generate electricity. These green materials also account for a significant percentage of organic compounds in industrial and municipal wastewaters [13]. There have been just a few reports on the use of these particle substrates in MFCs.

1.3.1. Physical Properties

The size of the agro-waste particles is usually irregular in shape and size, with some being needle, leave shaped, etc., with different surface areas. This influences the feeding rate along with fluidizing and mixing parameters during processing and pre-treatment. Additionally, the storage conditions of the agro-waste may also affect these processes. The efficiency of the conversion and energy requirements for these processes is associated with the beneficiation of the agro-waste, which can be affected by the variation in shape and size of the initial agro-waste, and the preceding processes generating it. Another characteristic is the length/diameter of the particulate matter constituting the agro-waste is an aspect ratio tending to unity, even when the finely granulated, i.e., converging to a spherical shape [14]. Such an aspect ratio was determined to be suitable for further processing of the agro-waste in bioenergy generating processes.

Other characteristic considerations include particle and averaged bulk density to determine the grindability of the waste, a known energy-consuming process, which is influenced by quality characteristic parameters of the waste such as moisture content, surface properties, shape, and size [15]. For instance, some agro-waste containing a high lignocellulosic content are difficult to grind due to the presence of fibrous cellulose and lignin. Applying the "Hard grove grindability index method" is usually performed to assess the grindability of such agro-waste. Generally, a particle size of 0.6-1.2 mm for the agro-waste is required to have a suitable grindability index. Similarly, fluid ability must also be considered as it seems to impact the operations associated with waste movement from one point to another during beneficiation and/or processing. Fluid ability is influenced by the biomass particle angle repose, cohesion coefficient, flow index, and compressibility index.

1.3.2. Chemical Properties

Proximate Analysis

Several analyses are performed for proximate analysis. This includes the internal and external amount of water content present in the waste sample, and it is expressed as a weight percent of the agro-waste. It is calculated by subjecting the agro-waste to thermal pre-treatment in a furnace usually operated at >105°C for at least 3 h until a constant weight is reached to get the exact amount of moisture within the initial un-pretreated agro-waste. The ash content can be obtained after the complete combustion of the waste when there is a specific amount of leftover residue in a process operated at >575°C for 3 h. The ash content can be determined by comparing the ash residue to the total amount of the feed agro-waste sample. For volatile matter, except for moisture, it is released when biomass is incinerated at high temperature (950°C for 7 min) in anaerobic conditions. The presence of a high amount of volatile matter indicates a high amount of liquids and gaseous by-products, which can be useful products. The total weight loss of the waste during such a thermal operation is estimated to be equivalent to the amount of volatile matter. After accounting for moisture, ash, and volatile matter, the amount of explosive residue, i.e., fixed carbon, can be determined.

Ultimate Analysis

For ultimate analysis, the total carbon and hydrogen in agro-wastes usually vary from 40-50% (w/w) and 4-6% (w/w). Overall, this analysis involves determining total carbon, hydrogen, nitrogen, and sulphur content in the agro-waste sample. The total oxygen can be calculated by subtracting the total amount of nitrogen, hydrogen, carbon, and sulphur from the known weight of the sample. Hence, this analysis can be carried out using a CHNS analyzer on a dry basis.

Similarly, compositional analysis can be performed using the "Van Soest" method, classified as the National renewable energy or Technical Association of pulp and paper industry method. Most agro-wastes are composed of cellulose, hemicellulose, and lignin; albeit, with a varying degree of composition in different waste samples. The degradation temperature of cellulose is around 240-360°C, leading to the production of liquid products after conversion. Hemicellulose, which surrounds the cellulose, comprises a short and heterogeneous branched chain of polymers. It also links cellulose with lignin. Lignin is the most complex and aromatic compound of higher molecular weight polymer with the crosslink made up of phenolic groups [16]. Other inorganic elements present in the biomass include Na, K, Mg, Cl, etc., and some components such as proteins, resins, gums, etc.

1.4. Pretreatment of Agricultural Wastes

The breakdown of complex molecular structures of agro-waste into simpler monomers is generally considered essential during the pretreatment process. Thus, it contributes to a high output after the conversion process [17]. Different technological approaches can be employed for biomass treatment; this includes physical (grinding, milling), thermal (e.g., steam explosion), biological (e.g., enzymatic), chemical (e.g., use of acids, alkalis) methods, and a combination of treatments such as thermochemical treatments [18]. These methods provide ease of accessibility to enzymes for hydrolysis, increasing the surface area while minimizing operational costs. For instance, the physical treatment of the waste enhances the surface area as it provides easy accessibility for microbial populations and enzymes during hydrolysis. On the other hand, the thermochemical method of pretreatment increases the rate of heat and mass transfer and facilitates the rate of uniform temperature distribution within the agro-waste particles, thus high efficiency for hydrolysate constituents' recovery during liquefaction. Similarly, thermochemical conversion involves two essential methods: drying and torrefaction [19]. The former involves moisture removal from the waste, which increases the efficiency of the process, while the latter involves the thermal treatment of waste at a temperature of 200-300 ° C, where sufficient oxygen is removed from the waste, including water.

For the biological conversion process, the waste can be treated at a temperature of 50-250 °C. This provides an efficient treatment process in terms of pathogen removal and biodegradability. However, in the biological pretreatment method, different types of enzymes and fungi are utilized; hence, it is considered the less energy-consuming method as it can be operated at both milder and economical temperature. However, it seems to be a very slow process as several days are required for the process to be completed. Therefore, various fungi are required for delignification of the agro-waste. It is carried out by inoculating the agro-waste with fungal spores or hydrolysis by a cocktail of enzymes [20]. In essence, ligninolytic enzymes play a role in the hydrolysis of the recalcitrant lignin. Simultaneously, the fungi (white-rot fungi) participate in lignin degradation with minimal holocellulose consumption [21].

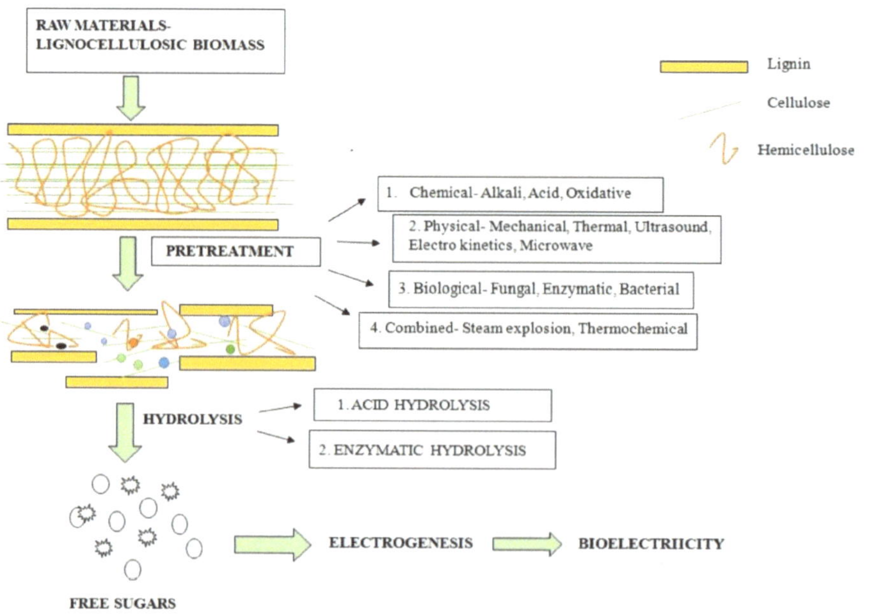

Figure 1. Effect of pretreatment on the biomass component.

The chemical pretreatment method involves using various chemicals such as acids and alkalis that contribute to the breakdown of organic components present in the agrowaste. This pretreatment method will break down the lignin-carbohydrate bond and crystalline cellulosic structure (Figure 1). Examples of the acid used during pretreatment include H_3PO4, H_2SO_4, HNO_3, HCl, etc., and alkalis such as NaOH and KOH. Many researchers have considered the use of liquid-ammonia-water mixture to treat the recalcitrant lignocellulosic constituents in the agro-waste.

1.4.1. Physical Pretreatment

Physical pretreatment techniques requiring mechanical processes such as chipping, milling, and grinding may decrease particle size, break down crystallinity, and increase the degree of polymerization, both of which significantly enhance the biodegradability of biomass in MFCs. Using a fermentation medium containing solid substrate resulted in a low PD attributable to the sluggish hydrolysis of the biodegradable materials, suggesting that particle size is a significant factor for optimum bioenergy production. Additional particle size reduction under 40 mesh has been reported to impact hydrolysis rates and yields, resulting in a significant amount of usable material in the biodegradation phase in MFCs [22]. Furthermore, various irradiation methods (such as ultrasonication, electron beams, X-rays, or gamma rays) may be used to pretreat biomass physically. Shen et al. (2018) studied the effects of ultrasonic pretreatment on electricity production in a dairy manure microbial fuel cell (DMMFC). At 600 W ultrasonic power, the pretreated DMMFC had a maximum PD of 102 mW/m^2, which was 241 percent higher than the untreated substrate [23]. According to Tao et al. (2013), ultrasonication may be an effective pretreatment technique for vegetable or grass wastes [24].

1.4.2. Acid Pretreatment

Of the numerous chemical pretreatment procedures, acid pretreatment is among the most widely utilized. Acid hydrolysis will boost enzymatic hydrolysis performance and increase the energy conversion efficiency of lignocellulosic biomass in MFCs. Concentrated mineral acid (CA), dilute mineral acid (DA), and dicarboxylic acid has been utilized to pretreat agro wastes. CAs like H_2SO_4 (Figure 2) and HCl are especially useful for agro wastes. These acids, however, are acidic, corrosive, and dangerous, necessitating the use of specialized reactors that can withstand corrosion. Meanwhile, lignocellulose hydrolysis of agrowastes displayed a strong reaction rate after pretreatment with dilute sulfuric acid. Initially, high temperature ($T > 160$ °C) and low temperature ($T < 160$ °C) dilute acid hydrolysis pretreatment methods were created [21]. In contrast, a high temperature throughout the DA hydrolysis is ideal for cellulose hydrolysis due to sugar decomposition. In an MFC inoculated with pure-culture, Wang et al. (2017) used diluted sulfuric acid pretreated corn straw as the substrate for direct power production. The maximum PD provided by this MFC was 17.2 ± 0.3 mW/m^2, demonstrating the viability of biomass hydrolysate as a source of power production in MFC. A high PD of 660 mW/m^2 from the hydrolysate with a pure-culture of *Shewanella oneidensis* MR-1 could also be obtained by integrating electrode alteration and electron shuttle attachment [25]. Ionic liquids have also been stated to be beneficial due to their thermal stability, low hydrophobicity, low toxicity, and increased electrochemical stability [26]. Ionic pretreatment of farm straw biomass substantially solubilizes cellulose and may recover 100% of the utilized liquid with high purity under moderate conditions.

Straw biomass is pretreated with 1-ethyl-3-methylimidazolium acetate in an ionic liquid (IL) at 120-140 °C (EmimAC). The materials are then washed with anti-solvent for a certain number of hours, resulting in cellulose regeneration. The cellulose is then separated, the lignin precipitated, and anti-solvent recycle, and IL is developed (Figure 3) [27]. Due to the self-evident intra-structure modifications and the disparity in crystallinity characteristics, the generated cellulose precipitate has a strong enzymatic digestibility compared to the rudimentary cellulose from straw waste [28].

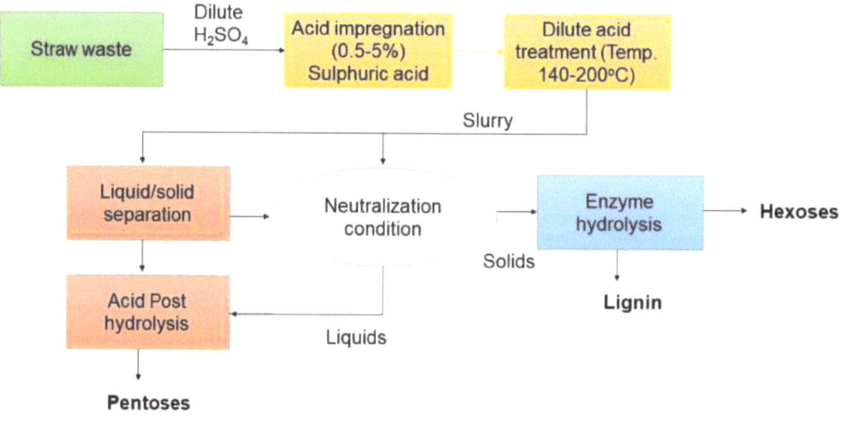

Figure 2. Schematic representation of sulphuric acid pretreatment.

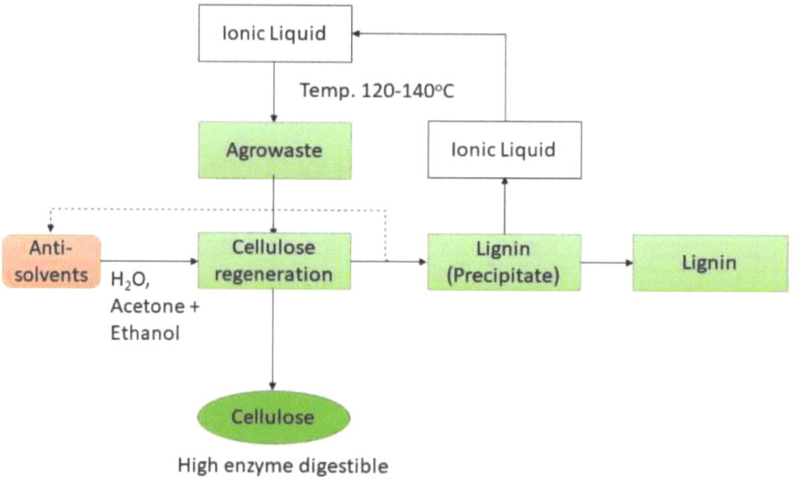

Figure 3. Schematic representation of ionic liquid pretreatment.

1.4.3. Alkali Pretreatment

Basic chemicals such as sodium hydroxide (NaOH), hydrazine, anhydrous ammonia, potassium hydroxide (KOH), or lime ($Ca(OH)_2$) (Figure 4) are used in alkali pretreatment. Even though this process can be used at room temperature, the reaction period is typically long, ranging from hours to days [29]. Song et al. (2018) showed that rice straw could be pretreated with sodium hydroxide (NaOH) for usage in a solid phase microbial fuel cell (SMFC). The SMFC with NaOH (5%) pretreated rice straw could maintain a maximal PD of 140 mW/m^2, which was 3.6 times that of the untreated SMFC [30]. The viability of alkaline pretreatment for sludge-fueled MFC was also verified by Xiao et al., which achieved a PD of 46.82-55.88 mW/m^2 with a quick alkaline procedure using concentrated sodium hydroxide [31].

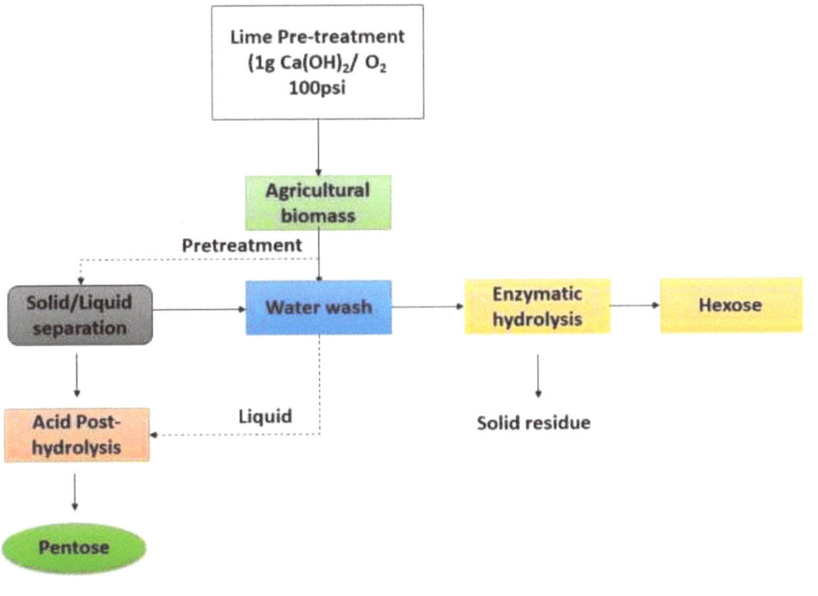

Figure 4. Schematic representation of agricultural biomass pretreatment using lime (Ca(OH^).

1.4.4. Biological Pretreatment

Biological pretreatment is a spectacular accomplishment that encourages the generation of minimal to no hazardous material, an environmentally sustainable procedure with low energy usage and moderate operating conditions. Cellulases generated by bacteria and fungi will hydrolyze and degrade the crystalline structure of lignocellulosic biomass, increasing sugar yields and improving MFC efficiency [32]. *Clostridium, Cellulomonas, Bacillus, Termomonospora, Ruminococcus, Bacteriodes, Erwinia, Acetovibrio, Microbispora,* and *Streptomyces* are among the bacteria that may generate cellulases [33]. The drawbacks of this approach include the need for a longer retention period of 10 to 14 days, close monitoring of growth conditions to prevent contamination, and a significant amount of room for biological pretreatment, both of which render it less economically feasible. Krishnaraj et al. (2015) used a novel three-chamber MFC to produce bioelectricity while simultaneously decaying lignocellulosic biomass (sugarcane bagasse and corncob). In the first compartment of the three-chamber MFCs, *Oscillatoria annae* degraded the LCB. Anodic inoculums of *Oscillatoria annae* and *Gluconobacter roseus* were used to produce electricity in MFCs utilizing decomposed substrates from the first chamber. For sugarcane bagasse and corncob as substrates, the maximum PD was 8.78 W/m^3 and 6.73 W/m^3, respectively [34].

1.5. Route for Conversion

At present, many technologies such as biochemical and thermochemical conversion techniques have been put in place for the proper utilization of agro-waste and beneficiation into valuable products. The biochemical conversion technique employs microbial consortia for the complete degradation of the agro-waste. On the other hand, the thermochemical conversion process usually requires the agro-waste with a minute amount of moisture content, which requires additional energy for drying.

1.5.1. Biochemical Conversion

Today, the biochemical conversion of agro-wastes into energy is a promising and emerging field of technology for sustainable development. Depending on the type and the nature of waste, different microbial consortia can play a crucial part in the conversion processes of such waste for energy generation. Two important processes, i.e., anaerobic digestion (AD) and fermentation, are coupled with biochemical conversion techniques.

AD is usually carried out in an oxygen-free environment where microorganisms help degrade or break down organic waste products into bioenergy. The four (4) main important stages in AD are known. These are hydrolysis, acidogenesis, acetogenesis, and methanogenesis. Each of the stages above is facilitated by different microbial populations that help convert one complex organic material to another. Most microorganisms associated with agro-waste biodegradation in AD processes include *Pseudomonas, Bacillus, Streptococcus, Clostridium, Methanococcus,* and *Methanobacteria* spp. These are mostly employed when handling waste with a high moisture content of about 80-90%. Equation (1) summarizes the stoichiometric relationship between agro-waste biodegradation by microorganisms in AD for biogas production, a renewable energy source.

$$\text{Agricultural waste} + \text{Microorganisms} \rightarrow \text{Biogas} + \text{Digestate} \quad (1)$$

In the fermentation process, which also works in the absence of oxygen, the production of valuable products such as alcohol, organic acids, and a mixture of gases due to microorganisms' action is observed. It is believed that the fermentation of agro-waste is difficult and time-consuming due to the presence of long-chain polymeric molecules and requires acid or enzymatic hydrolysis before fermentation to produce valuable endproducts. For instance, bio-butanol production can be achieved with the help of a bacteria called *Clostridium* spp., coupled with sugar production from various types of agro-waste. The process comprises two main steps, i.e., acidogenesis and solventogenesis, which is referred to as acetone, butanol, and ethanol (ABE) fermentation. Hence, it is a promising technology, but it is costlier and time-consuming.

1.5.2. Thermochemical Conversion

This conversion process consists of pyrolysis, gasification, and combustion. In this process, the treatment of agro-waste into valuable and important products such as biochar, bio-oil, biofuels, etc., usually requires a high temperature. Pyrolysis refers to the thermal depolymerization of agro-waste in an atmosphere with a constant supply of heat. Among the sources of feedstock used for pyrolysis is agro-waste such as rice husk, corn stover, wheat straw, etc., woody biomass (redwood, teak, etc.), and energy crops like bamboo, sorghum, etc., and municipal solid wastes [35]. As a result of the constant and rapid heating of such agro-waste leads to the production of vapor made up of various hydrocarbons coupled with condensation to yield an organic liquid called bio-oil [36]. Moreover, the product obtained from pyrolysis depends primarily upon the composition of the agrowaste used and the interaction between the produced liquefaction products influenced by different parameters such as temperature, heating rate, inert flow rate, and particle size, and conversion time. Due to the influence of these parameters, pyrolysis can be classified as fast, slow, or flash. Pyrolysis is a flash when it operates at a lower temperature, with a lower heating rate and longer vapor formation time. In comparison, fast pyrolysis tends to work at a higher temperature, higher heating rate, and short vapor formation time. Hence, the primary end-products of slow pyrolysis are biochar, bio-oil, and pyrolyzed gas with varying percentages of 35-40% for biochar and 30-35% for pyrolyzed gas. Similarly, flash and fast pyrolysis produce an end-product of biochar (12%) and pyrolyzed gas (13%), while bio-oil is about 75% of the end-product (Equation (2)) [37].

$$\text{Agricultural waste} + \text{heat} + \text{inert} \rightarrow \text{Bio-oil} + \text{Biochar} + \text{pyrolytic gas} \qquad (2)$$

Another vital thermochemical process is the gasification of the agro-waste that works on the principle of a partial oxidative atmosphere at some specific high temperature between 800-1000°C. It employs a similar feedstock (e.g., agricultural waste) to that which is used in pyrolysis but produces an important end-product, i.e., syngas, made up of 85% of carbon monoxide (CO)and hydrogen gas (H_2), with some proportion of tar (5%) and biochar (10%) [38,39]. The gases produced can also be used in a turbine or engines as fuel as they contain a high calorific value. Studies have shown that, for the gasification to work, it depends on two different modes of processing, i.e., fixed-or fluidized-bed processing. Gases with a lower calorific value of 4-6 MJ/NM^3 are seen in the fixed-bed processes. In contrast, fluidized-bed gasification is mostly seen in the provision of uniform temperature distribution, usually in the gasification zone [40]. For combustion, a standardized oxidative- high temperature process is used for the feedstock. As such, it is said to be a heat-based degradation process involving the conversion of chemical energy of biomass to yield heat and power in addition to carbon dioxide and water [41]. The generated energy from combustion can be used in turbines and boilers, among other processes, albeit the moisture content of the waste to be combusted should be below 50%.

1.6. Up-Gradation of End-Products

Recent studies have laid much emphasis on agro-waste-AD systems, in which a mixture of 40-65% of methane (CH_4), 35-55% of carbon dioxide (CO_2), some traces of hydrogen sulphide (H2S), nitrogen gas(N2), H2, water vapor and other components (e.g., volatile hydrocarbons, chlorinated hydrocarbons, etc.) are produced as raw biogas. Similarly, the removal of contaminants, mainly H_2S, CO_2, and water vapor, in addition to some other toxic components from the biogas stream, is termed biogas up-gradation. This is usually performed to obtain a methane-rich gas of >96% CH_4. In the biogas up-gradation process, three main techniques are usually used, i.e., pressure swing adsorption, absorption (physical and chemical), and membrane separation [42]. Pressure swing adsorption is carried out based on the molecular size to adsorb unwanted CO2, H2S, N2, and O2 from the biogas stream, and, as such, 96-98% of pure methane is obtained. The most commonly used adsorbent materials during biogas up-gradation techniques are activated carbon and zeolites. Another technique employed in biogas purification is the physical water scrubbing method based on the increased solubility of CO_2 and H_2S in the water compared to CH_4. Other separation techniques used to remove CO_2 and H_2S from the biogas stream include amine scrubbing, caustic scrubbing, and amino acid salt solution usage. Examples of commonly used amines for chemical absorption are monoethanolamine, aminoethoxy ethanol, etc. On the other hand, the membrane separation technique involves using permeable membranes, which helps in trapping some other biogas constituents. In fermentation, the conventional method used for liquid biofuels up-gradation is distillation, which operates on the principle of the volatile nature of the substances in a mixture. The separation can be carried out primarily based on the less heavy products. Other classes of distillation that can be used for the product up-gradation include extractive, conventional, azeotropic, and molecular distillation. Hence, when an end-product undergoes an up-gradation process using the above processes, it can be utilized efficiently in various technology fields and serve as a promising alternative for renewable energy process development.

2. AGRICULTURAL WASTE USAGE IN MICROBIAL FUEL CELL TECHNOLOGY

The technological approach of MFC in electricity generation fulfills numerous requirements. It allows the recovery of electricity from liquefied agro-waste and the removal of pollutants when wastewater is used. Therefore, MFC refers to the system of bioelectrochemical components that aids in converting organic matter to energy from a large source of complex carbon-based compounds. This has been achieved through the action of microorganisms on waste to produce electrical energy (Figure 5) [43].

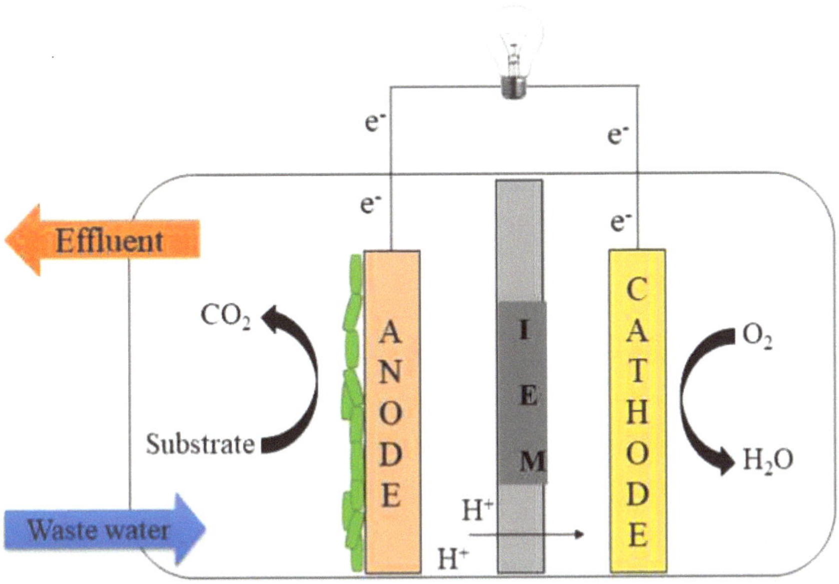

Figure 5. A schematic representation of an experimental set-up of a dual-chambered MFC.

However, within the MFC, the bacteria facilitate oxidation processes that oxidize the organic substrates, leading to the production of electrons transferred by several different enzymes within some essential cells. At the terminal section, electrons are released in the cathode compartment leading to a reduction in oxygen. Today, MFCs have developed two emerging solutions, which are of significance, contributing to environmental concerns' mitigation, i.e., the production of an abundance of pollutant-free and hygienic water while generating the required power at some stage of wastewater treatment in the MFC. In essence, MFC is a promising technology that can achieve the simultaneous production of energy and the treatment of wastewater [6].

The COD removal efficiency was reported in MFC technology for wastewater treatment using such technology being associated with an emergence of a need for renewable energy sources. Although MFC technology still needs further improvement that can make them economically viable and attractive on the international market, it can be another system of organic matter removal from the effluent of different industries. The organic matter removal rate of the MFC compared to the other wastewater treatment systems was estimated, with the result indicating that the removal rate was up to 7 kg COD/m^3-day. In comparison, a range of 0.5-2 kg COD/m^3-day was determined for generic wastewater treatment systems, with studies reporting 8-20 kg COD/m^3-day removal being directly associated with AD [44,45].

The inadequate production of power and current cannot be the sole measure of MFC's practical application and implementation on an industrial scale for electricity generation. For example, in comparison with AD, the gain in electricity generation in using MFCs was seen to be very low with reduced capital investment and operational costs, respectively [46].

In previous studies, the use of MFC has played a vital role in wastewater treatment. It provided various alternatives as a secondary means of energy production and a promising way for technological upscaling in wastewater treatment, particularly whereby agro-waste is to be oxidized [47]. Therefore, the use of wastewater from agro-based industries, in particular, seems to be promising, as such wastewater is constituted by a high content of oxidizable organic matter with its biodegradability, i.e., BOD/COD ratio, being greater than 60% [48-50].

Overall, a typical MFC consists of two chambers, an anode and cathode for oxidationreduction reactions, respectively, with the chambers usually separated by an anion exchange membrane (IEM) (Figure 4). Electrons are usually produced after an anodic oxidation reaction which leads to the production of electric current. In contrast, protons, on the other hand, travel through an IEM as they are utilized for cathodic reduction reactions to generate water [51]. Other studies have shown that the oxidation-reduction reactions from both anodic and cathodic sides from organic matter-containing wastewater using electron acceptors can be attached bio-electrochemically in an MFC [52].

In a typical mediator-less MFC, the extracellular electrons are transferred via electroactive bacteria (EAB). These microbes are dissimilatory metal or sulphate-reducing bacteria. In the presence of an anode, they donate extracellular electrons to the anode to continue anaerobic respiration. These EABs capture electrons released by the oxidized organic matter and transport it directly to the anode. This form of direct electron transfer is further divided into three pathways: Cytochrome mediated, nanowire, and electron shuttle or soluble mediators. In electron transport, cytochrome C (CTC) plays a critical function. It is a heme-containing protein that is found in both archaebacteria and eubacteria. Electricity harvesting is aided by Cytochrome C. CymA, whose N-terminal is connected to the inner membrane. At the same time, the C-terminal is exposed to the periplasm, is a good example of CTC. Because it links the inner membrane to the periplasmic region, CymA is an essential electron route. It is important in anaerobic respiration and interacts with a variety of terminal reductases, including nitrate and fumarate reductases. Microbial nanowires are one of the most recent methods for transporting electrons. These nanowires are the bacterium's pilus, which are electrically conductive and were found by reducing iron oxide using G. *sulfurreducens* bacteria. Other bacteria also have an electrically conducting pilus, indicating the presence of bacterial appendages in the environment. The electron shuttles, also known as electron mediators, are gram-negative bacteria secretions that assist power generation in MFCs. Ideally, these mediators should be soluble, stable, reusable, and environmentally benign, with a redox potential between the bacterial membrane protein and anodic substance. Endogenously generated flavins by *Shewanella* species are a well-known electron shuttle in MFCs. Riboflavin (RF) and flavin mononucleotide (FMN) are the most common, as described in Savla et al. [53]. As previously stated, MFC uses two types of bacteria: mediator-dependent and mediator-independent. *Actinobacillus succinogenes, Proteus mirabilis* and *Pseudomonas fluorescens* are among the bacteria that need mediators, according to the National Institutes of Health. There is a growing interest in bacteria that do not need mediators, such Shewanella putrefaciens [54], Rhodoferaxferrireducens, and D. desulfurcans [55]. Various materials utilized in the MFC components have been illustrated in Table 1.

Liquefied agro-waste is considered one of the most promising substrates for microbial oxidation in the anodic chamber of the MFC. It contains a high amount of carbohydrates, organic matter, and other nutrients [56].

Table 1. Various materials utilized in MFC configuration and construction.

MFC Configuration	MFC vol. (L)	Type of Operation	Anode Material	Cathode Material	Power Output	References
Single chamber	20	Continuous	Activated Carbon	Carbon cloth	0.35-0.9 W/m^3	[57]
	250	Continuous	Carbon brush	Carbon mesh	0.47 W/m^3	[58]

Two chamber	40	Continuous	Carbon cloth		0.44 W/m^3	[59]
	50	Batch	Activated semicoke		43.1 W/m^3	[60]
Stack	10	Continuous	Carbon felt	MEA	6 W/m^3	[61]
	60		Granular graphite		4 W/m^3	[62]
	72		Activated carbon		50.9 W/m^3	[63]
	94		Stainless steel mesh		2 W/m^3	[62]
	200		Carbon brush	Carbon cloth	0.009 W/m^3	[64]
	1000		Activated carbon		7-60 W/m^3	[65]

Microbial Fuel Cells Used in Laboratory Studies for Scale-Up Purposes

Several MFCs have been used in the laboratory for scale-up studies. These include single or dual-chambered cylindrical and cubic MFCs. Similarly, tubular or flat-plate designs have been employed for most scale-up studies. The MFC configuration mostly used in scale-up studies includes a tubular anode surrounded by a separator to isolate the anode from the cathode [66] electrically. Moreover, the MFCs, usually tubular or a product of cylindrical construction materials, can easily be upscaled. Most of the materials that are used as support materials to scale-up reactors include polyvinyl chloride, cylindrical glass, polypropylene, measuring cylinders, cation exchange membranes produced in a tubular shape, and nylon tubing. Most recent studies have shown the mechanism of operation of tubular designs in a continuous flow mode. This resulted in further opportunities for scale-up because of the tube length extension. As such, this culminated in the extension and additional tubular MFC modules to form an MFC stack. However, when considering the flat-plate and tubular designs, multiple MFCs configured this way can only be hydraulically controlled in parallel or series. In another arrangement in series, the effluent flow goes through each of the MFC modules sequentially; on the other hand, each MFC module obtains the same influent when a parallel connection is used. Overall, this means modules of MFCs can be connected in parallel or series to increases voltage and current generation, respectively [53].

3. AGRO-INDUSTRIAL WASTEWATER AS A SUBSTRATE FOR MICROBIAL FUEL CELLS

Generally, agro-food processing waste is comprised of a large number of organic constituents, which can be either in a solid, liquid, or gaseous state. For example, carbohydrates, fat, etc., present in wastewater, indicate the need for the maximum oxygen demand for biodegradation [67]. Various solids in the wastewater are known to halt or reduce the MFCs efficiency. Severe pollution challenges can occur to the environment when there is an absolute lack of proper treatment or management of such agro-industrial wastewater [66,68]. Conventionally, agricultural activities lead to the production of different amounts of food debris and wastes from either man or animal due to various activities derived directly from the human or animal population, which contribute significantly to effluent discharged into receiving streams [69]. The agricultural production streams that usually produce agricultural residues and wastewater require several treatments to avoid water pollution, which enormously varies in pollutant composition and concentration. Some effluent from various agro-industries and their suitability for use in MFCs is discussed in subsequent subsections.

3.1. Palm Oil Mill Effluent

Palm oil is a digestible and high nutritive oil manufactured in some regions globally, mainly for food and energy production in some countries. The palm oil industry usually produces two main products, crude palm oil and solid palm kernels [70]. Moreover, residual waste of different types is usually produced from palm oil-agro industrial processes [71]. Many waste types, including palm oil mill effluent (POME), are normally obtained from different extraction processes. The treatment of such waste can confer advantageous attributes to MFC processes, resulting in high energy generation [72,73]. POME inoculated with an anaerobic sludge has been treated as reported by some researchers with the aid of a simple two-chambered MFC. This process demonstrated a higher power output (Pd_{max}) of 45 mW/m^2 to 304 mW/m^2; albeit, achieving a considerably lower coulombic efficiency (CE) percentage of 0.8% and 45% COD removal [72].

3.2. Mustard Tuber and Molasses Wastewater

Mustard tuber processing is often known to generate a large volume of effluent, with the wastewater being of high strength and salinity. A case study was observed when a two-chambered MFC was employed to treat mustard tuber wastewater, recording a Pd_{max} of about 246 mW/m^2, including 67% and 85% of CE and COD removal [74]. On the other hand, molasses are broadly employed in many research fields and are usually derived from sugarcane mills. It is usually considered a rich source of sugar and minerals such as Ca, including vitamins. It contains a high COD concentration, which varies from 60-100 g/L and thus acts as a major pollutant from the sugarcane processing factories. When the diluted molasses wastewater is used in MFCs, a 62 mW/m^2 power density was recorded, whereas 81% of COD was removed using a mixed inoculum [75]. A generated bioelectricity of 0.18 W/m^2 was recorded from an MFC treating sugarcane molasses whereby *Brevibacillus borstelensis* STRII was used [76]; additionally, there was an increase in power density when the sugarcane molasses concentration was increased. This demonstrates a promising way to manage the substrates in wastewater and energy generation whereby MFC integration in dark fermentation processes can result in positive environmental outcomes [77]. Table 2 summarizes some of the agricultural product processing wastewater used in MFC technology for its treatment and CE.

Table 2. MFC types and performance using different agricultural product processing wastewater.

Wastewater Type	MFC Type	Feeding Mode	Volume (mL)	COD Removal (%)	CE (%)	Reference
Cassava mill wastewater	Two-chamber MFC	Continuous	1500	72	20	[78]
Cereal processing wastewater	Dual-chamber MFC	Batch	310	95	40.5	[79]
Mustard tuber wastewater	Dual-chamber MFC	Batch	150	57.1	67.7	[74]
Olive mill wastewater mixed with domestic wastewater (1:14)	Air-cathode single-chamber MFC	Batch	28	60	29	[80]
Starch extract (potatoes)	Mediator-less two-chamber MFC	Batch	100	61	18.5	[81]
Raw corn stover	Bottle-type air cathode MFC	Batch	250	42 ± 8	3.6	[8]
Rice milling	Earthen pot MFC	Batch	400	(cellulose) 17% ± 7 (hemicellulose)	21	[82]
Steam exploded corn stover	Batch	Batch	250	96.5	1.6	[8]
Rice straw hydrolysate	Air-cathode single-chambered	Batch	220	60 ± 4 (cellulose) 15 ± 4 (hemicellulose)	8.5-17	[83]
Steam exploded corn stover	Single-chambered air-cathode MFC	Batch	28	49-72 60-70	20-30	[84]

3.3. Brewery Wastewater

Brewery wastewater is another high-strength wastewater that can be used in energy generation. It is generated in large volumes during beer production, whereby several processing steps, such as fermentation, saccharification, etc., are undertaken [59]. It has been studied that the MFCs can carry out brewery wastewater treatment [85,86]. A 20 L MFC was tested for one year [59], with the MFC being operated in different modes and/or phases, with phase A producing an external resistance *(Rext)* of 10 ohms and phase B producing R_{ext} of 3.7 ohms, and up to R_{ext} of 5.3 ohms in other phases, with the overall COD removal efficiency of up to 94.6% being achieved; albeit, the energy obtained seemed to be low, i.e., 1.61 mW/m^2 [59].

3.4. Winery Wastewater

Winery wastewater with different compositions was tested using MFCs made up of two single-chamber air cathodes [80]. A wire made up of titanium along with carbon fiber was put in place of generic anodes, while on the other hand, cathodes were made up of platinum-coated carbon cloth of 0.4 mg Pt/cm^2. The white wine wastewater resulted in less promising results with an energy generation capacity of 263 mW/m^2; albeit a significant quantity of COD (90%) and BOD (95%) removal was achieved with a CE of 15%. Comparatively, red wine wastewater had 111 W/m^2, with a recorded 27% COD removal, whereas a maximum BOD removal of 83% was achieved; hence, a lower CE of 9% was observed. These experiments indicated that different wastewater, even from the same industry, can have different usable substrates with differentiated compositional characteristics, which determines the MFC's power output. This was facilitated by diverse and different microbial populations found in the anode; besides, the negative influence on MFC performance was due to a high concentration of polyphenolic compounds in some winery wastewater.

3.5. Other Agricultural Activity Effluents and Waste

Most agro-industrial activities in the food, plant, and animal processing industries generate a high quantity of wastewater. Most agricultural wastewater has been demonstrated as suitable for use in MFCs, i.e., either used in a minute and/or large quantities [87]. One of the recent studies has shown that sugar beet processing wastewater at a concentration of 2.56 g COD/L was able to generate a power of 1.41 W/m^2 which in turn contributed to the complete removal of suspended solids up to 97% of organic matter [87]. However, the hydraulic retention time was up to 40 days. Similarly, up to 93% can be removed from coconut husk retting wastewater containing phenols (potent toxicants) with a concentration of up to 320 mg/m^{3,} with 91% COD removal being achieved in a dual-chamber MFC [88]. Similarly, the use of crude starch extracts from potato processing has shown promising results when used in a dual-chamber mediator less MFC, culminating in 18.5% CE and a COD removal of 61% [81]. Furthermore, olive mill wastewater (OMW) is usually produced during the processing of olives and is considered the most pollutant containing wastewater constituted by up to 100,000 mg/L COD. This type of wastewater is characterized by a strong, intensive black color, acidic pH, a strong odor, and toxicant concentration in the range of 200-800 mg/L in the form of polyphenols [89] while possessing differentiated values of electro-conductivity [90].

It has been reported that an MFC made up of a 12 mL inner volume and a singlechambered air-cathode can be used with diluted OMW in a ratio of 1:10 with OMW acting as a sole source of carbon for bioenergy generation. A COD of 65% was removed for this type of set-up, and a total phenolic content of up to 49% was removed while reaching a maximum voltage generation of 381 mV. In a similar study, an MFC made up of a singlechambered air cathode was used; albeit, with a cathode equipped with platinum-coating (0.5 mg Pt/cm^2) enveloped with a carbon cloth; a configuration which was demonstrated as being efficient for OMW treatment [91]. However, OMW is not a promising substrate for energy production and as such, is not usually considered as an alternative source of substrates in MFC technology development. However, a mixture of any other wastewater combined with OMW generated high-power density results whereby COD was removed with a low yield in CE [56]. This combination was determined to produce a virtuous power output and has shown promising results, particularly for OMW treatment [91].

4. AGRICULTURAL RESIDUES

Agricultural residues are considered one of the most prominent substrates in renewable energy production and carbon source content due to their availability as a cheap renewable energy feedstock.

Moreover, any microbial community cannot directly utilize the agricultural residues in MFC to produce electricity. To generate fermentable sugar hydrolysates easily, either acidic or enzymatic pre-treatments is required [6]. Some of the important agricultural residues that contribute to bioelectricity generation include wheat straw, corn stover, rice straw, cassava mill effluent, plant and flower waste, vegetable waste, etc.

4.1. Wheat Straw and Corn Stover

Wheat straw is a known agricultural residue containing cellulose of about 34-40% of the total organic carbon of the waste. Hemicellulose is 21-26% of the organic carbon, while lignin is 11-23% of the total organic carbon content. Due to hydrolysis, the formation of a hydrolysate rich in carbohydrates can be achieved [7]. Some studies have shown wheat straw as an alternative means of carbon source provisioning in MFCs to generate electricity. The hydrolysate is formed after converting the solid residue into a carbohydrate-rich liquid, which can be used as a substrate in MFC to obtain a maximum power density of up to 123 mW/m^2 when the initial concentration of the substrate was 1 g/L. However, the reported energy output seemed to be on the lower side. Overall, wheat straw showed a high efficiency as a substrate in MFC. However, corn stover, another agricultural residue containing 70% cellulose and hemicellulose, can undergo conversion processes through cellulosic enzymatic treatment or steam explosion into sugar hydrolysates containing a similar profile of sugar content to other agricultural residue hydrolysates obtained [6]. In another experimental setup, the substrate as a hydrolysate from "raw corn stover" employed in the production of electricity in an MFC, generated a considerable amount of low power output, unlike in control MFC whereby glucose was employed [8]. This means an improvement is required in producing a hydrolysate from wheat straw and corn stover that is suitable for use in MFC technology development.

4.2. Rice Straw

Rice straw consists of mainly lignocellulosic biomass with varying compositions of organic carbon. The electricity production can be carried out using this agricultural residue, with its hydrolysate being observed to be suitable to serve as a substrate. Comparatively, industrial wastewater has been recorded to generate maximum energy of 2.3 mW/m^3, while achieving a 96.5% COD reduction; albeit, a pot MFC was used [82]. However, in the case of a carbohydrate-rich hydrolysate from rice straw in which 400 mg COD/L removal was observed, the recorded maximum energy output of 137 mW/m^2 was obtained. Still, when the conductivity of the solution was increased to about 17 mS/cm, about 293 mW/m^2 power density was reported [83]. In a two-chambered MFC to produce electricity without a pretreatment process, the powdered rice was applied directly to the anode side of the MFC in the presence of a mixed culture containing bacteria capable of breaking down the cellulose in the straw, culminating in a 54.3% increase in energy generation [76]. In other studies, the highest generated power was 190 W/m^3 when the utilized substrate underwent no pretreatment process; a mixed culture containing bacteria capable of breaking down cellulose in the MFC was used [92]. In this regard, MFC has shown a promising and convenient channel of treatment of wastewaters containing rice straw for the effective management of the wastewater to minimize pollution in the environment, which will simultaneously generate electricity.

4.3. Cassava Mill Effluents

In the processing of cassava to produce starch, a large number of effluents rich in COD and total solids with high acidic pH, including a minute concentration of cyanide, are released to freshwater streams can be alternatively redirected for energy generation in MFC technology. The starch processing industry effluent is rich in carbohydrates. It usually consists of a high organic content, which ranges from 10-16 g/L, thus making it suitable for use in MFCs. Some studies have demonstrated the feasibility and biodegradability of such effluent in MFCs for the treatment of cassava mill effluent, with a high percentage (88%) of COD being removed, while 1.7 W/m^2 of power was generated [78,93]. An increase in energy recovery of 22.19 W/m^3 in a single-chambered MFC was also recorded after adding a buffer solution for pH correction to within the range 5-9 [68]. Many microbial species such as *Pseudomonas aeruginosa, Bacillus cereus, Bacillus subtilis, Escherichia coli, Saccharomyces cerevisiae, Aspergillus niger, Aspergillus flavus,* and *Rhizopus* sp., were all found to be in the anodic biofilms in the MFC treating cassava mill effluent [94].

4.4. Vegetable Waste

Vegetable waste can be regarded as another promising substrate that can generate energy from various MFC designs. It is usually generated during the washing and cutting of vegetables from various vegetable markets, restaurants, and some vegetable packaging industries. The electrogenic population in vegetable waste MFC tended to utilize the slurry form of the waste better during hydrolysis. In other studies, when the proportion of cooked and uncooked potato substrate was used in an MFC, increasing the coulombic yield culminated in 86.3% of COD removal (Du 2017; Du et al., 2018). An average current density of about 72.2-100.2 mA/m^2 was recorded, and 15.6-17.3% COD removal was achieved using vegetable waste containing effluent in combination with MFC; however, a diverse microbial consortium was needed, with some organisms such as *Firmicutes, Proteobacteria,* and *Geobacter* sp. proliferating in the anodic solution of the MFC. These organisms were the most dominant when using potato wastewater as a substrate, conferring the characteristics of suitable electrogens for electron transfer (Du, 2017). Moreover, a U-shaped MFC generated a current density of 314 mA/m^2 at a resistance of 123 ohms when a vegetable waste extract was applied as a substrate, demonstrating a higher power density output than the dual-chambered MFCs [95].

4.5. Fruit Waste

To date, the biodegradation of fruit waste effluent is a challenge due to monosaccharides, disaccharides, and polysaccharides which can facilitate an exponential proliferation of disease-causing organisms when such effluent is release into rivers untreated. As observed indifferent MFCs configurations, a proportion of fruit wastes can generate about 330 mV during conversion, as observed in some biotransformation of effluent from fruit processing (Table 2) [96]. In another study, a voltage of 0.563 V and 0.492 V in MFC was generated when an orange and banana peel effluent was used with no chemical pretreatment. The residual total reducible sugars were a source of carbon for the microbial consortium [97]. Different fruit processing effluent containing residues and soluble components from orange, lemon, grape, and mixed fruit processing were observed for their performance in MFCs to assess the generation of power output compared to conventional MFCs. Improved performance resulting from highly fermentable carbohydrates was observed; albeit, the concentration of organic acids such as citric acid from lemon fruit processing, might have been detrimental for the MFC performance [98]. Lemon processing effluent was considered a source of energy in which the electrogenic population in the dual-chambered MFC led to an electron recovery of 0.99 A/m^2 with 32.3% CE [99]. Therefore, fruit waste and peel extract containing effluent can be considered an alternative source of energy-rich support for electrochemical oxidation in MFCs and possibly can invigorate an emerging renewable energy technology development field.

4.6. Plant and Yard Waste

Plant and yard waste residue contain a relatively high concentration of cellulose, hemicellulose, and lignin. This can undergo hydrolysis by providing a pretreatment step to make it easily biodegradable during microbial oxidation in MFCs. When a hydrolysate was generated from plant and yard waste, an energy output of 1.02 W/m^2 with COD removal efficiency of 76% and CE of 69% was recorded using an air cathode MFC [100]. Some aquatic plants have been demonstrated to have similar attributes to those observed for generic plant and yard waste hydrolysates with *Canna indica* (Canna)—rich in cellulose and hemicellulose. Lignin is observed to be suitable to generate a hydrolysate with a consortium maintenance capability. For the use of hydrolysates from these plant- and yard-based hydrolysates using an air cathode MFC, about 0.45 W/m^3 of volumetric power density can be generated [101].

5. TREATMENT OF ANIMAL DEBRIS WASTE AND WASTEWATER IN MICROBIAL FUEL CELLS

5.1 Slaughterhouse and Animal Debris Containing Waste

Generally, slaughterhouses and animal manure are usually derived from the livestock industry, which generates a large amount of wastewater containing suspended solids and high organic matter content. The release of such wastewater into municipal wastewater treatment works can cause major environmental odor problems if released untreated. The wastewater produced from slaughterhouses consists of different substances that the action of microorganisms can break down. The wastewater also has many suspended nutrients such as proteins, carbohydrates, minerals, and fat, all of which are also present in animal blood [102]. Wastewaters derived from abattoirs are usually discharged in several different channels due to a lack of monitoring for such discharges.

Previously for bioelectricity generation, both the slaughterhouse and animal carcass cleaning wastewater were employed in MFC technology, with a generated power of 578 mW/m^2 being recorded [103]. Similarly, the generation of bioelectricity using animal debris containing wastewater as a substrate achieved a maximum power density of 2.19 W/m^3 in an up-flow tubular MFC made up of an air-cathode, recording a COD removal of 50.66%. In contrast, a low CE (0.25%) was recorded elsewhere [104].

5.2. Livestock Compost Wastewater

Livestock compost wastewater is also one of the effluents most produced in the livestock industry. Livestock compost is described as an important source of some organic and inorganic components [105]. Some of these components can be easily broken down into simpler molecules, which in turn can provide a source of easily fermentable constituents for consortium support in MFCs [106]. Overall, a complex organic substrate may assist in the propagation of different species of microorganisms. Generally, exoelectrogenic bacteria possess a limited ability to utilize complex substrates. Many different microbial populations are needed for the wastewater to undergo the required oxidation processes and with microbial species undergoing directed evolution to decompose semi-biodegradable carbon-based compounds [107]. For example, when wastewater treatment was carried out in an air-cathode MFC using cattle manure sludge as a substrate with and without any mediators, increases in power density up to 200% were observed when methylene blue was used as a mediator [108]. In another study, a maximum power density of 16.3 W/m^3 was recorded using suspended cattle manure as a substrate. This was achieved when a cassette-electrode MFC configuration operating in a batch mode was used, with 41.9% COD removal being reported in the first ten days of the MFC operation [106]. Some studies have demonstrated that using a small number of substances derived from livestock waste via fractionation in combination with compost wastewater as a substrate showed promising results when fed into an MFC. Generally, 67-215 mW/m^2 power density was generated by livestock waste and compost wastewater which was greater than when supplied as a liquefied feedstock in MFCs. When livestock compost was employed and the substrate is halted in MFC, only 15.1 W/m^3 of power density was produced [109].

5.3. Swine Wastewater

A greater emphasis on the use of swine wastewater treatment is currently being advocated for, with some studies showing that a 110 L capacity MFC can achieve a maximum of 5 kg COD/m^3.day reduction, representing a 65% efficiency in terms of COD removal while generating 110 Wh/kg COD net energy [110]. In another case study, about 85.6%, 70.2% and 93.9% of ammonium nitrogen, total nitrogen, and total organic compounds were removed using *Chlorella vulgaris* algal-biocathode photosynthetic MFC, achieving a maximum power density of 3.7 W/m^3 with carbon dioxide sequestration. Others reported swine wastewater treatment while generating a maximum power density of 45 mW/m^2 in MFC made up of two-chambered aqueous cathodes [111]. In a further experiment of similar wastewater, a maximum power density of 261 mW/m^2 was generated in an MFC made up of a single-chambered air-cathode [112], whereas 382 mW/m^2 was recorded elsewhere [113].

5.4. Poultry Slaughterhouse Wastewater

Disposed poultry slaughterhouse waste and effluents containing excreta from birds, feed, feathers, hatchery waste, urine, feces, sawdust, etc., were determined to be suitable to generate an engineered biofilm in the anode of an MFC. The biofilm was constituted by *Escherichia coli, Enterobacter, Citrobacter, Geobacter, Klebsiella, Lactobacillus,* and *Pseudomonas* spp. [114]. Using MFC with rice husk charcoal as an electrode in combination with effluent from poultry slaughterhouses generated a volumetric power output of 6.9 W/m^3 while achieving a 40% dissolved organic carbon reduction [115]. Similarly, an energy generation harvesting rate of 278 mW/m^2 was observed with an effective 82% BOD removal efficiency in a continuous horizontal flow MFC [116].

5.5. Dairy Industry Wastewater

The most prominent characterization of dairy industrial wastewater is associated with its unique constituents attributed to differentially complex organics, including proteins, lipids, and polysaccharides. The hydrolysis of such wastewater can transform the wastewater components into organic acids, fatty acids, and sugars, respectively. The properties attributed to dairy industrial wastewater were seen as effective and efficient in an anolyte in MFC [117]. However, another important product rich in nutritional constituents from the dairy industry is cheese whey (CW) classified as milk casein obtained after the separation of milk constituents; hence, the diary industrial wastewater containing CW treatment using MFC was evaluated by many researchers and reported in several investigations regarding the bioelectrochemical recovery of electricity from such MFC operations. CW contains high organic carbon-based compounds that can be broken down into simpler constituents that are readily available to microorganisms [118]. Results indicated electron transfer variability using different materials in MFC with different designs, i.e., single, dual, and tubular chambered MFCs, and different anodic materials, e.g., carbon graphite, stainless steel, composites, etc. The highest CE of 37.2% was recorded using a catalyst-free and mediator-less MFC treating wastewater from the dairy industry [119]. The electrical performances of the MFC increased with an increase in organic matter loading rates (OLRs) [120]; albeit, it was noticed that a high acolyte's COD concentration of up to 2800 mg/L could lead to a reduction in electrical energy, and the flow rate of substances in the MFC IEM may be lowered. Overall, CW containing wastewater has shown a promising result with an MFC made up of an H-type-two-chambered system connected to a carbon paper anode and a platinum-coated (0.5 mg/cm^2) cathode, achieving the highest energy generation of up to18.4 mW/m² with 94% of COD removal being recorded with the said MFC; albeit, operated in a fed-batch mode [121]. A CE of 11.3% was also reported, further showing CW containing wastewater as a promising substrate in MFCs [122]. Another system of an MFC operated in a four-fed batch mode using a cylindrical cathode made from carbon brushes, and carbon powder was determined to serve as an example of a suitable electrode and catalyst configuration for dairy wastewater treatment in MFCs. A comparative account of the substrate used in MFCs and their performance is given in Table 3.

Table 3. MFC efficiency is based on various substrates.

Substrate	MFC Configuration	Volume (mL)	Power Density	CE (%)	COD (%)	References
For types of food industry wastewaters						
Brewery wastewater diluted with domestic wastewater	Single chambered MFC	100	30 mW/m²	--	90.4	[123]
Dairy wastewater	Single chambered MFC	480	1.1 W/m3 (~36 mW/m²)	7.5	95.49	[120]
Dairy wastewater	Annular single chamber MFC	90	20.2 W/m²	26.87	91	[124]
Dairy wastewater	Dual Chambered MFC	300	161 mW/m²	NA	90	[125]
Cheese whey	Dual chambered Tubular MFC	500	1.3 ± 0.5 W/m²	3.9 ± 1.7	59.0 ± 9.3	[126]
Chocolate industry wastewater	Dual Chambered MFC	400	1500 mW/m²	--	74.77	[127]
Molasses wastewater	Single chambered cuboid MFC	650	1410 mW/m²	-1	53,2	[128]
Distillery wastewater (Molasses based)	Single chambered MFC	400	124.35 mW/m²	Ft	72.84	[129]

Substrate	MFC Configuration	Volume (mL)	Power Density	CE (%)	COD (%)	References
Molasses wastewater mixedwith sewage	Single chambered MFC	800	382 mW/m^2	—	59	[130]
Palm oil mill effluent	Cylindrical MFC	2360	41.8 mW/m^2	—	-60	[72,73]
			44.6 mW/m^2	—	-90	
Vegetable waste	Single chambered MFC	400	57.38 mW/m^2	—	62.86	[131]
Fermented vegetable waste	Single chambered MFC	400	111.76 mW/m^2	—	80	[132]
Cereal-processing wastewater	Dual Chambered MFC	310	81 ± 7 mW/m^2	40.5	95	[79]
For types of agricultural wastes						
Dairy cow waste slurry	Air cathode Double chamber MFC	—	0.34 mW/m^2	0.22	84 (BOD)	[133]
Manure	Air cathode single chamber	—	67 mW/m^2	1.3-5.2	—	[134]
Manure wash water	Air cathode single chamber	—	215 mW/m^2	—	—	[134]
Soil organic matter	Solid-phase Soil MFC	—	0.72 mW/m^2	—	—	[135]
Bean residue, ground coffee waste and rice hull	Solid-phase Compost MFC	—	264 mW/m^2	—	—	[136]
Powdered rice straw	H type MFC	—	145 mW/m^2	54.3% to 45.3%	—	[137]
Cattle manure slurry	Air cathode Cassette-electrode microbial fuel cell	—	765 mW/m^2	28.8	41.9-56.7	[106]

Substrate	MFC Configuration	Volume (mL)	Power Density	CE (%)	COD (%)	References
Cow manure	Single chamber Compost MFC (Pt in cathode)	--	349 ± 39 mW/m^2	--	~50 (carbon)	[138]
Wheat straw hydrolysate	H-type double chamber MFC	--	123 mW/m^2	15.5-37.1	--	[139]
Diluted wheat straw hydrolysate	Double chamber MFC	--	148 mW/m^2	17±2	95% (xylan and glucan)	[140]
Steam exploded corn stover hydrolysate	Air cathode Single chamber MFC (Pt/C cathode)	--	371 ± 13 mW/m^2 (neutral) 367 ± 13mW/m^2 (acid)	20-30	± 2 (Neutral pH) ± 1 (Acidic pH)	[84]

6. COMPARISON OF RELATED WORKS

In comparison to related studies, the quantity of various important factors, including CE, COD removal, and maximum energy generated, was recorded using different agricultural residues, agro-industrial wastewater, and other by-products generated from the agricultural industry. These wastes can be used as substrates in MFC technology development. Among the agricultural waste that contributes to the highest power generation in MFC includes wheat straw effluent, rice straw hydrolysate (without pretreatment), and corn stover along with the application of glucose as a substrate; although, the percentage rate of COD removal was seen to be very effective in MFCs when rice straw was used. Similarly, the use of vegetable waste extract produces a high-power output in U-shaped MFCs than dual-chambered MFCs. Overall, some convincing results were obtained regarding power generation using slaughterhouse and animal carcass debris containing wastewater. In particular, cattle manure and manure wash wastewater were considered good substrates for bioelectricity production, with a high percentage of COD removal achieved. Other substrates such as disposed of poultry waste and swine wastewater can be used in MFCs, indicating a 65% COD removal, with plant, fruit, and cassava processing effluent achieving COD removal of up to 88%; although with low power production. A substantial amount of power density was observed with swine wastewater using an MFC made up of an air cathode in a single-chambered cell than in an algal bio-cathode in a photosynthetic MFC. Considering chicken manure treatment, 82% of BOD can be successfully removed with high energy production using a horizontal flow continuous MFC. Similarly, the treatment of mustard tuber wastewater in the dual-MFC has shown a good result with high energy recovery and a high percentage of COD removal. Other agro-industrial wastewaters containing agricultural activity by-products showed a poor performance in terms of power generation; this includes POME, brewery, and dairy wastewater, though it has proven that a high percentage of COD can be removed from these wastewaters with a moderate quantity of CE. For winery wastewater, only white wine wastewater with the less organic matter has shown a good result during power production with a high reduction of BOD and CE. Thus, the application of agricultural waste and its effluents to generate bioelectricity was demonstrated, with some adequate energy recovery. This can thus be considered as an alternative source of renewable energy technology, supported by different microbial communities largely found in the anodic solution of most MFCs; as these types of microbial communities confer characteristics of electrogens for efficient electron transfer, most especially to support redox reactions; therefore, this ultimately characterize the ability of these organisms to support AD.

It has further been shown that sugar beet processing wastewater at a concentration of 256 g COD/L could generate a power of 1.41 W/m^2. Even for coconut husk retting wastewater containing phenol, 91% of COD was successfully removed when employed a dual-chamber MFC. Similarly, the use of crude starch extract from potato processing wastewater has shown promising results when used in a dual-chambered MFC, with a substantial amount of energy recovered. In another comparative analysis, it has been observed that OMW is not usually considered a promising substrate for MFCs compared to sugar beet processing wastewater and coconut husk retting wastewater in terms of energy production. Thus, it is not usually considered an alternative substrate source in MFC technology unless combined with another substrate source. Conclusively, when various comparative analyses of related MFC-substrate studies were conducted, it is clear that substrates such as wheat straw effluent, rice straw hydrolysate, and corn stover are good substrate sources that can be utilized in MFCs for energy generation. Others include slaughterhouse and animal carcass debris containing wastewater, which provides a large quantity of energy. In contrast, substrates such as POME, brewery, plant and yard waste, and dairy wastewaters showed a poor performance in energy recovery using MFCs. To this end, other studies have shown better applicability of some substrates such as swine wastewater, livestock compost wastewater, fruit waste, vegetable waste, and cassava mill effluent to facilitate the generation of electricity using MFCs. However, all these studies have elucidated the fundamental MFC design approach in increasing power generation quantity.

However, among the physical factors affecting the performance of MFC include the type of electrode materials used (graphite rod, graphite fiber brush, carbon cloth, carbon mesh, carbon paper), the surface area of the electrode, and electrode-spacing, and characteristics of the catholyte. On the other hand, biological factors are considered as another key component that governs the overall MFC performance, which includes biocatalyst (mixed culture, monoculture) proliferation and activity, including their biofilm-forming ability and the complex organic matter degradation efficiency, whereas the operational factors affecting the working principle of MFC in terms of power generation include pH conditions, the nature and the type of anolyte and load configuration.

7. FACTORS AFFECTING THE PERFORMANCE OF MFC UTILIZING FOOD WASTE

7.1. pH

For ideal microbial growth, MFCs are usually controlled at pH nearby neutral environments. However, due to reduced ionic concentration at neutral pH, the internal resistance of MFCs is strong in comparison to chemical fuel cells that are using alkaline or acid as electrolytes. Unintended pH shift reduces the power generating potential of MFCs. Ghangrekar et al. [141] analyzed the pH-change effect on the overall efficiency of a two-chambered MFC. When the pH gap between the two chambers was high, they measured optimum current and voltage. Cathode alkalization and anode acidification have been documented to affect the efficiency of MFCs [142]. During the short- or long-term activity, the pH gradient is created at the membrane. Because the electrons aggregate at the anode, an equal amount of H^+ is released into the electrolyte and eventually travels into the cathode, where they are absorbed in cathodic reactions. However, the pH of the anodic compartment reduces because of inadequate or slow migration and diffusion of H+ via the membrane. On the other side, as a consequence of proton intake, the pH of the cathodic compartment decreases for the oxygen reduction reaction (ORR). The presence of H+ is the main element in evaluating the ORR efficiency of electrochemical water splitting devices [143] in the cathode chamber. In the anodic container, the performance of electron transfer and the function of neutrophilic biofilm microbes are decreased if pH is dropped too suddenly. Although alkaline pH decreases power production in the cathode chamber markedly. Zhang et al. [144] analyzed the role of initial pH on the anodic bacteria, biofilm, and MFC's efficiency in power generation. At acidic conditions, they achieved voltage output of 232-284 mV vs. 311-339 mV along with a power density of 95-116 mW/m^2 vs. 182-237 mW/m^2. Reduced and cracked biofilm at pH 5. Around pH 4, the MFCs were unable to obtain the optimum power around neutral pH. The findings indicated that the power supply corresponds to the output voltage and time-speed pH variance of the cathodic and anodic chambers of the MFCs. MFC's poor performance at pH 4 remained for a long time and could be irreversible; therefore, low pH conditions in MFCs should be avoided.

7.2. Substrate Concentration

The impact of substrate concentration on electricity output was explored by the dual chamber MFC (DCMFC) to treat domestic wastewater [145]. The performance of DCMFC in the removal of COD was analyzed at various organic loading levels varying from 435 to 870 mg COD/L-d. It can be said that the COD removal efficiency is greater than 90% as the organic loading rate rises from 435 to 720 mg COD/L-d. In contrast, the COD removal efficiency declined to about 70% at a lower loading rate (870 mg COD/L-d).

Various performance evaluation studies about pH and substrate concentration have been illustrated in Table 4.

Table 4. Effect of pH and substrate concentration on MFC performance.

Sr. No.	MFC Configuration	Substrate	Substrate Conc. (COD mg/L)	pH	COD (%)	Power Density	Ref.
1.	Dual Chambered MFC	Dairy wastewater	1600	7	91	2.7 mW/m^2	[125]
2.		Food waste leachate	39,048	6.3-7.6	84.5	5.591 mW/m^2	[146]
3.		Wastewater	1587	6.3	41	461 mW/m^2	[147]
4.	Single Chambered MFC		1000	9.5-11.50	91	20.2 mW/m^2	[124]
5.		Lactate		8	80	4.8 mW/m^2	[148]
6.	Single Chambered tubular MFC	Fruit and Vegetable slurry	48,320	3.0 ± 0.5	45	55 mW/m^2	[149]
7.	Sediment MFC	Aquaculture wastewater	170-185	8.5	96	4.52 mW/m^2	[150]

7.3. Temperature

The temperature influences the efficacy of MFCs because it affects ORR catalyzed by Pt on the cathode, bacterial kinetics, and the rate of mass transfer of protons through the liquid. MFC experiments are often conducted at about room temperature or somewhat higher (20-35°C). At low temperatures in the range of 4-30°C, MFC functionality requires a longer starting time to provide consistent power cycles and performance. MFCs could not generate significant electricity at temperatures below 15°C, even after a month of operation [151]. Researchers are now focusing their efforts on creating effective MFCs based on thermophilic bacteria as it has an advantage over agri-waste as it also promotes pre-treatment. Thermophilic microorganisms have a high rate of metabolic processes and electron generation, which may be advantageous for their use in MFCs operating at elevated temperatures [152]. Choi et al. (2004) constructed an efficient MFC using thermophilic microorganisms. The authors utilized thermophilic bacteria *(Bacillus licheniformis* and *Bacillus thermoglucosidasius)* to investigate various operational parameters in the MFC system, including redox mediators, temperature, pH, and carbon sources. The authors stated that they produced a significant quantity of power via the use of a redox mediator. Maximum performance was found at 50°C, and cell productivity remained constant at this temperature [153].

7.4. Salinity

Although approximately 5% of the earth's wastewaters are extremely saline, MFC may be more helpful in treating these wastewaters. Increased salinity improves power generation through increased conductivity. Increased conductivity promotes proton transport and therefore reduces the system's internal resistance. Lefebvre et al. demonstrated that adding up to 20 g/L of NaCl improved the cumulative efficiency of MFCs by decreasing internal resistance by 33% and boosting maximum power output by 30% [154].

8. STRATEGY TO ENHANCE THE EFFICIENCY OF MFC PERFORMANCE

The recent studies conducted by Nadafpour have revealed that to upsurge the current in MFCs that have unique characteristics, including strong electric conductivity stability in microbial cultures. In addition, vast surface area and oxidizing agents such as potassium permanganate have a great ability along with anode that is made up of carbon-containing material such as graphite rod carbon paper carbon cloth; graphite fiber brush carbon cloth reticulated vitreous carbon, and carbon flesh [155]. Along with increasing surface area, Nanoengineering material is being used as anode material instead of conventional material, improving the electronic transfer mechanism [44,45]. As well to improve the output electrical power conductive polymer along with modified carbon and metal-based anode are being used in other suitable matter where during operation of MFC system charged balanced must be maintained for unhindered migration of H, OH ions and attention must be paid to electrode stability [156] and at the same time between electrode compartment any kind of diffusion should be avoided, but significant losses in the performance of the bioelectrical microbial system as always happening due to the crossover process. A study conducted by Miyake et al. (2003) has shown that by using functionalized hydrocarbon polymer in polymer electrolyte fuel cell as proton conductive material an increase in conductivity of fuel cell under the humid and heated condition it is seen that long term stability and higher conductivity then 0.01 cm has been provided by the MFC system along with impermeably to hydrogen methanol and oxygen [157]

In a study conducted by Li et al. (2016) has shown that the characteristic of the substrate in food waste after MFC treatment to perceive information about how the organic material was biodegraded and transform during MFC treatment and the aromatic compound in the hydrophilic fraction in comparison to non-aromatic compound such as aliphatic compound tryptophan were far preferably removed along with average output voltage of 0.51 V and maximum power density of 5.6 W per meter cube was achieved [158]. For the power generation and routine electrical purposes, MFC is not the economical--method; it is incapable of producing as much electricity as is required, nothing less electric current merely [159] and the very first fuel cell ad produces 1/40 mW/m^2 energy, in addition, a mixed bacterial culture having carbon sources as glucose has been reported to produce power up to 3.6 W/m^2 microbial fuel cell, which is the higher power output of about 5 fold then the very first fuel cell.

For practical application, it's crucial to use cost-effective material for building the system. There has been ample focus to make high surface and low-cost electrode materials for high-performing systems. The surface area of an anode directly impacts power generation. Higher the surface area of anode could lead large accessible surface area for biofilms results in higher charge generation in the system.

9. TECHNO-ECONOMIC EVALUATION OF MICROBIAL FUEL CELL TECHNOLOGY

Trapero et al. year evaluated the techno-economic status of MFC utilizing juice industry wastewater in an aerobic system [160] through modifications conducted in the parameters like utilization of a dual-chambered reactor with carbon cloth as an anode along with the two types of cathodes; Pt coated carbon cloth and non-Pt coated carbon cloth, both in comparison to the conventional process utilizing activated sludge. This configuration has an effluent flow rate of 54 m^3/day with a COD of 15,000 mg/L. The removal of COD ranges from 40-90%, along with the Coulombic efficiency from 2-30% chosen for the techno- economic assessment relying mostly on the power efficiency and wastewater treatment. The initial investment in consideration of MFCs, including electrodes, DC/AC converter, membranes, pumps, and the fan, is much greater than the conventional treatment plant, requiring just pump fans and a biological tank. Various other investments need to be estimated, such as the costs for operating the treatment plant of 100 m^3

of the volume, including the labor cost of 35%, which is around EUR 3248/year, 19% of the investment for management of the sludge costing EUR 1763/year along with 34% of the investment costing EUR 3155/year for electricity, indicating that most of the investment will be carried out for providing labor and electricity for the plant. In comparison to MFCs, the overall cost of the investment can be reduced because of the automation and no necessity for aeration at the wastewater treatment plant. Based on this estimation, the overall operating cost of MFC ranges around EUR 1700-2300/year, which is very low compared to the conventional process. However, this estimation can be considered only if there is no requirement of replacing the electrodes or membranes. Thus, the construction of highly durable MFC parts is essential for making the process economically viable. Therefore, it is a practical implementation because the capital cost is high compared to the operating cost, which is directly in contrast to the conventional treatment system. In a general scenario, at 30% Coulombic Efficiency, there is 90% efficient COD removal creates a relatively better cash flow than the conventional process (EUR 2600-3400/year) [161].

10. MFC COMMERCIALIZATION

MFC is a well-established and contemplated technology and provides various functional benefits compared to the technologies used to generate energy from organic chemicals [162,163]. It has been exhibited that any compound, which can be used by microorganisms, transformed into electricity using microorganisms [164]. MFCs offer many alluring attractions, e.g., (a) direct conversion of chemical energy to electricity which prompts high transformation effectiveness; (b) the fuel to electricity transformation by MFCs is not constrained by the Carnot cycle since it does not include the change of energy into heat, rather straightforwardly into electricity and, hypothetically it is possible to accomplish higher transformation proficiency (70%)" (c) MFC operate at ambient temperature, due to involvement of microbes as a catalyst; (d) MFCs generate sustainable electricity; and (e) calm and safe execution of performance [165] and (f) no off-gas treatment is necessary because MFC usually generate carbon dioxide which has no useful content of energy [166]. It is hypothesized that MFCs can generate about half the power needed for a conventional treatment process involving aeration of the activated sludge [167]. Even if MFCs are still holding on to be completely commercialized, they are not confined to the laboratories alone. MFCs have ventured into a few smaller-scale applications, which mostly require long-haul, sustainable low power supply, viz. for sensors for small electronic devices, cell phones, robots, and urinals. From the industrial application point of view, several start-up companies have been already established and are trying to commercialize it as illustrated in Table 5. Such recent progress in MFC design highlighted the optimization and economic efficiency of operating conditions. However, practical MFC systems must be demonstrated in a step towards marketing, but they can present new challenges and limitations that must be tackled systematically in the years to come. Table 6 comprises various patents involving lignocellulosic biomass as substrates in MFC for the production of value-added products.

Table 5. Various MFC commercialization.

Sr. No	Head Quarter	Company Name	Website/ Information Link	Foundation Year	Services	Specific Product
1.	USA	Cambrian Innovation Inc.,	https://www.cambrianinnovation.com/ (accessed date—13 August 2021)	2006	wastewater treatment technology	EcoVolt; EcoVolt MBR
2.	Israel	Fluence Corporation Limited (earlier Emefcy)	https://www.fluencecorp.com/emefcy-and-rwl-water-merge-to-create-fluence/ (accessed date—13 August 2021)	2008	wastewater treatment	electrogenic bioreactors (EBR)
3.	USA	Zigco LLC,	http://www.zigcollc.com/ (accessed date—13 August 2021)		soil powered battery	-
4.		Magical microbes	https: //www.magicalmicrobes.com/ (accessed date—13 August 2021)	2010	educational kit	MudWatt; MudWatt Core Kit; MudWatt DeepDig Kit
5.	Canada	Pron-gineer	http://prongineer.com/ (accessed date—13 August 2021)		water and wastewater treatment technology	-
6.		CASCADE Clean Energy, Inc.	http://www.ccleanenergy.com/ (accessed date—13 August 2021)		clean energy production	Wastewater Works (WWW)

Table 6. Various Patents associated with MFC utilizing Agricultural biomass as substrate.

Patent No.	Description	Reference
US9716287B2	A fuel cell with an anode electrode, a cathode electrode, and a reference electrode that are all electronically connected to each other; a first biocatalyst with a consolidated bioprocessing organism; and a second biocatalyst with a consolidated bioprocessing organism. (e.g., a Cellulomonasor clostridium or related strains, like Cellulomonas uda (C. uda), C. lentocellum, A. cellolulyticus, C. cellobioparum, alcohol-tolerant C. cellobioparum, alcohol-tolerant C. uda, Clostridium cellobioparum (C. cellobioparum) and combinations thereof) capable of fermenting biomass (e.g., cellulosic biomass or glycerin-containing biomass) to produce a fermentation byproduct; and a second biocatalyst comprising an electricigen (e.g., Geobacter sulfurreducens) suitable for transferring nearly all the electrons in the fermentation byproduct (e.g., hydrogen, one or more organic acids, or a combination thereof) to the anode electrode to produce electricity is disclosed. A consolidated bioprocessing organism is also disclosed, as well as systems and methods relevant to it.	
EP3071517A1	A plant-derived nanocellulose material that consists of nanocellulose particles or fibers derived from a plant material with a hemicellulose content of 30% or more (w/w) (calculated as a weight percentage of the lignocellulosic components of the material). Aspect ratios of more than 250 are possible for nanocellulose. Plant materials with a C4 leaf morphology could be used to make the nanocellulose. Arid Spinifex is a good source of plant material. Mild processing conditions can be used to create nanocellulose.	
US20090017512A1	In other implementations, the invention relates to a method for generating ethanol and electricity or ethanol and hydrogen that involves supplying a microbial catalyst and a fuel source to a fermentation vessel in operable connection with a microbial fuel cell or a BEAMR device, where the microbial catalyst has cellulolytic, ethanologenic, and electricigenic operation, and the microbial catalyst has a cellulolytic, ethanologenic, and electricigenic activity. Compositions and apparatus for carrying out the invention are examples of other embodiments.	[170]

Patent No.	Description	Reference
US10686205B2	An electrochemical cell with an anode electrode, a cathode electrode, and a reference electrode that are all electronically connected; the first biocatalyst with a consolidated bioprocessing organism. (e.g., a *Cellulomonad* or *Clostridium* or related strains, such as *Cellulomonas uda (C. uda), Clostridium lentocellum (C. lentocellum), Acetivibriocelluloyticus* (*A. cellulolyticus*) *Clostridium cellobioparum (C. cellobioparum),* alcohol-tolerant *C. cellobioparum,* alcohol-tolerant *C. uda,* and combinations thereof) capable of fermenting biomass (e.g., cellulosic biomass or glycerin-containing biomass) to produce a fermentation byproduct; and a second biocatalyst comprising an electricigen (e.g., *Geobacter sulfurreducens*) capable of transferring substantially all the electrons in the fermentation byproduct (e.g., hydrogen, one or more organic acids, or a combination thereof) to the anode electrode to produce electricity is disclosed. A consolidated bioprocessing organism is also revealed by systems and methods relevant to it.	[171]

11. LIFE CYCLE ASSESSMENT

Distinguishing the decrease in energy and discharges from bioenergy production and use, an exhaustive assessment from "cradle to grave" is to be deliberately carried out [172]. Life cycle assessment (LCA) is a universally acknowledged way to deal with the climatic impact of a certain product over its whole production cycle. This structured outlook will uncover the genuine capability of the product assessed and recognize the situation dilemma in the product trials in the long run so that prudent advances can be proposed to lessen the cynical climatic impact [173]. However, LCA is a technique to characterize and decrease the ecological weights from a product, procedure, or activity by distinguishing and measuring energy along with usage of materials, in addition to waste releases, surveying the effects of these wastes on nature and opportunity evaluation for ecological refinements over the entire life cycle. LCA is, therefore, important to avoid unplanned outcomes of new technology or alleviation strategy. A cycle assessment study including MES ought to characterize the objective, purview, and practical unit as the essential strides of the investigation. The objective should be to evaluate the energy and financial flows related to the MES systems. There must be well characterized to certify its affinity with the objective. The extent of LCA can be assessed by various MESs and some novel systems for transforming the food waste to straightforwardly produce electricity or other chemical products. The practical unit is setting the correlation scale for at least two or more products giving the reference for which the sources and yields are standardized to establish the inventory. The basic role of the unit in functionality is to cite the information sources and yields connected and important to guarantee the likeness of results [174]. In the waste treatment plan's LCA, the practical unit is characterized regarding systems input which is waste. Thus, if MES is anticipated as a sewerage treatment apparatus, the operative system will likewise contrast in like manner.

12. CHALLENGES IN USING MICROBIAL FUEL CELLS

In recent years, the evolution of MFC technology has raised many concerns due to its unrealized potential for simultaneous bioenergy production and wastewater treatment. There is also a rising concern for environmental waste management and the amount of organic matter being released to the environment in the form of untreated effluent, which affects both terrestrial and marine life. However, these effluents and waste can be used for the production of energy. Some companies have launched MFC-based wastewater treatment systems; although, there are numerous challenges still associated with the use of such technology. One of the paramount shortcomings of MFC technology is the problem of high operating costs and low power output. Other challenges associated with MFC technology are the selection of appropriate and suitable substrate thus wastewater and the complexity of the molecular structure of the agro-waste identified as a suitable substrate for MFCs and its resistance to oxidation, which will, in turn, affect its treatability and organic constituents' removal which will affect the MFC working principle. However, the pH of the substrate and its sudden alteration while performing some remediation activities must be considered for substrate conditioning as the changes in pH may lower the activity of microorganisms if the optimum is altered, which will, in turn, affect the MFC performance and quality of the wastewater being treated.

Another factor affecting the performance of the MFC system is the CE when the MFC is alternatively fed with low strength wastewaters, which will affect electrode performance due to the diversion of electrons into non-exoelectrogenic growth when using both plant and animal waste materials, resulting in the rapid depletion of the substrate along with the process of metabolism, which will lower processes such as methanogenesis thus low electron transfer efficiency. Scaling-up is another challenging factor that needs to be considered, which requires an economic evaluation with appropriate safeguards for a simple wastewater treatment-MFC set-up that can be maintained effectively and easily to generate a high-power output. Considering the capital cost of MFC based on simplified designs and their configuration and agro-waste treatment capability, this technology it is said to be more promising for long energy security than the sole use of conventional wastewater treatment systems for domestic wastewater [117]. Most of the expensive electrode materials used, such as catalyst and membrane materials, may result in the high capital and operational cost implications of MFC [175]. A high potential loss has been observed at the surface of the electrodes, leading to a reduction in current density when the upscaling of the MFC technology from a few milliliters to hectoliters is considered [6]. Furthermore, the inability of the MFC to recover heavy metal ions is a great challenge, also affecting its selection as a preferred technology. Overall, the bio-toxicity of certain heavy metals would negatively affect the performance of MFCs, imparting low energy production rates and limited wastewater remediation efficiencies [176].

13. CONCLUSIONS AND FUTURE DIRECTION

The use of agricultural waste in MFC has been critical in the renewable energy industry, contributing significantly to the production of bioenergy. Similarly, the development of MFC technology has enabled the use of agricultural waste as a feedstock by various microbial communities in the anode compartment of the fuel cell. Due to the complex structure and crystallinity of agricultural biomass, biodegradation is limited. As a result, the agricultural waste's greater moisture content would assist in overcoming this barrier. By using a variety of wastewater resources, such as agricultural wastewater and fruit wastewater, the bacteria may easily break down the solid biomass. Conclusively, this review has demonstrated the use of various agricultural wastes for bioelectricity generation. Therefore, the use of different agricultural wastes and wastewater containing different industrial-by products for bioelectricity production in MFC seems to be a promising and alternative source of renewable energy generation. Moreover, it has been shown that different varieties of agricultural wastes and wastewater can be utilized using several different MFCs to enhance bioenergy production; thus, the conversion of agro-waste into bioenergy can be carried out by both biochemical and thermochemical MFC routes. Several papers report numerous experimental studies, whereby the use of various substrates from different agri-based industries and with different compositions for application in MFCs, has been demonstrated: most importantly, in terms of simultaneous wastewater treatment and energy recovery. Another attractive and fascinating trait of MFC technology development is the incorporation of wastewater treatment, which provides an alternative solution to wastewater management, pollutant removal, and the maintenance of a safe and eco-friendly environment in addition to energy production. Overall, it has been noted that MFC technology offers significant advantages such as low input energy cost and a low level of residual biosolid production. In essence, improvement has been made in the total bioenergy production arena in using—concentrated wastewater derived from various agro-waste, indicating that various microbial consortia of different origins play an important role during the oxidation-reduction reactions for bioenergy production using different anodes and cathodes. Overall, an effective pretreatment approach has been made to solve problems associated with agri-waste mitigation even when such waste has a different particle size, calorific value, etc. There is also a need to promote environmental sustainability in agricultural activities and the standard management of agro-wastes that will reduce the volume of wastes released into the environment and provide a channel for bioenergy generation. To this end, local governments and regulatory agencies should explore ways of generating bioelectricity from various agro-waste as there is an urgent need to disabuse the general public's minds that agro-waste is useless.

REFERENCES

1. Bhagchandanii, D.D.; Babu, R.P.; Sonawane, J.M.; Khanna, N.; Pandit, S.; Jadhav, D.A.; Khilari, S.; Prasad, R. A Comprehensive Understanding of Electro. *Fermentation* **2020**, 6, 92. [CrossRef]
2. Azam, M.; Jahromy, S.S.; Raza, W.; Raza, N.; Lee, S.S.; Kim, K.-H.; Winter, F. Status, Characterization and Potential Utilization of Municipal Solid Waste as Renewable Energy Source: Lahore Case Study in Pakistan. *Environ. Int.* **2020**,*134*,105291. [CrossRef] [PubMed]
3. Savla, N.; Pandit, S.; Khanna, N.; Mathuriya, A.S.; Jung, S.P Microbially Powered Electrochemical Systems Coupled with Membrane-Based Technology for Sustainable Desalination and Efficient Wastewater Treatment. *J. Korean Soc. Environ. Eng.* **2020**, *42,* 360-380. [CrossRef]
4. Kumbhar, P.; Savla, N.; Banerjee, S.; Mathuriya, A.S.; Sarkar, A.; Khilari, S.; Jadhav, D.A.; Pandit, S. Chapter 26—Microbial Electrochemical Heavy Metal Removal: Fundamental to the Recent Development. In *Wastewater Treatment;* Elsevier: Amsterdam, The Netherlands, 2021; pp. 521-542. [CrossRef]
5. Yang, Y.; Liew, R.K.; Tamothran, A.M.; Foong, S.Y.; Yek, P.N.Y.; Chia, P.W.; Van Tran, T.; Peng, W.; Lam, S.S. Gasification of Refuse-Derived Fuel from Municipal Solid Waste for Energy Production: A Review. *Env. Chem Lett.* **2021**, 19, 2127-2140. [CrossRef] [PubMed]
6. ElMekawy, A.; Srikanth, S.; Bajracharya, S.; Hegab, H.M.; Nigam, P.S.; Singh, A.; Mohan, S.V.; Pant, D. Food and Agricultural Wastes as Substrates for Bioelectrochemical System (BES): The Synchronized Recovery of Sustainable Energy and Waste Treatment. *Food Res. Int.* **2015**, 73, 213-225. [CrossRef]
7. Khan, T.S.; Mubeen, U. Wheat Straw: A Pragmatic Overview. *Curr. Res. J. Biol. Sci.* **2012**, 4, 673-675.
8. Wang, X.; Feng, Y.; Wang, H.; Qu, Y.; Yu, Y.; Ren, N.; Li, N.; Wang, E.; Lee, H.; Logan, B.E. Bioaugmentation for Electricity Generation from Corn Stover Biomass Using Microbial Fuel Cells. *Env. Sci. Technol.* **2009**, 43, 6088-6093. [CrossRef]

9. Lin, H.; Ying, Y. Theory and Application of near Infrared Spectroscopy in Assessment of Fruit Quality: A Review. *Sens. Instrum. Food Qual. Saf.* **2009**, 3,130-141. [CrossRef]

10. Jin, B.; van Leeuwen, H.J.; Patel, B.; Yu, Q. Utilisation of Starch Processing Wastewater for Production of Microbial Biomass Protein and Fungal a-Amylase by Aspergillus Oryzae. *Bioresour. Technol.* **1998**, 66, 201-206. [CrossRef]

11. Shah, A.V.; Srivastava, V.K.; Mohanty, S.S.; Varjani, S. Municipal Solid Waste as a Sustainable Resource for Energy Production: State-of-the-Art Review. *J. Environ. Chem. Eng.* **2021**, *9*,105717. [CrossRef]

12. Bustillo-Lecompte, C.F.; Mehrvar, M. Slaughterhouse Wastewater Characteristics, Treatment, and Management in the Meat Processing Industry: A Review on Trends and Advances. *J. Environ. Manag.* **2015**,161, 287-302. [CrossRef]

13. Rezaei, F.; Richard, T.L.; Logan, B.E. Analysis of Chitin Particle Size on Maximum Power Generation, Power Longevity, and Coulombic Efficiency in Solid-Substrate Microbial Fuel Cells. *J. Power Sour.* **2009**,192, 304-309. [CrossRef]

14. *Engineering and Science of Biomass Feedstock Production and Provision;* Shastri, Y.; Hansen, A.; Rodriguez, L.; Ting, K.C. (Eds.) Springer-Verlag: New York, NY, USA, 2014; ISBN 978-1-4899-8013-7.

15. Nagarajan, D.; Varjani, S.; Lee, D.-J.; Chang, J.-S. Sustainable Aquaculture and Animal Feed from Microalgae—Nutritive Value and Techno-Functional Components. *Renew. Sustain. Energy Rev.* **2021**,150,111549. [CrossRef]

16. Lebo, S.E.; Gargulak, J.D.; McNally, T.J. Lignin. In *Kirk-Othmer Encyclopedia of Chemical Technology;* American Cancer Society: Atlanta, GA, USA, 2001; ISBN 978-0-471-23896-6.

17. Kan, T.; Strezov, V.; Evans, T.J. Lignocellulosic Biomass Pyrolysis: A Review of Product Properties and Effects of Pyrolysis Parameters. *Renew. Sustain. Energy Rev.* **2016**, 57,1126-1140. [CrossRef]

18. Prajapati, P.; Varjani, S.; Singhania, R.R.; Patel, A.K.; Awasthi, M.K.; Sindhu, R.; Zhang, Z.; Binod, P.; Awasthi, S.K.; Chaturvedi, P. Critical Review on Technological Advancements for Effective Waste Management of Municipal Solid Waste—Updates and Way Forward. *Environ. Technol. Innov.* **2021**,23,101749. [CrossRef]

19. Varjani, S.; Shah, A.V.; Vyas, S.; Srivastava, V.K. Processes and Prospects on Valorizing Solid Waste for the Production of Valuable Products Employing Bio-Routes: A Systematic Review. *Chemosphere* **2021**,282,130954. [CrossRef]

20. Lloyd, T.A.; Wyman, C.E. Combined Sugar Yields for Dilute Sulfuric Acid Pretreatment of Corn Stover Followed by Enzymatic Hydrolysis of the Remaining Solids. *Bioresour. Technol.* **2005**, 96,1967-1977. [CrossRef] [PubMed]

21. Sun, Y.; Cheng, J. Hydrolysis of Lignocellulosic Materials for Ethanol Production: A Review. *Bioresour. Technol.* **2002**, 83,1-11. [CrossRef]

22. Agbor, V.B.; Cicek, N.; Sparling, R.; Berlin, A.; Levin, D.B. Biomass Pretreatment: Fundamentals toward Application. *Biotechnol. Adv.* **2011**, 29, 675-685. [CrossRef] [PubMed]

23. Shen, J.; Wang, C.; Liu, Y.; Hu, C.; Xin, Y.; Ding, N.; Su, S. Effect of Ultrasonic Pretreatment of the Dairy Manure on the Electricity Generation of Microbial Fuel Cell. *Biochem. Eng. J.* **2018**,129, 44-49. [CrossRef]

24. Tao, K.; Quan, X.; Quan, Y. Composite vegetable degradation and electricity generation in microbial fuel cell with ultrasonic pretreatment. *Env. Eng. Manag. J.* **2013**,12,1423-1427. [CrossRef]

25. Wang, Y.-Z.; Shen, Y.; Gao, L.; Liao, Z.-H.; Sun, J.-Z.; Yong, Y.-C. Improving the Extracellular Electron Transfer of Shewanella Oneidensis MR-1 for Enhanced Bioelectricity Production from Biomass Hydrolysate. *Rsc Adv.* **2017**, 7, 30488-30494. [CrossRef]

26. Cao, Y.; Zhang, R.; Cheng, T.; Guo, J.; Xian, M.; Liu, H. Imidazolium-Based Ionic Liquids for Cellulose Pretreatment: Recent Progresses and Future Perspectives. *Appl. Microbiol. Biotechnol.* **2017**,101, 521-532. [CrossRef]

27. Yang, B.; Qin, X.; Hu, H.; Duan, C.; He, Z.; Ni, Y. Using Ionic Liquid (EmimAc)-Water Mixture in Selective Removal of Hemicelluloses from a Paper-Grade Bleached Hardwood Kraft Pulp. *Cellulose* **2020**, 27, 9653-9661. [CrossRef]

28. Wahlstrom, R.M.; Suurnakki, A. Enzymatic Hydrolysis of Lignocellulosic Polysaccharides in the Presence of Ionic Liquids. *Green Chem.* **2015**,17, 694-714. [CrossRef]

29. Amin, F.R.; Khalid, H.; Zhang, H.; Rahman, S.U.; Zhang, R.; Liu, G.; Chen, C. Pretreatment Methods of Lignocellulosic Biomass for Anaerobic Digestion. *AMB Express* **2017**, 7, 72. [CrossRef] [PubMed]

30. Song, T Production of Electricity from Rice Straw with Different Pretreatment Methods Using a Sediment Microbial Fuel Cell. *Int. J. Electrochem. Sci.* **2018**, 461-471. [CrossRef]

31. Xiao, B.; Yang, F.; Liu, J. Evaluation of Electricity Production from Alkaline Pretreated Sludge Using Two-Chamber Microbial Fuel Cell. *J. Hazard. Mater.* **2013**, *254-255*, 57-63. [CrossRef]

32. Wagner, A.O.; Lackner, N.; Mutschlechner, M.; Prem, E.M.; Markt, R.; Illmer, P. Biological Pretreatment Strategies for Second- Generation Lignocellulosic Resources to Enhance Biogas Production. *Energies* **2018**,11,1797. [CrossRef]

33. Galbe, M.; Wallberg, O. Pretreatment for Biorefineries: A Review of Common Methods for Efficient Utilisation of Lignocellulosic Materials. *Biotechnol. Biofuels* **2019**,12, 294. [CrossRef]

34. Krishnaraj, R.N.; Berchmans, S.; Pal, P. The Three-Compartment Microbial Fuel Cell: A New Sustainable Approach to Bioelectricity Generation from Lignocellulosic Biomass. *Cellulose* **2015**,22, 655-662. [CrossRef]

35. Zamri, M.F.M.A.; Hasmady, S.; Akhiar, A.; Ideris, F.; Shamsuddin, A.H.; Mofijur, M.; Fattah, I.M.R.; Mahlia, T.M.I. A Comprehensive Review on Anaerobic Digestion of Organic Fraction of Municipal Solid Waste. *Renew. Sustain. Energy Rev.* **2021**,137,110637. [CrossRef]

36. Gupta, G.K.; Mondal, M.K. Bio-Energy Generation from Sagwan Sawdust via Pyrolysis: Product Distributions, Characterizations and Optimization Using Response Surface Methodology. *Energy* **2019**,170, 423-437. [CrossRef]

37. Hossain, A.K.; Davies, P.A. Pyrolysis Liquids and Gases as Alternative Fuels in Internal Combustion Engines—A Review. *Renew. Sustain. Energy Rev.* **2013**, 21,165-189. [CrossRef]

38. Bridgewater, A.V. Biomass Fast Pyrolysis. *Therm. Sci.* **2004**, 8, 21-50. [CrossRef]

39. Ram, M.; Mondal, M.K. Comparative Study of Native and Impregnated Coconut Husk with Pulp and Paper Industry Waste Water for Fuel Gas Production. *Energy* **2018**,156,122-131. [CrossRef]

40. Cai, J.; Zeng, R.; Zheng, W.; Wang, S.; Han, J.; Li, K.; Luo, M.; Tang, X. Synergistic Effects of Co-Gasification of Municipal Solid Waste and Biomass in Fixed-Bed Gasifier. *Process. Saf. Environ. Prot.* **2021**,148,1-12. [CrossRef]

41. Chand Malav, L.; Yadav, K.K.; Gupta, N.; Kumar, S.; Sharma, G.K.; Krishnan, S.; Rezania, S.; Kamyab, H.; Pham, Q.B.; Yadav, S.; et al. A Review on Municipal Solid Waste as a Renewable Source for Waste-to-Energy Project in India: Current Practices, Challenges, and Future Opportunities. *J. Clean. Prod.* **2020**, 277,123227. [CrossRef]

42. Miltner, M.; Makaruk, A.; Harasek, M. Review on Available Biogas Upgrading Technologies and Innovations towards Advanced Solutions. *J. Clean. Prod.* **2017**,161,1329-1337. [CrossRef]

43. Capodaglio, A.; Molognoni, D.; Dallago, E.; Liberale, A.; Cella, R.; Longoni, P.; Pantaleoni, L. Microbial Fuel Cells for Direct Electrical Energy Recovery from Urban Wastewaters. *Sci. World J.* **2013,** *2013,* 634738. [CrossRef]

44. Logan, B.E.; Hamelers, B.; Rozendal, R.; Schroder, U.; Keller, J.; Freguia, S.; Aelterman, P.; Verstraete, W.; Rabaey, K. Microbial Fuel Cells: Methodology and Technology. *Env. Sci. Technol.* **2006**, 40, 5181-5192. [CrossRef] [PubMed]

45. Molognoni, D.; Puig, S.; Balaguer, M.D.; Liberale, A.; Capodaglio, A.G.; Callegari, A.; Colprim, J. Reducing Start-up Time and Minimizing Energy Losses of Microbial Fuel Cells Using Maximum Power Point Tracking Strategy. *J. Power Sour.* **2014**,269,403-411. [CrossRef]

46. Rozendal, R.A.; Hamelers, H.V.M.; Rabaey, K.; Keller, J.; Buisman, C.J.N. Towards Practical Implementation of Bioelectrochemical Wastewater Treatment. *Trends Biotechnol.* **2008**, 26, 450-459. [CrossRef] [PubMed]

47. Savla, N.; Khilari, S.; Pandit, S.; Jung, S.P. Effective Cathode Catalysts for Oxygen Reduction Reactions in Microbial Fuel Cell. In *Bioelectrochemical Systems: Vol. 1 Principles and Processes;* Kumar, P., Kuppam, C., Eds.; Springer: Singapore, 2020; ISBN 9789811568725. pp. 189-210.

48. Callegari, A.; Cecconet, D.; Molognoni, D.; Capodaglio, A.G. Sustainable Processing of Dairy Wastewater: Long-Term Pilot Application of a Bio-Electrochemical System. *J. Clean. Prod.* **2018**,189, 563-569. [CrossRef]

49. Capodaglio, A.G.; Molognoni, D.; Pons, A.V. A Multi-Perspective Review of Microbial Fuel-Cells for Wastewater Treatment: Bio-Electro-Chemical, Microbiologic and Modeling Aspects. *Aip. Conf. Proc.* **2016**,*1758*, 030032. [CrossRef]

50. Cercado-Quezada, B.; Delia, M.-L.; Bergel, A. Testing Various Food-Industry Wastes for Electricity Production in Microbial Fuel Cell. *Bioresour. Technol.* **2010**,101, 2748-2754. [CrossRef]

51. Jung, S.P.; Pandit, S. Chapter 3.1—Important Factors Influencing Microbial Fuel Cell Performance. In *Microbial Electrochemical Technology;* Mohan, S.V., Varjani, S., Pandey, A., Eds.; Biomass, Biofuels and Biochemicals; Elsevier: Amsterdam, The Netherlands, 2019; pp. 377-406. ISBN 978-0-444-64052-9.

52. Pandit, S.; Das, D. Principles of Microbial Fuel Cell for the Power Generation. In *Microbial Fuel Cell: A Bioelectrochemical System that Converts Waste to Watts;* Das, D., Ed.; Springer International Publishing: Berlin, Germany, 2018; pp. 21-41. ISBN 978-3-319-66793-5.

53. Savla, N.; Anand, R.; Pandit, S.; Prasad, R. Utilization of Nanomaterials as Anode Modifiers for Improving Microbial Fuel Cells Performance. *J. Renew. Mater.* **2020**, 8,1581-1605. [CrossRef]

54. Kim, H.J.; Park, H.S.; Hyun, M.S.; Chang, I.S.; Kim, M.; Kim, B.H. A Mediator-Less Microbial Fuel Cell Using a Metal Reducing Bacterium, Shewanella Putrefaciens. *Enzym. Microb. Technol.* **2002**, 30,145-152. [CrossRef]

55. Ieropoulos, I.A.; Greenman, J.; Melhuish, C.; Hart, J. Comparative Study of Three Types of Microbial Fuel Cell. *Enzym. Microb. Technol.* **2005**, 37, 238-245. [CrossRef]

56. Roy, S.; Pandit, S. 1.2—Microbial Electrochemical System: Principles and Application. In *Microbial Electrochemical Technology;* Mohan, S.V., Varjani, S., Pandey, A., Eds.; Biomass, Biofuels and Biochemicals; Elsevier: Amsterdam, The Netherlands, 2019; pp. 19-48. ISBN 978-0-444-64052-9.

57. Jiang, D.; Curtis, M.; Troop, E.; Scheible, K.; McGrath, J.; Hu, B.; Suib, S.; Raymond, D.; Li, B. A Pilot-Scale Study on Utilizing Multi-Anode/Cathode Microbial Fuel Cells (MAC MFCs) to Enhance the Power Production in Wastewater Treatment. *Int. J. Hydrog. Energy* **2011**, 36, 876-884. [CrossRef]

58. Feng, Y.; He, W.; Liu, J.; Wang, X.; Qu, Y.; Ren, N. A Horizontal Plug Flow and Stackable Pilot Microbial Fuel Cell for Municipal Wastewater Treatment. *Bioresour. Technol.* **2014**,156,132-138. [CrossRef]

59. Lu, M.; Chen, S.; Babanova, S.; Phadke, S.; Salvacion, M.; Mirhosseini, A.; Chan, S.; Carpenter, K.; Cortese, R.; Bretschger, O. Long-Term Performance of a 20-L Continuous Flow Microbial Fuel Cell for Treatment of Brewery Wastewater. *J. Power Sour.* **2017**, *356*, 274-287. [CrossRef]

60. Liang, P.; Wei, J.; Li, M.; Huang, X. Scaling up a Novel Denitrifying Microbial Fuel Cell with an Oxic-Anoxic Two Stage Biocathode. *Front. Env. Sci. Eng.* **2013**, 7, 913-919. [CrossRef]

61. Zhuang, L.; Yuan, Y.; Wang, Y.; Zhou, S. Long-Term Evaluation of a 10-Liter Serpentine-Type Microbial Fuel Cell Stack Treating Brewery Wastewater. *Bioresour. Technol.* **2012**,123, 406-412. [CrossRef]

62. Vilajeliu-Pons, A.; Puig, S.; Salcedo-Davila, I.D.; Balaguer, M.; Colprim, J. Long-Term Assessment of Six-Stacked Scaled-up MFCs Treating Swine Manure with Different Electrode Materials. *Environ. Sci. Water Res. Technol.* **2017**, 3, 947-959. [CrossRef]

63. Wu, S.; Li, H.; Zhou, X.; Liang, P.; Zhang, X.; Jiang, Y.; Huang, X. A Novel Pilot-Scale Stacked Microbial Fuel Cell for Efficient Electricity Generation and Wastewater Treatment. *Water Res.* **2016**, 98, 396-403. [CrossRef] [PubMed]

64. Ge, Z.; He, Z. Long-Term Performance of a 200 Liter Modularized Microbial Fuel Cell System Treating Municipal Wastewater: Treatment, Energy, and Cost. *Environ. Sci. Water Res. Technol.* **2016**, 2, 274-281. [CrossRef]

65. Liang, P.; Duan, R.; Jiang, Y.; Zhang, X.; Qiu, Y.; Huang, X. One-Year Operation of 1000-L Modularized Microbial Fuel Cell for Municipal Wastewater Treatment. *Water Res.* **2018**,141,1-8. [CrossRef]

66. Prasertsan, P.; Prasertsan, S.; Kittikun, A.H. Recycling of Agro-Industrial Wastes Through Cleaner Technology. *Biotechnology* **2010**, *10*, 1-11.

67. Mohanty, S.S.; Koul, Y.; Varjani, S.; Pandey, A.; Ngo, H.H.; Chang, J.-S.; Wong, J.W.C.; Bui, X.-T. A Critical Review on Various Feedstocks as Sustainable Substrates for Biosurfactants Production: A Way towards Cleaner Production. *Microb. Cell Factories* **2021**, 20,120. [CrossRef]

68. Kim, G.; Kim, W.; Yun, C.; Son, D.; Chang, D.; Bae, H.; Lee, Y.; Sunwoo, Y.; Hong, K. Agro-Industrial Wastewater Treatment by Electrolysis Technology. *Int. J. Electrochem. Sci.* **2013**, *8*, 9835-9850.

69. Vinayak, V.; Khan, M.J.; Varjani, S.; Saratale, G.D.; Saratale, R.G.; Bhatia, S.K. Microbial Fuel Cells for Remediation of Environmental Pollutants and Value Addition: Special Focus on Coupling Diatom Microbial Fuel Cells with Photocatalytic and Photoelectric Fuel Cells. *J. Biotechnol.* **2021**, 338, 5-19. [CrossRef]

70. Garcia-Nunez, J.A.; Ramirez-Contreras, N.E.; Rodriguez, D.T.; Silva-Lora, E.; Frear, C.S.; Stockle, C.; Garcia-Perez, M. Evolution of Palm Oil Mills into Bio-Refineries: Literature Review on Current and Potential Uses of Residual Biomass and Effluents. *Resour. Conserv. Recycl.* **2016**,110, 99-114. [CrossRef]

71. Ng, W.P.Q.; Lam, H.L.; Ng, F.Y.; Kamal, M.; Lim, J.H.E. Waste-to-Wealth: Green Potential from Palm Biomass in Malaysia. *J. Clean. Prod.* **2012**, 34, 57-65. [CrossRef]

72. Baranitharan, E.; Khan, M.R.; Yousuf, A.; Teo, W.F.A.; Tan, G.Y.A.; Cheng, C.K. Enhanced Power Generation Using Controlled Inoculum from Palm Oil Mill Effluent Fed Microbial Fuel Cell. *Fuel* **2015**,143, 72-79. [CrossRef]

73. Cheng, J.; Zhu, X.; Ni, J.; Borthwick, A. Palm Oil Mill Effluent Treatment Using a Two-Stage Microbial Fuel Cells System Integrated with Immobilized Biological Aerated Filters. *Bioresour. Technol.* **2010**,101, 2729-2734. [CrossRef] [PubMed]

74. Guo, F.; Fu, G.; Zhang, Z.; Zhang, C. Mustard Tuber Wastewater Treatment and Simultaneous Electricity Generation Using Microbial Fuel Cells. *Bioresour. Technol.* **2013**,136, 425-430. [CrossRef] [PubMed]

75. Zhang, Y.; Sun, C.; Liu, X.; Han, W.; Dong, Y.; Li, Y. Electricity Production from Molasses Wastewater in Two-Chamber Microbial Fuel Cell. *Water Sci. Technol.* **2013**, 68, 494-498. [CrossRef]

76. Hassan, S.H.A.; el Nasser, A.; Zohri, A.; Kassim, R.M.F. Electricity Generation from Sugarcane Molasses Using Microbial Fuel Cell Technologies. *Energy* **2019**,178, 538-543. [CrossRef]

77. Pandit, S.; Balachandar, G.; Das, D. Improved Energy Recovery from Dark Fermented Cane Molasses Using Microbial Fuel Cells. *Front. Chem. Sci. Eng.* **2014**, 8, 43-54. [CrossRef]

78. Kaewkannetra, P.; Chiwes, W.; Chiu, T.Y. Treatment of Cassava Mill Wastewater and Production of Electricity through Microbial Fuel Cell Technology. *Fuel* **2011**, 90, 2746-2750. [CrossRef]

79. Oh, S.; Logan, B.E. Hydrogen and Electricity Production from a Food Processing Wastewater Using Fermentation and Microbial Fuel Cell Technologies. *Water Res.* **2005**, 39, 4673-4682. [CrossRef]

80. Pepe Sciarria, T.; Merlino, G.; Scaglia, B.; D'Epifanio, A.; Mecheri, B.; Borin, S.; Licoccia, S.; Adani, F. Electricity Generation Using White and Red Wine Lees in Air Cathode Microbial Fuel Cells. *J. Power Sour.* **2015**,274, 393-399. [CrossRef]

81. Herrero-Hernandez, E.; Smith, T.J.; Akid, R. Electricity Generation from Wastewaters with Starch as Carbon Source Using a Mediatorless Microbial Fuel Cell. *Biosens. Bioelectron.* **2013**, 39,194-198. [CrossRef] [PubMed]

82. Behera, M.; Jana, P.S.; More, T.T.; Ghangrekar, M.M. Rice Mill Wastewater Treatment in Microbial Fuel Cells Fabricated Using Proton Exchange Membrane and Earthen Pot at Different PH. *Bioelectrochemistry* **2010**, 79, 228-233. [CrossRef]

83. Wang, Z.; Lee, T.; Lim, B.; Choi, C.; Park, J. Microbial Community Structures Differentiated in a Single-Chamber Air-Cathode Microbial Fuel Cell Fueled with Rice Straw Hydrolysate. *Biotechnol. Biofuels* **2014**, 7, 9. [CrossRef]

84. Zuo, Y.; Maness, P.-C.; Logan, B.E. Electricity Production from Steam-Exploded Corn Stover Biomass. *Energy Fuels* **2006**,20,1716-1721. [CrossRef]

85. Feng, Y.; Wang, X.; Logan, B.E.; Lee, H. Brewery Wastewater Treatment Using Air-Cathode Microbial Fuel Cells. *Appl. Microbiol. Biotechnol.* **2008**, 78, 873-880. [CrossRef]

86. Wen, Q.; Wu, Y.; Zhao, L.; Sun, Q. Production of Electricity from the Treatment of Continuous Brewery Wastewater Using a Microbial Fuel Cell. *Fuel* **2010**, *89*,1381-1385. [CrossRef]

87. Rahman, A.; Borhan, M.S.; Rahman, S. Evaluation of Microbial Fuel Cell (MFC) for Bioelectricity Generation and Pollutants Removal from Sugar Beet Processing Wastewater (SBPW). *Water Sci. Technol.* **2018**, *77*, 387-397. [CrossRef] [PubMed]

88. Jayashree, C.; Arulazhagan, P.; Adish Kumar, S.; Kaliappan, S.; Yeom, I.T.; Rajesh Banu, J. Bioelectricity Generation from Coconut Husk Retting Wastewater in Fed Batch Operating Microbial Fuel Cell by Phenol Degrading Microorganism. *Biomass Bioenerg.* **2014**, *69*, 249-254. [CrossRef]

89. Azbar, N.; Bayram, A.; Filibeli, A.; Muezzinoglu, A.; Sengul, F.; Ozer, A. A Review of Waste Management Options in Olive Oil Production. *Crit. Rev. Environ. Sci. Technol.* **2004**, 34, 209-247. [CrossRef]

90. Ochando-Pulido, J.M.; Martinez-Ferez, A. On the Purification of Agro-Industrial Wastewater by Membrane Technologies: The Case of Olive Mill Effluents. *Desalination* **2017**. [CrossRef]

91. Sciarria, T.P.; Tenca, A.; D'Epifanio, A.; Mecheri, B.; Merlino, G.; Barbato, M.; Borin, S.; Licoccia, S.; Garavaglia, V.; Adani, F. Using Olive Mill Wastewater to Improve Performance in Producing Electricity from Domestic Wastewater by Using Single-Chamber Microbial Fuel Cell. *Bioresour. Technol.* **2013**,147, 246-253. [CrossRef] [PubMed]

92. Gurung, A.; Oh, S.-E. Rice Straw as a Potential Biomass for Generation of Bioelectrical Energy Using Microbial Fuel Cells (MFCs). *Energy Sour. Part A Recovery Util. Environ. Eff.* **2015**, 37, 2625-2631. [CrossRef]

93. Kaewkannetra, P; Imai, T.; Garcia-Garcia, F.J.; Chiu, T.Y. Cyanide Removal from Cassava Mill Wastewater Using Azotobactor Vinelandii TISTR 1094 with Mixed Microorganisms in Activated Sludge Treatment System. *J. Hazard. Mater.* **2009**,172, 224-228. [CrossRef] [PubMed]

94. Agarry, S.E.; Oghenejobor, K.M.; Solomon, B.O. Bioelectricity production from cassava mill effluents using microbial fuel cell technology. *Niger. J. Technol.* **2016**, 35, 329-336. [CrossRef]

95. Javed, M.; Nisar, M.A.; Ahmad, M.; Muneer, B. Production of Bioelectricity from Vegetable Waste Extract by Designing a U-Shaped Microbial Fuel Cell. *Pak. J. Zool.* **2017**, 49, 711-716. [CrossRef]

96. Logrono, W.; Ramirez, G.; Recalde, C.; Echeverria, M.; Cunachi, A. Bioelectricity Generation from Vegetables and Fruits Wastes by Using Single Chamber Microbial Fuel Cells with High Andean Soils. *Energy Procedia* **2015**, 75, 2009-2014. [CrossRef]

97. Toding, O.S.L.; Virginia, C.; Suhartini, S. Conversion Banana and Orange Peel Waste into Electricity Using Microbial Fuel Cell. *IOP Conf. Ser. Earth Env. Sci.* **2018**, 209, 012049. [CrossRef]

98. Khan, A.M.; Obaid, M. Comparative Bioelectricity Generation from Waste Citrus Fruit Using a Galvanic Cell, Fuel Cell and Microbial Fuel Cell. *J. Energy South. Afr.* **2015**,26, 90-99. [CrossRef]

99. Miran, W.; Nawaz, M.; Jang, J.; Lee, D.S. Sustainable Electricity Generation by Biodegradation of Low-Cost Lemon Peel Biomass in a Dual Chamber Microbial Fuel Cell. *Int. Biodeterior. Biodegrad.* **2016**,106, 75-79. [CrossRef]

100. Wang, M.; Yan, Z.; Huang, B.; Zhao, J.; Liu, R. Electricity Generation by Microbial Fuel Cells Fuelled with Enteromorpha Prolifera Hydrolysis. *Int. J. Electrochem. Sci.* **2013**, 8, 2104-2111.

101. Zang, G.-L.; Sheng, G.-P.; Tong, Z.-H.; Liu, X.-W.; Teng, S.-X.; Li, W.-W.; Yu, H.-Q. Direct Electricity Recovery from Canna Indica by an Air-Cathode Microbial Fuel Cell Inoculated with Rumen Microorganisms. *Env. Sci. Technol.* **2010**, 44, 2715-2720. [CrossRef] [PubMed]

102. Sunder, G.C.; Satyanarayan, S. Efficient Treatment of Slaughter House Wastewater by Anaerobic Hybrid Reactor Packed with Special Floating Media. Available online: https://www.semanticscholar.org/paper/Efficient-Treatment-of-Slaughter-House- Wastewater-Sunder-Satyanarayan/10da3a3b 26f8ccc5c53a1421085e591d490649c6 (accessed on 28 August 2021).

103. Katuri, K.P.; Enright, A.-M.; O'Flaherty, V.; Leech, D. Microbial Analysis of Anodic Biofilm in a Microbial Fuel Cell Using Slaughterhouse Wastewater. *Bioelectrochemistry* **2012**, 87,164-171. [CrossRef] [PubMed]

104. Li, X.; Zhu, N.; Wang, Y.; Li, P.; Wu, P.; Wu, J. Animal Carcass Wastewater Treatment and Bioelectricity Generation in Up-Flow Tubular Microbial Fuel Cells: Effects of HRT and Non-Precious Metallic Catalyst. *Bioresour. Technol.* **2013**,128,454-460. [CrossRef] [PubMed]

105. Rout, P.R.; Shahid, M.K.; Dash, R.R.; Bhunia, P.; Liu, D.; Varjani, S.; Zhang, T.C.; Surampalli, R.Y. Nutrient Removal from Domestic Wastewater: A Comprehensive Review on Conventional and Advanced Technologies. *J. Environ. Manag.* **2021**, 296, 113246. [CrossRef]

106. Inoue, K.; Ito, T.; Kawano, Y.; Iguchi, A.; Miyahara, M.; Suzuki, Y.; Watanabe, K. Electricity Generation from Cattle Manure Slurry by Cassette-Electrode Microbial Fuel Cells. *J. Biosci. Bioeng.* **2013**,116, 610-615. [CrossRef]

107. Kiely, P.D.; Cusick, R.; Call, D.F.; Selembo, P.A.; Regan, J.M.; Logan, B.E. Anode Microbial Communities Produced by Changing from Microbial Fuel Cell to Microbial Electrolysis Cell Operation Using Two Different Wastewaters. *Bioresour. Technol.* **2011**,102, 388-394. [CrossRef]

108. Scott, K.; Murano, C. Microbial Fuel Cells Utilising Carbohydrates. *J. Chem. Technol. Biotechnol.* **2007**, 82, 92-100. [CrossRef]

109. Zhang, G.; Zhao, Q.; Jiao, Y.; Wang, K.; Lee, D.-J.; Ren, N. Biocathode Microbial Fuel Cell for Efficient Electricity Recovery from Dairy Manure. *Biosens. Bioelectron.* **2012**, 31, 537-543. [CrossRef] [PubMed]

110. Babanova, S.; Jones, J.; Phadke, S.; Lu, M.; Angulo, C.; Garcia, J.; Carpenter, K.; Cortese, R.; Chen, S.; Phan, T.; et al. Continuous Flow, Large-Scale, Microbial Fuel Cell System for the Sustained Treatment of Swine Waste. *Water Environ. Res.* **2020**, 92, 60-72. [CrossRef]

111. Min, B.; Kim, J.; Oh, S.; Regan, J.M.; Logan, B.E. Electricity Generation from Swine Wastewater Using Microbial Fuel Cells. *Water Res.* **2005**, 39, 4961-4968. [CrossRef] [PubMed]

112. Kim, J.R.; Dec, J.; Bruns, M.A.; Logan, B.E. Removal of Odors from Swine Wastewater by Using Microbial Fuel Cells. *Appl. Env. Microbiol.* **2008**, 74, 2540-2543. [CrossRef]

113. Cheng, C.Y.; Li, C.C.; Chung, Y.C. Continuous Electricity Generation and Pollutant Removal from Swine Wastewater Using a Single-Chambered Air-Cathode Microbial Fuel Cell. Available online: https://www.scientific.net/AMR.953-954.158 (accessed on 23 August 2020).

114. Nyoyoko, V. Generation of Electricity from Poultry Waste Using Microbial Fuel Cell. *J. Microbiol. Biotechnol.* **2015**, 5, 31-37.

115. Oyiwona, G.E.; Ogbonna, J.C.; Anyanwu, C.U.; Okabe, S. Electricity Generation Potential of Poultry Droppings Wastewater in Microbial Fuel Cell Using Rice Husk Charcoal Electrodes. *Bioresour. Bioprocess.* **2018**, 5,13. [CrossRef]

116. Jaeel, A.J. Electricity Production from Dual Chambers Microbial Fuel Cell Fed with Chicken Manure-Wastewater. *Wasit J. Eng. Sci.* **2015**, 3, 9-18. [CrossRef]

117. He, L.; Du, P.; Chen, Y.; Lu, H.; Cheng, X.; Chang, B.; Wang, Z. Advances in Microbial Fuel Cells for Wastewater Treatment. *Renew. Sustain. Energy Rev.* **2017**, 71, 388-403. [CrossRef]

118. Demirer, G.; Duran, M.; Erguder, T.H.; Guven, E.; Ugurlu, O.; Tezel, U. Anaerobic Treatability and Biogas Production Potential Studies of Different Agro-Industrial Wastewaters in Turkey. *Biodegradation* **2000**,11, 401-405. [CrossRef]

119. Mansoorian, H.J.; Mahvi, A.H.; Jafari, A.J.; Khanjani, N. Evaluation of Dairy Industry Wastewater Treatment and Simultaneous Bioelectricity Generation in a Catalyst-Less and Mediator-Less Membrane Microbial Fuel Cell. *J. Saudi Chem. Soc.* **2016**, 20, 88-100. [CrossRef]

120. Venkata Mohan, S.; Mohanakrishna, G.; Velvizhi, G.; Babu, V.L.; Sarma, P.N. Bio-Catalyzed Electrochemical Treatment of Real Field Dairy Wastewater with Simultaneous Power Generation. *Biochem. Eng. J.* **2010**, 51, 32-39. [CrossRef]

121. Antonopoulou, G.; Stamatelatou, K.; Bebelis, S.; Lyberatos, G. Electricity Generation from Synthetic Substrates and Cheese Whey Using a Two Chamber Microbial Fuel Cell. *Biochem. Eng. J.* **2010**, 50,10-15. [CrossRef]

122. Tremouli, A.; Antonopoulou, G.; Bebelis, S.; Lyberatos, G. Operation and Characterization of a Microbial Fuel Cell Fed with Pretreated Cheese Whey at Different Organic Loads. *Bioresour. Technol.* **2013**,131, 380-389. [CrossRef] [PubMed]

123. Scott, K.; Rimbu, G.A.; Katuri, K.P.; Prasad, K.K.; Head, I.M. Application of Modified Carbon Anodes in Microbial Fuel Cells. *Process. Saf. Environ. Prot.* **2007**, 85, 481-488. [CrossRef]

124. Mahdi Mardanpour, M.; Nasr Esfahany, M.; Behzad, T.; Sedaqatvand, R. Single Chamber Microbial Fuel Cell with Spiral Anode for Dairy Wastewater Treatment. *Biosens. Bioelectron.* **2012**, 38, 264-269. [CrossRef] [PubMed]

125. Elakkiya, E.; Matheswaran, M. Comparison of Anodic Metabolisms in Bioelectricity Production during Treatment of Dairy Wastewater in Microbial Fuel Cell. *Bioresour. Technol.* **2013**,136, 407-412. [CrossRef] [PubMed]

126. Kelly, P.T.; He, Z. Understanding the Application Niche of Microbial Fuel Cells in a Cheese Wastewater Treatment Process. *Bioresour. Technol.* **2014**,157,154-160. [CrossRef]

127. Patil, S.A.; Surakasi, V.P.; Koul, S.; Ijmulwar, S.; Vivek, A.; Shouche, Y.S.; Kapadnis, B.P. Electricity Generation Using Chocolate Industry Wastewater and Its Treatment in Activated Sludge Based Microbial Fuel Cell and Analysis of Developed Microbial Community in the Anode Chamber. *Bioresour. Technol.* **2009**,100, 5132-5139. [CrossRef]

128. Zhang, B.; Zhao, H.; Zhou, S.; Shi, C.; Wang, C.; Ni, J. A Novel UASB-MFC-BAF Integrated System for High Strength Molasses Wastewater Treatment and Bioelectricity Generation. *Bioresour. Technol.* **2009**,100, 5687-5693. [CrossRef]

129. Mohanakrishna, G.; Venkata Mohan, S.; Sarma, P.N. Bio-Electrochemical Treatment of Distillery Wastewater in Microbial Fuel Cell Facilitating Decolorization and Desalination along with Power Generation. *J. Hazard. Mater.* **2010**,177, 487-494. [CrossRef]

130. Sevda, S.; Dominguez-Benetton, X.; Vanbroekhoven, K.; De Wever, H.; Sreekrishnan, T.R.; Pant, D. High Strength Wastewater Treatment Accompanied by Power Generation Using Air Cathode Microbial Fuel Cell. *Appl. Energy* **2013**,105,194-206. [CrossRef]

131. Venkata Mohan, S.; Mohanakrishna, G.; Sarma, P.N. Composite Vegetable Waste as Renewable Resource for Bioelectricity Generation through Non-Catalyzed Open-Air Cathode Microbial Fuel Cell. *Bioresour. Technol.* **2010**,101, 970-976. [CrossRef]

132. Mohanakrishna, G.; Venkata Mohan, S.; Sarma, P.N. Utilizing Acid-Rich Effluents of Fermentative Hydrogen Production Process as Substrate for Harnessing Bioelectricity: An Integrative Approach. *Int. J. Hydrog. Energy* **2010**, 35, 3440-3449. [CrossRef]

133. Yokoyama, H.; Ohmori, H.; Ishida, M.; Waki, M.; Tanaka, Y. Treatment of Cow-Waste Slurry by a Microbial Fuel Cell and the Properties of the Treated Slurry as a Liquid Manure. *Anim. Sci. J.* **2006**, 77, 634-638. [CrossRef]

134. Zheng, X.; Nirmalakhandan, N. Cattle Wastes as Substrates for Bioelectricity Production via Microbial Fuel Cells. *Biotechnol. Lett.* **2010**, 32,1809-1814. [CrossRef] [PubMed]

135. Deng, H.; Wu, Y.-C.; Zhang, F.; Huang, Z.-C.; Chen, Z.; Xu, H.-J.; Zhao, F. Factors Affecting the Performance of Single-Chamber Soil Microbial Fuel Cells for Power Generation. *Pedosphere* **2014**, 24, 330-338. [CrossRef]

136. Wang, C.-T.; Liao, F.-Y.; Liu, K.-S. Electrical Analysis of Compost Solid Phase Microbial Fuel Cell. *Int. J. Hydrog. Energy* **2013**, *38*, 11124-11130. [CrossRef]

137. Hassan, S.H.A.; Gad El-Rab, S.M.F.; Rahimnejad, M.; Ghasemi, M.; Joo, J.-H.; Sik-Ok, Y.; Kim, I.S.; Oh, S.-E. Electricity Generation from Rice Straw Using a Microbial Fuel Cell. *Int. J. Hydrog. Energy* **2014**, 39, 9490-9496. [CrossRef]

138. Wang, X.; Tang, J.; Cui, J.; Liu, Q.; Giesy, J.P.; Hecker, M. Synergy of Electricity Generation and Waste Disposal in Solid- State Microbial Fuel Cell (MFC) of Cow Manure Composting. *Int. J. Electrochem. Sci.* **2014**, 9,14.

139. Zhang, Y.; Min, B.; Huang, L.; Angelidaki, I. Generation of Electricity and Analysis of Microbial Communities in Wheat Straw Biomass-Powered Microbial Fuel Cells. *Appl. Env. Microbiol.* **2009**, 75, 3389-3395. [CrossRef]

140. Thygesen, A.; Poulsen, F.W.; Angelidaki, I.; Min, B.; Bjerre, A.-B. Electricity Generation by Microbial Fuel Cells Fuelled with Wheat Straw Hydrolysate. *Biomass Bioenerg.* **2011**, 35, 4732-4739. [CrossRef]

141. Jadhav, G.S.; Ghangrekar, M.M. Performance of Microbial Fuel Cell Subjected to Variation in PH, Temperature, External Load and Substrate Concentration. *Bioresour. Technol.* **2009**,100, 717-723. [CrossRef]

142. Picioreanu, C.; van Loosdrecht, M.C.M.; Curtis, T.P.; Scott, K. Model Based Evaluation of the Effect of PH and Electrode Geometry on Microbial Fuel Cell Performance. *Bioelectrochemistry* **2010**, 78, 8-24. [CrossRef]

143. Zhang, L.; Li, J.; Zhu, X.; Ye, D.; Liao, Q. Effect of Proton Transfer on the Performance of Unbuffered Tubular Microbial Fuel Cells in Continuous Flow Mode. *Int. J. Hydrog. Energy* **2015**, 40, 3953-3960. [CrossRef]

144. Zhang, L.; Li, C.; Ding, L.; Xu, K.; Ren, H. Influences of Initial PH on Performance and Anodic Microbes of Fed-Batch Microbial Fuel Cells. *J. Chem. Technol. Biotechnol.* **2011**, 86,1226-1232. [CrossRef]

145. Ye, Y.; Ngo, H.H.; Guo, W.; Chang, S.W.; Nguyen, D.D.; Liu, Y.; Nghiem, L.D.; Zhang, X.; Wang, J. Effect of Organic Loading Rate on the Recovery of Nutrients and Energy in a Dual-Chamber Microbial Fuel Cell. *Bioresour. Technol.* **2019**, 281,367-373. [CrossRef]

146. Li, X.M.; Cheng, K.Y.; Wong, J.W.C. Bioelectricity Production from Food Waste Leachate Using Microbial Fuel Cells: Effect of NaCl and PH. *Bioresour. Technol.* **2013**,149, 452-458. [CrossRef]

147. Madani, S.; Gheshlaghi, R.; Mahdavi, M.A.; Sobhani, M.; Elkamel, A. Optimization of the Performance of a Double-Chamber Microbial Fuel Cell through Factorial Design of Experiments and Response Surface Methodology. *Fuel* **2015**, 150, 434-440. [CrossRef]

148. Pandit, S.; Khilari, S.; Roy, S.; Pradhan, D.; Das, D. Improvement of Power Generation Using Shewanella Putrefaciens Mediated Bioanode in a Single Chambered Microbial Fuel Cell: Effect of Different Anodic Operating Conditions. *Bioresour. Technol.* **2014**,166, 451-457. [CrossRef]

149. Jannelli, N.; Anna Nastro, R.; Cigolotti, V.; Minutillo, M.; Falcucci, G. Low PH, High Salinity: Too Much for Microbial Fuel Cells? *Appl. Energy* **2017**,192, 543-550. [CrossRef]

150. Sajana, T.K.; Ghangrekar, M.M.; Mitra, A. Effect of Operating Parameters on the Performance of Sediment Microbial Fuel Cell Treating Aquaculture Water. *Aquac. Eng.* **2014**, 61,17-26. [CrossRef]

151. Pandit, S.; Savla, N.; Jung, S.P 16—Recent advancements in scaling up microbial fuel cells. In *Integrated Microbial Fuel Cells for Wastewater Treatment;* Abbassi, R., Yadav, A.K., Khan, F., Garaniya, V., Eds.; Butterworth-Heinemann: Oxford, UK, 2020; pp. 349-368. ISBN 978-0-12-817493-7.

152. Yadav, M.; Sehrawat, N.; Singh, M.; Kumar, V.; Sharma, A.K.; Kumar, S. Chapter 11—Thermophilic microbes-based fuel cells: An eco-friendly approach for sustainable energy production. In *Bioremediation for Environmental Sustainability;* Kumar, V., Saxena, G., Shah, M.P., Eds.; Elsevier: Amsterdam, The Netherlands, 2021; pp. 235-246. ISBN 978-0-12-820318-7.

153. Choi, Y.-J.; Jung, E.-K.; Park, H.-J.; Paik, S.R.; Jung, S.-H.; Kim, S.-H. Construction of Microbial Fuel Cells Using Thermophilic Microorganisms, Bacillus Licheniformis and Bacillus Thermoglucosidasius. *Bull. Korean Chem. Soc.* **2004**, 25, 813-818. [CrossRef]

154. Lefebvre, O.; Tan, Z.; Kharkwal, S.; Ng, H.Y. Effect of Increasing Anodic NaCl Concentration on Microbial Fuel Cell Performance. *Bioresour. Technol.* **2012**,112, 336-340. [CrossRef] [PubMed]

155. Najafpour, G.; Rahimnejad, M.; Ghoreshi, A. The Enhancement of a Microbial Fuel Cell for Electrical Output Using Mediators and Oxidizing Agents. *Energy Sour. Part A Recovery Util. Environ. Eff.* **2011**, 33, 2239-2248. [CrossRef]

156. Niessen, J.; Schroder, U.; Rosenbaum, M.; Scholz, F Fluorinated Polyanilines as Superior Materials for Electrocatalytic Anodes in Bacterial Fuel Cells. *Electrochem. Commun.* **2004**, 6, 571-575. [CrossRef]

157. Miyatake, K.; Chikashige, Y.; Watanabe, M. Novel Sulfonated Poly(Arylene Ether): A Proton Conductive Polymer Electrolyte Designed for Fuel Cells. *Macromolecules* **2003**, 36, 9691-9693. [CrossRef]

158. Li, Y.; Khanal, S.K. *Bioenergy: Principles and Applications;* John Wiley & Sons: New York, NY, USA, 2016; ISBN 978-1-118-56831-6.

159. Bond, D.R.; Holmes, D.E.; Tender, L.M.; Lovley, D.R. Electrode-Reducing Microorganisms That Harvest Energy from Marine Sediments. *Science* **2002**, 295, 483-485. [CrossRef]

160. Trapero, J.R.; Horcajada, L.; Linares, J.J.; Lobato, J. Is Microbial Fuel Cell Technology Ready? An Economic Answer towards Industrial Commercialization. *Appl. Energy* **2017**,185, 698-707. [CrossRef]

161. Logan, B.E. Scaling up Microbial Fuel Cells and Other Bioelectrochemical Systems. *Appl. Microbiol. Biotechnol.* **2010**, 85,1665-1671. [CrossRef] [PubMed]

162. Abourached, C.; Catal, T.; Liu, H. Efficacy of Single-Chamber Microbial Fuel Cells for Removal of Cadmium and Zinc with Simultaneous Electricity Production. *Water Res.* **2014**, 51, 228-233. [CrossRef]

163. Adekunle, A.; Raghavan, V.; Tartakovsky, B. On-Line Monitoring of Heavy Metals-Related Toxicity with a Microbial Fuel Cell Biosensor. *Biosens. Bioelectron.* **2019**,132, 382-390. [CrossRef] [PubMed]

164. Ghadge, A.N.; Ghangrekar, M.M. Performance of Low Cost Scalable Air-Cathode Microbial Fuel Cell Made from Clayware Separator Using Multiple Electrodes. *Bioresour. Technol.* **2015**,182, 373-377. [CrossRef] [PubMed]

165. Ghadge, A.N.; Jadhav, D.A.; Ghangrekar, M.M. Wastewater Treatment in Pilot-Scale Microbial Fuel Cell Using Multielectrode Assembly with Ceramic Separator Suitable for Field Applications. *Environ. Prog. Sustain. Energy* **2016**, 35,1809-1817. [CrossRef]

166. Hiegemann, H.; Herzer, D.; Nettmann, E.; Lubken, M.; Schulte, P.; Schmelz, K.-G.; Gredigk-Hoffmann, S.; Wichern, M. An Integrated 45L Pilot Microbial Fuel Cell System at a Full-Scale Wastewater Treatment Plant. *Bioresour. Technol.* **2016**, 218,115-122. [CrossRef]

167. Huang, L.; Yang, X.; Quan, X.; Chen, J.; Yang, F A Microbial Fuel Cell-Electro-Oxidation System for Coking Wastewater Treatment and Bioelectricity Generation. *J. Chem. Technol. Biotechnol.* **2010**, 85, 621-627. [CrossRef]

168. Reguera, G.; Speers, A.; Young, J. Biofuel and Electricity Producing Fuel Cells and Systems and Methods Related to Same. 2017. Available online: https://patents.google.com/patent/US9716287B2/en?oq=US9716287B2 (accessed on 28 August 2021).

169. Martin, D.J.; Annamalai, P.K.; Amiralian, N. Nanocellulose. 2016. Available online: https://patents.google.com/patent/EP30715 17A1/en?oq=lignocellulosic+biomass+%2b+MFCs (accessed on 31 March 2021).

170. May, H.D.; Shimotori, T. Apparatus and Methods for the Production of Ethanol, Hydrogen and Electricity. 2009. Available online: https://patents.google.com/patent/US20090017512A1/en?q=lignocellulosic+biomass+%2b+MFC&oq=lignocellulosic+biomass+%2b+MFC (accessed on 31 March 2021).

171. Reguera, G.; Speers, A.; Young, J. Biofuel and Electricity Producing Fuel Cells and Systems and Methods Related to Same. 2020. Available online: https://patents.google.com/patent/US10686205B2/en?q=lignocellulosic+biomass+%2b+MFC&oq=lignocellulosic+biomass+%2b+MFC (accessed on 1 April 2021).

172. Singh, A.; Pant, D.; Korres, N.E.; Nizami, A.-S.; Prasad, S.; Murphy, J.D. Key Issues in Life Cycle Assessment of Ethanol Production from Lignocellulosic Biomass: Challenges and Perspectives. *Bioresour. Technol.* **2010**,101, 5003-5012. [CrossRef]

173. Lam, M.K.; Lee, K.T.; Mohamed, A.R. Life Cycle Assessment for the Production of Biodiesel: A Case Study in Malaysia for Palm Oil versus Jatropha Oil. *Biofuels Bioprod. Biorefin.* **2009**, 3, 601-612. [CrossRef]

174. Vlasopoulos, N.; Memon, F.A.; Butler, D.; Murphy, R. Life Cycle Assessment of Wastewater Treatment Technologies Treating Petroleum Process Waters. *Sci. Total Environ.* **2006**, 367, 58-70. [CrossRef] [PubMed]

175. Oliveira, V.B.; Simoes, M.; Melo, L.F.; Pinto, A.M.F.R. Overview on the Developments of Microbial Fuel Cells. *Biochem. Eng. J.* **2013**, 73, 53-64. [CrossRef]

176. Pandit, S.; Chandrasekhar, K.; Kakarla, R.; Kadier, A.; Jeevitha, V. Basic Principles of Microbial Fuel Cell: Technical Challenges and Economic Feasibility. In *Microbial Applications Vol. 1: Bioremediation and Bioenergy*; Kalia, V.C., Kumar, P., Eds.; Springer International Publishing: Cham, Switzerland, 2017; pp. 165-188, ISBN 978-3-319-52666-9.

CHAPTER

5

Optimization of Yeast, Sugar and Nutrient Concentrations for High Ethanol Production Rate Using Industrial Sugar Beet Molasses and Response Surface Methodology

1. INTRODUCTION

Over the years, climate changes and air pollution attributed to the use of fossil fuels has been an increased concern all around the world [1]. One of the main reasons related to the actual environmental issues is the high amount of greenhouse gases (GHG) continuously released into the atmosphere due to anthropogenic activities (industries, land use, transportation, etc.) [2]. In this context, the use of biofuels such as bioethanol has emerged as a sustainable alternative to fossil fuels due to numerous advantages, including reduced emission of GHG, hydrocarbons and nitrogen oxides during both bioethanol production and combustion [3]. In this context, many countries have been adopting new policies to promote the use of biofuels, and the market perspectives are a 2-fold increase in Brazil and the USA and a 4-fold increase in China and EU by 2040 [4].

Among biofuels, bioethanol production through anaerobic fermentation of carbohydrates by yeast microorganisms is the most developed and implemented process at an industrial scale [5]. Fermentable sugars are generally extracted from various plants such as corn, sugarcane and sugar beet, which represent the main feedstock used for the actual worldwide ethanol production [6]. However, even if part of the carbon released after the fuel combustion is captured by the crops during their growth, the production of bioethanol from conventional crops can still impact carbon neutrality [4].

One way to reach a carbon neutrality (or even negative carbon emissions) could involve reusing biomass residues such as agricultural and food wastes as alternative feedstocks [7]. These by-products represent a concrete solution to avoid the direct land use impact of using crops for the production of fuels, especially in light of life cycle assessment calculations that will determine the carbon impact of biofuels [8]. Recent studies reported the use of various residues such as food waste [9], corncobs [10] and algae waste [11] for the sustainable production of bioethanol. In the case of lignocellulosic biomass, several pre-treatment steps are often required to break down the complex lignin structure and depolymerise the crystalline cellulose to access fermentable sugars [12]. However, residues from the food industry often have the advantage of being rich in easily fermentable carbohydrates, which could simplify the overall process while reducing the energy requirements for the process and possibly, the carbon intensity of the downstream fuel [13].

Sugar beet and cane molasses are abundant liquid by-products from the sugar industry, which are generally found at high amount of total sugars (50.6-71.0% *w/w)* and traces of micronutrients such as minerals (Ca, Mg, Na and K), phosphate and nitrogen compounds [14]. These sugar-rich solutions does not require any major physical or chemical pretreatment (such as hydrolysis, filtration, sterilization, etc.) before fermentation, making them very appropriate for ethanol production. In this context, several studies already reported the efficient production of bioethanol using sugar beet and sugarcane molasses. For instance, Razmovski et al. compared the fermentation performances of immobilised and free yeast cells using sugar beet molasses and thick juice at three initial glucose concentrations of 100, 200 and 300 g/L [13]. For both substrates, immobilised yeasts resulted in higher ethanol yields than free cells condition, with the highest ethanol volumetric productivity of 1.257 and 1.422 g/L/h obtained with molasses and thick juice diluted at 150 g/L of initial sugar, respectively. In another study, the continuous production of ethanol was investigated using an immobilised yeast cell reactor and sugarcane molasses as low-cost fermentation substrate [15]. The highest ethanol production of 19.15 g/L was obtained with a hydraulic retention time of 15.63 h combined with an initial sugar concentration of 150 g/L. In a recent study, a rotary biofilm reactor was developed for long-term bioethanol production using non-sterilised sugar beet molasses [16]. By recycling 30% of the fermentation broth every 36 h, a stable production of 52.3 g/L of ethanol was achieved over a period of 60 days. In addition, molasses can also be used for the production of different alcoholic beverages such as rum, a spirit distillate with an ethanol content of 37-43% alcohol by volume [17]. In another study, the ethanol production performances of corn marsh feedstock were improved by the addition of sugarcane molasses [18]. Mixing 50% of corn mash with 50% of sugar cane molasses generated the highest ethanol concentration (8.2%).

To implement an efficient fermentation process using industrial by-products, several process parameters such as sugar and yeast concentrations must be optimised to ensure high ethanol productivity while keeping in mind the economic aspects of the process [19]. Nutrient supplementation is also an important parameter to take into consideration since an adequate amount of nutrient can significantly improve yeast viability and resistance to the medium, stimulating ethanol production performances. Since alcoholic fermentation is a complex biological process involving various operating factors, the use of the classical "one factor at a time" approach could be time-consuming due to the large number of experiments to perform. Hence, tools such as the statistical design of experiment (DoE) allow investigating the effect of several operational factors as well as their interactions on the overall process while considerably reducing the number of experimental tests [20]. For instance, a central composite design coupled with response surface methodology (CCD- RSM) represents a powerful and effective statistical tool that could commonly be used for the optimization of biotechnological processes such as fermentation [19,21,22]. Once developed, the CCD-RSM can be used to predict a process output, while imposing specific constraints based on economical and technical aspects.

In this context, the present work aims to optimise the production of ethanol from non-treated sugar beet molasses produced by a local sugar refinery. For this purpose, a CCD-RSM was designed and developed to investigate the effect of three fermentation process parameters (initial sugar, yeast and nutrient concentrations) on ethanol productivity while considering several operating parameters such as ethanol yield and sugar utilization rate. Then, the second-order mathematical model obtained through the CCD-RSM was tested to evaluate its ability to make accurate predictions based on specific desired process outputs. To the best of our knowledge, this is the first study reporting the use of a CCD- RSM statistical approach to maximise the production of ethanol from non-sterilised sugar beet molasses while scaling up the experimental results up to a 100 L bioreactor scale.

2. MATERIALS AND METHODS

2.1. Sugar Beet Molasses Preparation

The sugar beet molasses was collected from a local sugar refinery located in Coaticook (Quebec, Canada) and stored at 4°C until further use. The sugar beet crops were cultivated and harvested by the same company before being processed to produce sugar. The fresh molasses was initially characterised for pH, density as well as for sugars and other metabolites concentrations (see Table 1).

Table 1. Chemical and physical properties of the raw sugar beet molasses used for the RSM fermentation assays.

Parameters	Values
p[h]	6.36
Density (kg/L)	1.30
Total sugars (g/L)	655.18
Sucrose (g/L)	560.95
Glucose (g/L)	55.23
Fructose (g/L)	39.00
Acetic acid (g/L)	0.79
Lactic acid (g/L)	11.28

2.2. Statistical Experimental Design Methodology

The first part of this study aimed to optimise the ethanol fermentation of sugar beet molasses using a CCD-RSM. This statistical approach allows investigating specific operational parameters together with their related effects on the fermentation process with a limited number of experiments. Therefore, a CCD-RSM was implemented for three process variables: initial sugar concentration (Factor A) as well as yeast (Factor B) and nutrient (Factor C) concentrations. These investigated factors were studied at low, middle and high levels, coded -1, 0 and +1, respectively (Table 2). In addition, two centre points were added to check for curvature and verify the repeatability of the fermentation process. Finally, confirmation experiments were performed to verify the accuracy of the developed mathematical model and its capacity to predict process responses within the design space. The CCD-RSM was designed and analysed using the Design-Expert 12 software (Stat-Ease Inc., Minneapolis, MN, USA).

Factors	Symbols	-1	0	+1
Sugar concentration (g/L)	A	125	225	325
Yeast concentration (g/L)	B	0.2	0.6	1.0
Nutrient concentration (g/L)	C	0	2	4

The ethanol productivity was used as the main experimental process response (Y1) for the CCD-RSM and calculated using Equation (1).

$$\text{Ethanol productivity (g/L/d)} = EtOH_{max} / t_{fermentation} \quad (1)$$

where $EtOH_{max}$ represents the maximum ethanol concentration (g/L) and $tf_{ermentation}$ represents the time of incubation (d) required to achieve the $EtOH_{max}$.

In addition, ethanol yields were calculated based on the maximum theoretical ethanol concentration calculated for the three initial sugar concentrations. Using Equation (2), it can be assumed that 1 g of glucose produces 0.51 g of ethanol and 0.49 g of carbon dioxide (CO2).

$$C_6H_{12}O_6 \rightarrow 2C_2H_6O + 2CO_2 \quad (2)$$

Ethanol yields were expressed as % of the maximum theoretical ethanol concentration.

2.3. Fermentation Experiments

The fermentations assays were carried out according to the different experimental lines described by the DoE (Table 3). Therefore, sugar beet molasses was diluted to reach the required initial sugar concentration using tap water and the respective amount of nutrient (Fermaid O, Lallemand Biofuels & Distilled Spirits, Montreal, QC, Canada) was added to the solution. Fermaid O is a mix of inactivated yeast fractions, rich in organic nitrogen, generally used in the wine industry. This nutrient was not supplemented by any ammonia salts or micronutrients. Fermaid O was referred to as "nutrient" parameter in this study. After pH adjustment to 5.5 using an 11 M H_2SO_4 solution, fermentation broths were placed into 50 mL serum vials with a working volume of 25 mL. Yeast inoculum was prepared by rehydrating *Saccharomyces cerevisiae* dry yeast cells (Thermosacc® DRY, Lallemand Biofuels & Distilled Spirits, Canada) with tap water at 30°C, 140 rpm for a period of 15 min using a shaking incubator (Ecotron, Infors-HT Inc., Bottmingen, Switzerland). Three concentrated yeast suspensions were prepared to obtain the desired initial yeast concentration in the final fermentation media. After yeast inoculation, all the fermentation vials were capped with rubber septum stoppers and aluminium rings, before being flushed with N2 for 5 min to ensure an anaerobic environment. The fermentation vials were finally incubated at 30°C for 112 h under a 140 rpm stirring. Temperature and agitation parameters were not optimised in this work and their values were selected based on our previous study dealing with the fermentation of softwood residues [21]. Liquid samples were taken at the beginning and at the end of the experiment to quantify different metabolites and products such as sugars and ethanol concentrations. All conditions were carried out in triplicate and results are expressed as average value ± standard deviation.

Table 3. Experimental design matrix of the CCD-RSM study presenting the different experimental lines as well as the investigated response generated during the molasses fermentation optimization.

Run	Factor A	Factor B	Factor C	Response Y1
	Sugar Concentration g/L	Yeast Concentration g/L	Nutrient Concentration g/L	Ethanol Productivity ± SD g/L/d
1	125	0.2	0	5.7 ± 0.1
2	125	0.6	2	15.0 ± 0.0
3	325	0.2	4	12.8 ± 0.1
4	325	0.6	2	12.8 ± 0.0
5	125	1	4	18.3 ± 0.1
6	225	0.6	4	19.8 ± 0.8
7	325	0.2	0	2.6 ± 1.0
8	125	0.2	4	17.9 ± 0.2
9	325	1	4	15.8 ± 0.1
10	225	0.2	2	11.4 ± 0.1
11	325	1	0	10.6 ± 0.8
12	225	0.6	2	15.4 ±0.2
13	125	1	0	9.8 ± 0.1
14	225	0.6	0	10.3 ± 0.1
15	225	1	2	17.2 ± 0.6
16	225	0.6	2	15.3 ± 0.3

The optimal conditions obtained during the CCD were used to design and perform scale-up assays using 7.5 and 100 L bioreactors with working volumes of 5 and 50 L, respectively (Figure 1). The 7.5 L glass bioreactor (Infors-HT Inc., Bottmingen, Switzerland) was equipped with a double envelope to control the temperature at 30°C, meanwhile, a mechanical stirrer provided a continuous agitation at 250 rpm. I addition, a water- cooled condenser was attached to the reactor to prevent evaporation. The 100 L stainless bioreactor was equipped with a 2000 W heating element to maintain the fermentation media temperature to 30°C. Constant stirring at 70 rpm was achieved using a 2-blades impeller system powered by a 12 Volt DC electric motor.

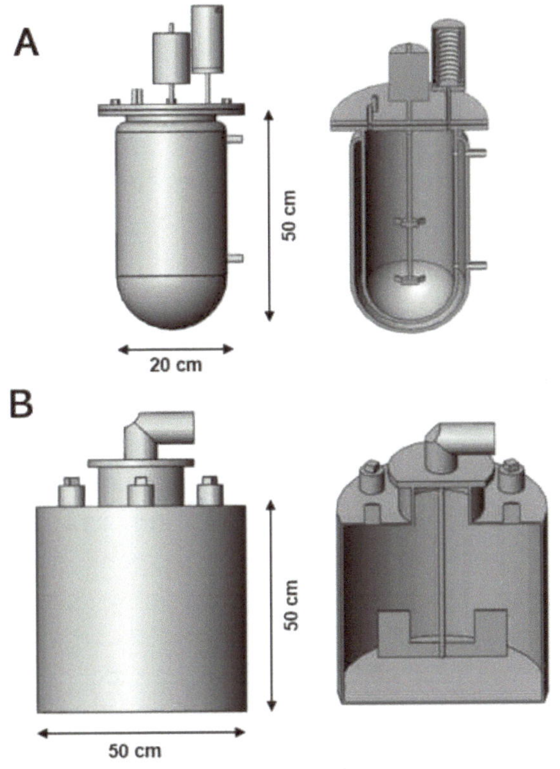

Figure 1. Schematic view of the 7.5 L (A) and 100 L (B) stirred-tank bioreactors used to perform the fermentation scale-up experiments. The optimal operating conditions determined during the CCD-RSM were utilised to maximise the production of ethanol from non-treated sugar beet molasses.

2.4. Analytical Procedures

2.4.1. Gravimetric Analysis

To monitor the ethanol production of a large number of fermentation media during the CCD-RSM, a gravimetric methodology was developed to quantify the amount of CO2 released during the fermentation procedure. Considering the alcoholic fermentation reaction (Equation (2)) that theoretically gives two moles of ethanol and two moles of CO_2 per initial mole of glucose, the ethanol concentration could be indirectly calculated by following the production of CO2. This gravimetric technique is a simple and efficient tool to follow ethanol production kinetics, with the advantage to be less time- and resourceconsuming than conventional methods. Moreover, this technique assumes that only CO_2 and ethanol are produced during fermentation.

To validate this technique, a calibration curve was prepared by plotting "the ethanol calculated by mass loss" versus "the ethanol quantified by HPLC" (details in Section 2.4.2), which were referred to as $E_m l$ and $Ehpl_c$, respectively (Figure S1). The linear regression showed a coefficient of determination (R^2) of 0.97 with Equation (3).

$$E_{hplc} = 0.84 \times E_{ml} - 9.95 \tag{3}$$

All the ethanol results shown in this study, except for the ones obtained during the scale-up experiments, were calculated using the latest calibration curve.

2.4.2. Sugars and Ethanol Quantification

Although the gravimetric analysis is an alternative option to follow the fermentation development, it is still necessary the quantify monomeric sugars and other fermentation- related metabolites to ensure accurate results. To this purpose, a Dionex ICS-5000+ ion chromatography system was used but more specifically for sugars quantification. The system was equipped with a KOH eluent generator to provide a proper eluent concentration, an analytical gradient pump, a thermostatic AS-AP auto-sampler and an electrochemical detector. A 200 mM KOH post-injection with a Dionex GP 50 gradient pump was implemented to ensure signal stability. For this project, a Dionex CarboPac Sa10-4 gM column was used while the oven was set at 45°C and the electrochemical detector was at 30°C. The injection volume was 0.4 gL and elution was made with an aqueous solution of KOH at 1.25 mL/min at the following concentrations: 1 mM for 12 min, 10 mM for 5 min and 1 mM for 1 min. The calibration curve involved a concentration of standards varying from 10 ppm to 1000 ppm and involved the following standards: fructose 99%, glucose 99% and sucrose 99% which were all purchased from Sigma-Aldrich.

An Agilent 1100 series HPLC (Agilent Technologies Inc., Colorado Springs, CO, USA) equipped with a G1362A Refractive Index Detector was used to quantify ethanol. The system was operated at 40°C with an isocratic elution method (2.5 mM). The HPLC set-up also had a G1322A Degasser and a G1311A Quaternary Pump. A G1313A Autosampler injected 40 gL of sample and the column used was a ROA-Organic Acid H+ (8%) at 65°C. The elution was performed at a constant flow of 0.08 mL/min of a 0.01 M H_2SO_4 solution. A calibration curve from 10 ppm to 1000 ppm was performed using the following standards: L-lactic 99% (Alfa Aesar), acetic acid 99.9% (Sigma Aldrich), ethanol 99% (Sigma Aldrich) and glycerol 99% (Sigma Aldrich).

3. RESULTS AND DISCUSSION

3.1. Ethanol Production Kinetics Overview

One of the main objectives of this study was to elucidate the effect of different operating fermentation parameters on the ethanol production process to develop a cost-effective and efficient strategy to valorise sugar beet molasses. In this sense, a CCD-RSM was designed and developed using three experimental process parameters (initial sugar, yeast and nutrient concentrations) investigated at three levels, using ethanol productivity as the main process response.

Ethanol production was strongly affected by the three investigated process parameters, which resulted in different fermentation kinetics among the 16 experiments of the CCD- RSM (Figure 2). For all the conditions, a short lag phase of approximately 24 h was noticed before measuring any production of ethanol. This period is probably due to the adaptation of yeast cells after their inoculation into the sugar-rich media. By the end of the incubation time, only three conditions with an initial sugar concentration of 125 g/L, various yeast (0.2, 0.6 and 1.0 g/L) and nutrient (2 and 4 g/L) concentrations reached a plateau with a constant ethanol concentration of 54 g/L. The highest ethanol production of 92 g/L was observed when the sugar beet molasses was diluted to 225 g/L of initial sugars combined with the use of 0.6 g/L and 4.0 g/L of yeast and nutrient, respectively. The positive effect of nutrient supplementation was highlighted by observing very slow fermentation kinetics for media not supplemented with nutrients compared with the ones with nutrients.

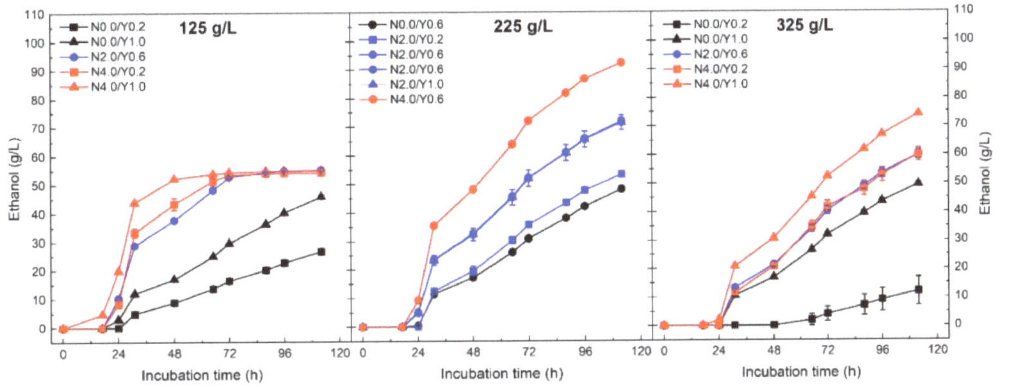

Figure 2. Ethanol production for the different initial sugar concentration of sugar beet molasses (125,225 and 325 g/L) using Saccharomyces cerevisiae yeast. Colours are attributed to the initial nutrient concentration: black (0.0 g/L), blue (2.0 g/L) and red (4.0 g/L). Symbols are related to the initial yeast concentration: square (0.2 g/L), circle (0.6 g/L) and triangle (1.0 g/L).

3.2. Design Matrix and Ethanol Productivity Analysis

The fermentation results obtained during the CCD-RSM were analysed by calculating the ethanol productivity of each experimental line, based on the maximum ethanol concentration achieved after a specific incubation time (Table 3). To quantify the fermentation performances and perform an in-depth comparison of the results, ethanol yields expressed as a percentage of the maximum theoretical ethanol yield were calculated and discussed in the following section.

As observed for the fermentation kinetics, various responses were detected among the experimental lines of the CCD-RSM, with ethanol productivity values ranging from 2.6 to19.8g/L/d with average productivity of13.2g/L.A total of 8 experiments achieved ethanol productivities higher than 15.0 g/L by the end of incubation time.

However, ethanol yields calculated for all the 16 experimental runs revealed that only a few numbers of conditions reached high ethanol yields (Figure 3). In fact, the initial sugar concentration had an important effect on the fermentation performances since only the experiments with sugar concentrations comprised between 125 and 225 g/L reached ethanol yields above 60%. In addition, the highest ethanol yields of 82% were generated for three experimental conditions (run 8, 5 and 2), where a minimum of 2 g/L of nutrient and 0.2 g/L of yeast were used, combined with an initial sugar concentration of 125 g/L. Finally, the lower ethanol yield of 7.0% was obtained for the fermentation media incubated without any nutrient supplementation combined with a low amount of yeast (0.2 g/L) and a high concentration of initial sugar (325 g/L).

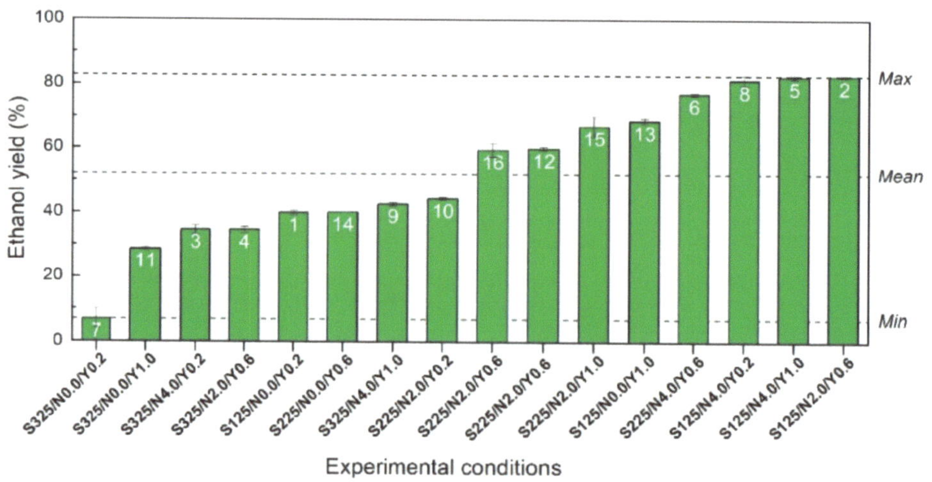

Figure 3. Ethanol yields expressed as a percentage of the maximum theoretical ethanol concentration for all the investigated runs of the CCD-RSM (white label). Yields were sorted in ascending order for a better interpretation of the ethanol performances. Dashed lines show the lowest (Min), the highest (Max) and the average (Mean) ethanol yields.

The high ethanol volumetric productivity values obtained in our study, ranging from 15.0 to 19.8 g/L/d, were in agreement with the fermentation results previously presented by Razmovski et al. using sugar beet molasses [13]. The use of non-immobilised yeast cells generated ethanol volumetric productivities of 20.09,19.82 and 13.51 g/L/d when using initial sugar concentration of 100,150 and 300 g/L, respectively. Similarly, Vucurovic et al. noticed a significant reduction in ethanol performances including sugar utilization, ethanol productivity and yield when fermenting sugar beet molasses with sugar concentration higher than 175 g/L, using free yeast cells [23]. These observations were correlated with lower cell yeast viability, attributed to the presence of high sugar concentration as well as non-sugar compounds such as mineral salts which might affect yeast metabolism.

The analysis of experimental ethanol yields was useful to correlate the ethanol productivity values with actual ethanol performances based on the maximum theoretical ethanol yield. The latest parameter is often calculated considering 100% sugar utilization by yeast cells during fermentation. Such operational parameter must be evaluated when implementing a fermentation process in order to take full advantage of the initial sugars present in the feedstock for the production of ethanol. In this sense, sugar concentration between 125 and 225 g/L and the addition of nutrient should be carefully considered to achieve high ethanol yields while producing at least 15 g/L/d of ethanol.

3.3. Mathematical Model Analysis

The ethanol productivity results presented in the last section were subjected to an analysis of variance to evaluate if the mathematical model was significant (Table 4). β-values lower than 0.0500 showed that the model is significant while a Fit-value of 107.40 indicates there is only a 0.01% of chance that a Fit-value this large could occur due to noise. Similarly, A, B, C, AB, BC, A^2, B^2 model terms were statistically significant with β-values lower than 0.0006. Regarding the pure error, the Lack of Fit was considered as not significant with a Fit-value of 57.66 and a β-value of 0.0996. Thus, there is only a 9.96% chance that a Lack of Fit-value this large could occur due to noise.

In addition, the accuracy of the model and its ability to predict in the design space was verified by the close relationship between the predicted and the experimental values obtained for the ethanol productivity main response. The Predicted R^2 of 0.9436 was found to be in reasonable agreement with the Adjusted R^2 of 0.9846 since the difference was less than 0.2. A correlation between these two parameters was also confirmed by a coefficient of determination (R^2) of 0.9938 when conducting a linear regression of the predicted and the experimental data. The signal to noise ratio of 36.9494 ("Adeq Precision") was significantly higher than 4, indicating an adequate signal to navigate the design space. In addition, the very similar ethanol productivity of 15.4 and 15.3 g/L/d obtained for both centre points (run 12 and 16, respectively) confirmed the stability and the reproducibility of the results (Table 3).

Table 4. ANOVA table of CCD-RSM used to optimise the fermentation of diluted sugar beet molasses using S. cerevisiae yeast.

Source	Sum of Squares	df	Mean Square	F-Value	p-Value[1]
Model	325.15	9	36.13	107.40	<0.0001
A—Sugar concentration	14.78	1	14.78	43.94	0.0006
B—Yeast concentration	44.81	1	44.81	133.20	<0.0001
C—Nutrient concentration	209.71	1	209.71	623.45	<0.0001
AB	5.55	1	5.55	16.51	0.0066
AC	3.58	1	3.58	10.65	0.0172
BC	9.79	1	9.79	29.10	0.0017
A^2	8.37	1	8.37	24.87	0.0025
B^2	5.16	1	5.16	15.34	0.0078
C^2	1.15	1	1.15	3.40	0.1146
Residual	2.02	6	0.3364		
Lack of Fit	2.01	5	0.4023	57.66	0.0996
Pure Error	0.0070	1	0.0070		
Cor Total	327.17	15			

[1] *β-values < 0.0500 represent significant model terms.*

3.4. RSM Optimization

The RSM analysis highlighted the combined effect of the three fermentation process parameters on the ethanol productivity response (Figure 4). As observed previously, nutrient supplementation was one of the most influential variables, strongly affecting the fermentation performances. Optimal regions for maximizing the overall ethanol productivity (>15 g/L/d) were detected for a nutrient concentration of 4 g/L combined with an initial sugar concentration lower than 225 g/L and a minimum of 0.2 g/L of yeast (Figure 4C). In the absence of nutrient, ethanol production rates were very slow (<10.6 g/L/d), suggesting that the molasses feedstock alone did not offer a sufficient level of nutrient for *S. cerevisiae* yeast metabolism. These findings are slightly different from other studies reporting the efficient production of ethanol from sugar beet molasses without nutrient supplementation [13,23]. However, yeast cells used in these studies were previously grown in a nutrient-rich solution before inoculation, which was very different from our simple rehydration yeast preparation protocol.

In addition, ethanol productivity performances were negatively affected when increasing the amount of initial sugar present in the fermentation system, which was probably due to an increase in osmotic stress. This behaviour was similarly reported by Vucurovic et al. when measuring *S. cerevisiae* yeast viability at the end of sugar beet molasses fermentation assays. In the absence of nutrient, the increase of initial sugar concentration from 100 to 300 g/L significantly decreased the viability of free cells from around 100 to 63.9% [17].

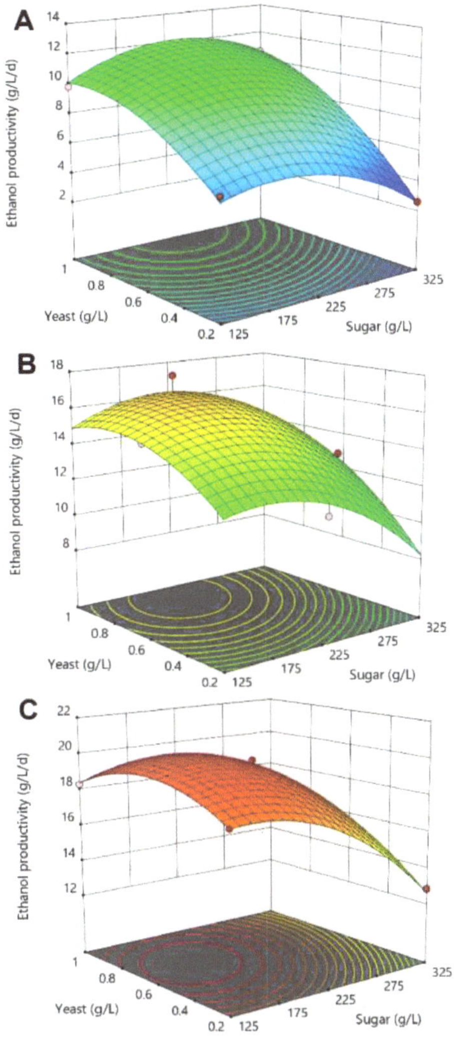

Figure 4. 3D-response curves of the ethanol productivity as a function of yeast and sugar concentrations for the three investigated nutrient concentrations of 0 (A), 2 (B) and 4 g/L (C). Untreated sugar beet molasses, diluted at several initial sugar concentration, was used as the fermentation feedstock. Red dots within the response curves represent experimental design points obtained during the CCD-RSM while contour lines show prediction outputs calculated by the mathematical model. 3.5. Validation Experiments

To confirm that the CCD-RSM model can be used to make accurate process predictions, validation experiments were performed at lab-scale using 50 mL vials based on the results presented in the first part of this study. Two process scenarios called respectively "scenario A" and "scenario B" were designed to maximise the ethanol productivity while taking into consideration specific operational process constraints. For both scenarios, the initial sugar concentration was kept between 125 and 225 g/L at a specific concentration of,170 g/L to trigger high ethanol yield and productivity responses. In addition, scenario B was designed to minimise the quantity of initial yeast while still trying to maximise the ethanol productivity response. Finally, the nutrient content was predicted within the design space with 3.9 and 4.0 g/L of initial nutrient concentration for scenario A and B, respectively.

Isoresponses curves predicting the ethanol productivity give valuable information regarding the combinatory effect of nutrient and yeast concentration on the process output, at a specific sugar concentration of 170 g/L (Figure 5A). Ethanol productivity performances higher than 19 g/L/d could be achieved by using a minimum of 3.5 g/L of nutrient during the fermentation. Interestingly, the quantity of yeast initially introduced in the system could be significantly lower to 0.2 g/L while still generating ethanol at a high production rate.

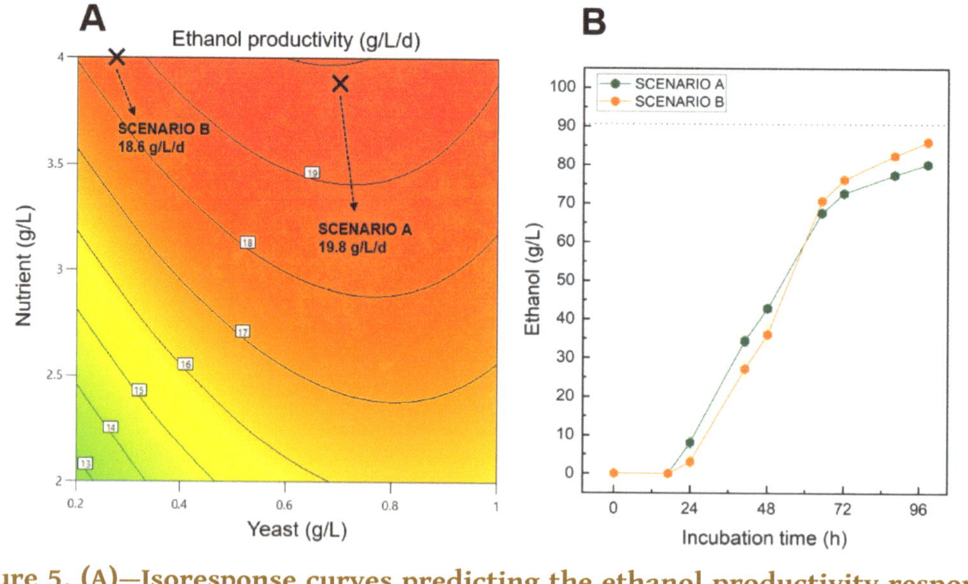

Figure 5. (A)—Isoresponse curves predicting the ethanol productivity responses based on nutrient and yeast concentrations using non-treated sugar beet molasses diluted at 170 g/L of initial sugar. (B)—Experimental fermentation kinetics of the two investigated scenarios for an initial sugar concentration of 170 g/L. Both scenarios A and B were designed for maximizing ethanol productivity with a high (0.7 g/L) and low concentration (0.27) of yeast, respectively. In addition, 3.9 and 4.0 g/L of nutrient were respectively used for scenario A and B. The dashed line represents the theoretical maximum ethanol yield of both scenarios.

Confirmation experiments showed that after a first lag phase of 24 h, both scenarios continuously produced ethanol before reaching more than 70 g/L of ethanol after 72 h of incubation (Figure 5B). By the end of the fermentation, scenario A and B reached final ethanol concentrations of 80.2 and 86.0 g/L, which represent theoretical ethanol yields of 88.5% and 94.9%, respectively. In addition, high ethanol productivities were achieved during the confirmation experimental assays with 19.5 and 21 g/L/d obtained for scenario A and B, respectively.

The mathematical model developed in this study was able to accurately predict the ethanol productivity response with an error lower than 12.9%, which we considered as statistically acceptable based on the challenges of reproducibility associated with the implementation of biotechnological processes such as the fermentation one. In addition, the confirmation experimental results highlighted the fact that the quantity of yeast can be considerably reduced while obtaining high ethanol productivity of 18.6 g/L/d (scenario B). This might have a major impact on the operating expenses of the fermentation process, reducing by 2.6-fold the cost related to the use of yeast.

3.6. Scale-Up Assays

Scaling up biological processes such as the alcoholic fermentation process often requires further investigation at an intermediate scale to realise a smooth transition from bench-scale to pilot-scale. In this context, experimental results obtained in the first part of this study were used to design and perform fermentation assays at bioreactor-scale. Diluted sugar beet molasses, with an initial total sugar concentration of 150 g/L, were used as feedstock to perform batch fermentation using two bioreactors with total volumes of 7.5 and 100 L, respectively. Fermentation broths were regularly sampled to monitor the sugar and ethanol concentration throughout the experiment.

During the first 24 h of incubation, the diminution of sucrose concentration was correlated with the production of 22.5 g/L of ethanol as well as an increase of glucose and fructose concentrations, for both the 7.5 and 100 L bioreactors (Figure 6). The latter phenomena can certainly be attributed to the hydrolysis of sucrose by *S. cerevisiae's* invertase enzyme [24]. It is worth mentioning that glucose was preferentially utilised by yeast cells, with a higher consumption rate than fructose. Then, the concentration of both reducing sugars continuously decreased over the fermentation and after 64 h of reaction, could not be detected anymore.

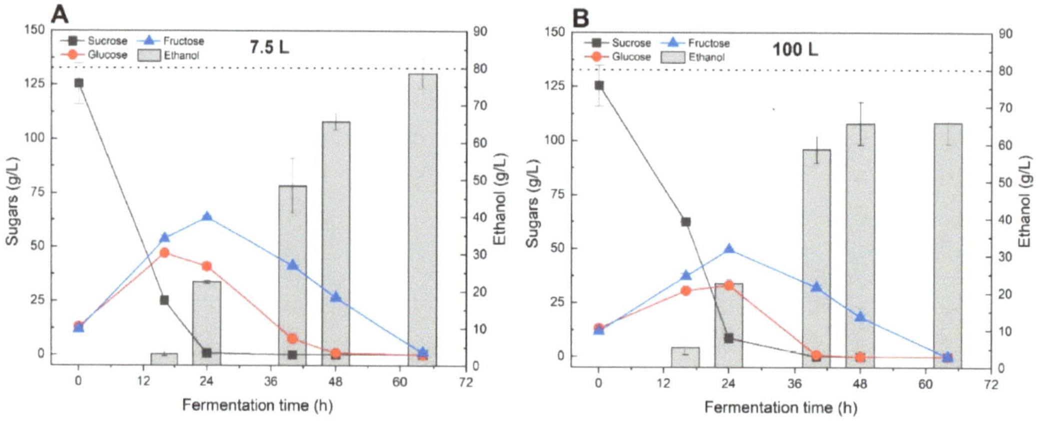

Figure 6. Ethanol and sugar concentration variations during fermentation scale-up assays performed in 7.5 L (A) and 100 L (B) batch bioreactors. Non-treated diluted sugar beet molasses with a total initial sugar concentration of 145 g/L was used for this experiment with 4.0 g/L of nutrient and 0.27 g/L of yeast. The dashed line refers to the theoretical maximum ethanol yield.

The two bioreactors presented slightly different behaviours in term of ethanol productivity as well as sugar utilization rate. For the 7.5 L bioreactor, ethanol concentration increased from 65.6 to 79.6 g/L between 48 and 64 h of fermentation, corresponding to a variation of ethanol yields from 81.9% and 99.5%, respectively. Interestingly, the ethanol concentration measured in the 100 L bioreactor remained similar between 48 and 64 h of incubation (around 65 g/L of ethanol), even if the fructose concentration decreased from 18.4 to 0 g/L, after 16 h. The latest observation suggests that some ethanol might have been produced during the last hours of incubation but directly escaped the fermentation broth due to evaporation. This situation could be explained by the fact that the 100 L bioreactor was not equipped with a water-cooled condenser system as compared to the 7.5 L bioreactor.

In terms of ethanol productivity, the 7.5 L bioreactor generated 29.9 g/L/d of ethanol compared with 23.7 g/L/d for the 100 L bioreactor when calculating the productivity after 64 h of incubation (Table 5). The lower fermentation performances observed for the 100 L bioreactor highlights the risk of ethanol evaporation during the scale-up of fermentation processes. In addition, analysis of the final fermentation broths showed very low levels of organic compounds such acetic and lactic acids, which confirmed that the scale-up experiments were not subjected to any major microbial contamination.

Table 5. Fermentation performances and by-products concentrations measured after 64 h of incubation at 7.5 and 100 L bioreactors scale using diluted sugar beet molasses (170 g/L of initial sugar) together with 0.27 g/L of Saccharomyces cerevisiae and 4.0 g/L of nutrient.

Parameters	7.5 L	100 L
Process performances		
Ethanol (g/L)	79.6 ± 0.7	63.2 ± 1.1
Ethanol productivity (g/L/d)	29.9 ± 0.4	23.7 ± 0.4
Ethanol yield (%)	99.5 ± 0.9	78.9 ± 1.4
Sugar utilization (%)	99.2 ± 0.8	100.0 ± 0.0
By-products		
Glycerol (g/L)	8.7 ± 0.2	6.5 ± 0.2
Lactic acid (g/L)	3.8 ± 0.1	3.4 ± 0.2
Acetic acid (g/L)	0.8 ± 0.1	0.2 ± 0.0

The ethanol productivity values obtained during the scale-up assays were found to be significantly higher than the ones obtained at lab-scale using the 50 mL vials, probably due to the difference of scale and reactor configuration. The most important difference between the incubator and bioreactor scales might be related to the mixing conditions, impacting the mass transfer between the yeast cells and the fermentation broth. The agitation performed in the incubator with the 50 mL fermentation vials was indirectly using an orbital shaker, which was certainly less efficient than the direct agitation maintained by a mechanical impeller, in the case of the 7.5 and 100 L reactors. In the case of the bioreactors, the microorganisms were continuously in suspension which could have significantly increased their contact with the molasses, resulting in a higher sugar utilization rate and ethanol productivity. In a recent study, the production of ethanol using raw cassava starch was investigated from a 5 L laboratory and a 200 L pilot scale up to a 3000 L industrial-scale system [25]. Although the technology and dimension of the three latter bioreactors were different, very comparable ethanol productivities and kinetics parameters were obtained for the three scales by adjusting the power input, also known as the energy dissipation rate per unit mass, to a specific value of 0.10 W/kg. Consequently, the agitation rates of the 5, 200 and 3000 L bioreactors were respectively fixed at 200,125 and 55 rpm based on the fermenter dimension, impeller type and size.

In addition to the mixing differences, the fermentation vials used during the CCD- RSM were sealed and flushed with nitrogen while the bioreactors were let at atmospheric pressure without nitrogen sparging. Thus, the build-up of pressure accumulated within the small fermentation vials could have affected the yeast metabolism, lowering the ethanol production performances. Similar findings were reported when investigating the effect of different process parameters on the fermentation of apple pomace by co-culturing filamentous fungi with yeast [26]. At an inoculation rate of 4.0% (v/v) of *S. cerevisiae* combined with 4.0% *(w/v)* of *T. harzianum* and *A. sojae,* increasing the agitation speed from 0 to 200 rpm showed a significant increase in the ethanol production when employing non-sealed reactor vessels.

These results highlighted the challenges associated with the scale-up of fermentation processes from lab to bioreactor scales, especially regarding the impact of agitation speed and reactor dimension. In another study, the scale-up of bioethanol production from soybean molasses raised different problems, especially with the formation of foam and the risk of bacterial contamination at pilot and industrial scales [27].

4. CONCLUSIONS

This work highlighted the valorisation of agro-industrial by-products for the efficient production of ethanol by yeast fermentation. The CCD-RSM methodology was employed as a powerful statistical tool to optimise ethanol productivity while considering the specific techno-economical process constraints such as the cost of *S. cerevisiae* yeast. Experimental results suggested that the raw sugar beet molasses should be diluted between 125-250 g/L of initial sugar and supplemented with a high concentration of nutrient (4 g/L) to achieve high ethanol productivity and yield responses. In addition, the quantity of yeast present in the fermentation media was significantly reduced from 0.70 to 0.27 g/L while achieving high ethanol production performances. Assays performed in 7.5 and 100 L bioreactors suggested that the fermentation process has the potential to be implemented at pilot and industrial scales for the production of bioethanol in the context of the circular economy. In addition, a detailed techno-economic evaluation of the whole fermentation process could provide valuable information for the potential commercialization of this integrated biorefinery approach.

REFERENCES

1. Heede, R. Tracing anthropogenic carbon dioxide and methane emissions to fossil fuel and cement producers, 1854-2010. *Clim. Chang.* **2014**,122, 229-241. [CrossRef]
2. Boden, T.A.; Andres, R.J.; Marland, G. *Global, Regional, and National Fossil-Fuel CO2 Emissions;* Carbon Dioxide Information Analysis Center, Oak Ridge National Laboratory, US Department of Energy: Oak Ridge, TN, USA, 2017. [CrossRef]

3. Kumar, S.; Dheeran, P.; Singh, S.P.; Mishra, I.M.; Adhikari, D.K. Kinetic studies of ethanol fermentation using *Kluyveromyces* sp. IIPE453. *J. Chem. Technol. Biotechnol.* **2013**, 88,1874-1884. [CrossRef]

4. Sydney, E.B.; Letti, L.A.J.; Karp, S.G.; Sydney, A.C.N.; Vandenberghe, L.P.S.; de Carvalho, J.C.; Woiciechowski, A.L.; Medeiros, A.B.P.; Soccol, V.T.; Soccol, C.R. Current analysis and future perspective of reduction in worldwide greenhouse gases emissions by using first and second generation bioethanol in the transportation sector. *Bioresour. Technol. Rep.* **2019**, 7,100234. [CrossRef]

5. Mohd Azhar, S.H.; Abdulla, R.; Jambo, S.A.; Marbawi, H.; Gansau, J.A.; Mohd Faik, A.A.; Rodrigues, K.F. Yeasts in sustainable bioethanol production: A review. *Biochem. Biophys. Rep.* **2017**,10,52-61. [CrossRef]

6. Joelsson, E.; Erdei, B.; Galbe, M.; Wallberg, O. Techno-economic evaluation of integrated first- and second-generation ethanol production from grain and straw. *Biotechnol. Biofuels* **2016**, 9,1-16. [CrossRef] [PubMed]

7. Ro, K.S.; Dietenberger, M.A.; Libra, J.A.; Proeschel, R.; Atiyeh, H.K.; Sahoo, K.; Park, W.J. Production of ethanol from livestock, agricultural, and forest residuals: An economic feasibility study. *Environments* **2019**, 6, 97. [CrossRef]

8. Tomei, J.; Helliwell, R. Food versus fuel? Going beyond biofuels. *Land Use Policy* **2015**, 56, 320-326. [CrossRef]

9. Prasoulas, G.; Gentikis, A.; Konti, A.; Kalantzi, S.; Kekos, D.; Mamma, D. Bioethanol production from food waste applying the multienzyme plasma at laboratory and bench-scale levels and its application as a starter culture in a meat product. *Fermentation* **2020**, 6, 39. [CrossRef]

10. Zheng, T.; Yu, H.; Liu, S.; Jiang, J.; Wang, K. Achieving high ethanol yield by co-feeding corncob residues and tea-seed cake at high-solids simultaneous saccharification and fermentation. *Renew. Energy* **2020**,145, 858-866. [CrossRef]

11. Alfonsin, V.; Maceiras, R.; Gutierrez, C. Bioethanol production from industrial algae waste. *Waste Manag.* **2019**, 87, 791-797. [CrossRef] [PubMed]

12. Naik, S.N.; Goud, V.V.; Rout, P.K.; Dalai, A.K. Production of first and second generation biofuels: A comprehensive review. *Renew. Sustain. Energy Rev.* **2010**,14, 578-597. [CrossRef]

13. Razmovski, R.; Vucurovic, V. Bioethanol production from sugar beet molasses and thick juice using *Saccharomyces cerevisiae* immobilized on maize stem ground tissue. *Fuel* **2012**, 92,1-8. [CrossRef]

14. Palmonari, A.; Cavallini, D.; Sniffen, C.J.; Fernandes, L.; Holder, P.; Fagioli, L.; Fusaro, I.; Biagi, G.; Formigoni, A.; Mammi, L. Short communication: Characterization of molasses chemical composition. *J. Dairy Sci.* **2020**,103, 6244-6249. [CrossRef]

15. Ghorbani, F.; Younesi, H.; Esmaeili Sari, A.; Najafpour, G. Cane molasses fermentation for continuous ethanol production in an immobilized cells reactor by *Saccharomyces cerevisiae*. *Renew. Energy* **2011**, 36, 503-509. [CrossRef]
16. Roukas, T.; Kotzekidou, P Rotary biofilm reactor: A new tool for long-term bioethanol production from non-sterilized beet molasses by *Saccharomyces cerevisiae* in repeated-batch fermentation. *J. Clean. Prod.* **2020**,257,120519. [CrossRef]
17. Mangwanda, T.; Johnson, J.B.; Mani, J.S.; Jackson, S.; Chandra, S.; McKeown, T.; White, S.; Naiker, M. Processes, Challenges and Optimisation of Rum Production from Molasses—A Contemporary Review. *Fermentation* **2021**, 7, 21. [CrossRef]
18. Alcantara, G.U.; Nogueira, L.C.; de Stringaci, L.A.; Moya, S.M.; Costa, G.H.G. Brazilian "Flex Mills": Ethanol from Sugarcane Molasses and Corn Mash. *Bioenergy Res.* **2020**,13, 229-236. [CrossRef]
19. Darvishi, F.; Moghaddami, N.A. Optimization of an industrial medium from molasses for bioethanol production using the Taguchi statistical experimental-design method. *Fermentation* **2019**, 5,14. [CrossRef]
20. Keskin Gundogdu, T.; Deniz, I.; Caliskan, G.; Sahin, E.S.; Azbar, N. Experimental design methods for bioengineering applications. *Crit. Rev. Biotechnol.* **2016**, 36, 368-388. [CrossRef]
21. Boboescu, I.; Gelinas, M.; Beigbeder, J.; Lavoie, J. A two-step optimization strategy for 2nd generation ethanol production using softwood hemicellulosic hydrolysate as fermentation substrate. *Bioresour. Technol.* **2017**, 244, 708-716. [CrossRef]
22. Pattanakittivorakul, S.; Lertwattanasakul, N.; Yamada, M.; Limtong, S. Selection of thermotolerant *Saccharomyces cerevisiae* for high temperature ethanol production from molasses and increasing ethanol production by strain improvement. *Antonie Leeuwenhoek Int. J. Gen. Mol. Microbiol.* **2019**,112, 975-990. [CrossRef] [PubMed]
23. Vucurovic, V.M.; Puskas, V.S.; Miljic, U.D. Bioethanol production from sugar beet molasses and thick juice by free and immobilised *Saccharomyces cerevisiae*. *J. Inst. Brew.* **2019**,125,134-142. [CrossRef]
24. Marques, W.L.; Raghavendran, V.; Stambuk, B.U.; Gombert, A.K. Sucrose and *Saccharomyces cerevisiae:* A relationship most sweet. *FEMS Yeast Res.* **2015**,16,1-16. [CrossRef]
25. Krajang, M.; Malairuang, K.; Sukna, J.; Rattanapradit, K.; Chamsart, S. Single-step ethanol production from raw cassava starch using a combination of raw starch hydrolysis and fermentation, scale-up from 5-L laboratory and 200-L

pilot plant to 3000-L industrial fermenters. *Biotechnol. Biofuels* **2021**,14,1-15. [CrossRef] [PubMed]

26. Evcan, E.; Tari, C. Production of bioethanol from apple pomace by using cocultures: Conversion of agro-industrial waste to value added product. *Energy* **2015**, *88*, 775-782. [CrossRef]

27. Siqueira, P.F.; Karp, S.G.; Carvalho, J.C.; Sturm, W.; Rodriguez-Leon, J.A.; Tholozan, J.L.; Singhania, R.R.; Pandey, A.; Soccol, C.R. Production of bio-ethanol from soybean molasses by *Saccharomyces cerevisiae* at laboratory, pilot and industrial scales. *Bioresour. Technol.* **2008**, 99, 8156-8163. [CrossRef] [PubMed]

CHAPTER 6

The Use of Life Cycle Assessment in the Support of the Development of Fungal Food Products from Surplus Bread

1. INTRODUCTION

In food production, extensive resources such as water, land, energy and nutrients are used with significant impact to the environment. It is forecasted that by 2050, the demand for agricultural products will increase by 35-50% due to a growing population and rising incomes, leading to higher pressure on the environment [1]. The problem is aggravated by the significant wastage of food throughout the supply chain. The wastage of food represents not only the loss of the product itself, but also all the resources used in food production, transportation and packaging.

The provision of animal-based protein generally has a higher environmental impact compared with vegetable or fungi-based proteins [2]. Concerns regarding the ethical and environmental implications of meat consumption have increased the demand for meat substitutes, such as those based on legumes and fungi [3]. Recently, the use of filamentous fungi as a commercial food product has gained considerable attention, due to its high protein content, the presence of essential amino acids and easy digestibility [4,5]. Filamentous fungi have traditionally been used by different societies as food, e.g., tempeh and oncom are indigenous staple foods in Indonesia [6,7]. One example of a product currently sold in the European market is Quorn™, which is made from the filamentous fungus Fusarium venenatum cultivated in synthetic medium with glucose, ammonium and mixed with egg albumen [8]. One problem with this product is the high cost of the cultivation medium [4]. Moreover, Smetana et al. [9] performed an environmental assessment of a similar product based on mycoprotein cultivated in molasses and concluded that the substrate is an environmental hotspot in the system. Therefore, it is relevant to consider other types of substrates to decrease production costs and environmental impacts.

One alternative substrate for fungal cultivation is surplus bread. The valorisation of food surplus into high-value products has gained interest as part of the strategies of transition to a bio-based economy grounded on a sustainable use of resources. Bread is rich in carbohydrates and its porosity makes it an ideal substrate for fungal cultivation [6,10]. Moreover, bread is a product with high waste generation, particularly at the level of the consumer and at the supplier-retailer interface. In the UK, it was estimated that 10% of all food waste are bakery products [11]. Brancoli et al. [12] studied food waste from retailers in Sweden and concluded that bread is a product category with large environmental impacts and economic costs [12]. In Sweden, the quantity of bread waste generated was estimated to be 80,410 tons/year, of which 28,200 tons/year are generated at the retail level [13]. The large waste volumes of uniform composition and quality, and the distribution model in which bread is sold makes it a good substrate for manufacturers who strive for stable delivery of raw materials.

There are mainly two different categories of bread, bake-off and pre-packaged. The technology proposed in this article can be used for both categories. Bake-off are products baked from pre-made dough in supermarkets or by a bakery. Pre-packaged bread is produced by the bakeries, and in some countries, it is often sold to retailers under takeback agreements (TBAs) [13-15]. In such agreements, the bakeries collect the bread that is not sold in the supermarkets, and here, bread waste is segregated from other food waste fractions from the retailers. This enables alternative pathways for the waste management and valorisation of the unsold bread. These pathways should preferably be located at higher levels in the food waste hierarchy, such as the production of fungal biomass for food, ethanol or animal feed [16,17]. Nevertheless, it is necessary to ensure that such technologies are sustainable and that they minimize the use of natural resources and the generation of waste as well as emissions of pollutants over their life cycles.

The life cycle assessment (LCA) methodology can be used to assess the environmental performance of a product to ensure its sustainability. The methodology can also be applied to support the early design stages in the development of the product by assessing the implication of design choices on the environmental performance of the technology [18]. Such studies are crucial since they can identify and prevent environmental impacts before the technology has been embedded by path dependency and lock-in [19]. Decisions made in the early stages of product development can have a significant influence on its subsequent environmental impacts. McAloone and Bey [20] estimated that 80% of the environmental performance of a product is determined by decisions made in the early stages of the technology development. For this reason, the European Union [21] recommends the use of LCA in product development to guarantee its sustainability.

Research on life cycle assessment of valorisation of bread waste is primarily focused on the lower stages of the food waste hierarchy. Few studies address bread waste valorisation into high-value products. Moreover, most of the research focuses on technologies with high technology readiness levels. Previous studies have investigated the environmental performance of bread waste management and valorisation alternatives, such as anaerobic digestion and incineration [22], conversion into biofuels [17], animal feed [17,23], and value-added chemicals [24]. There is a large body of research on the development of new technologies to valorise bread to high value products [6,25,26], but most of this research does not integrate lifecycle thinking into the early stages of the process development, increasing the risk for sub optimization of the technology.

The aim of this study was to integrate environmental considerations in the early stages of the development of a solid-state fermentation (SSF) process using bread as feedstock for the production of a protein-rich food product. The SSF process is proposed to be implemented as a stand-alone business, or on-site in small-scale bakeries to recover their otherwise discarded surplus bread. The technology can also be implemented in supermarkets that have a fresh bakery department in house, i.e., supermarkets that bake their own bread. The food product can be sold as a ready-to-eat meal directly to customers.to reduce waste while gener ating addition al income from the fungal product and provide a new sustainable food alternative that can replace other less sustainable options.

The goal of this study is to suppore the technology developers during the design and development phases by assessing the environmental sustainability of a range- of possible scenarios in which the technology can operate. This will enable a better understanding of the relation between design choices and the environmental performance of tha technology, cis well as a llow the steering of the technolagy towards solutions with a lewer environmental impace. Moreover, research on fungal growth paiterns in solid-state cultures are limited and this study investigates whether morphological differences of the inoculation culture influence the performance in the subsequent SSF step.

2. MATERIALS AND METHODS

The process development was iterative and used the concept of proxy eechnology transfer-proaess [27], whish aonsiders that emerging tepimologies are not often based on completely new procefses. instead, it relies on a new application of existing technologies. The development of the process started with the assessment of an incumbent Sechnology, namely thf production of mycoprotein cultivated in molasses to identify hotspots in tire system. Thp first results were then communicated to the researchers involved en the process development, starting an iterative epproach wherein the hotspots in then system are identffeed and alternatives were proposed and assessed. In total, four different scenarios for the inoculation were proposed and comparedin relation to the environmental performance an d sharscteristics of the final product, suc h as protein content and morphology. A detailed description of the scenarios is available in Section 2.1.

2.1. Process Description

The basic steps in fhe process are the (drying and grinding hf the bread, mixing with water and the staster spores (inoculum), and fermentation. The later processes after fermentation, such as frying, and freezing; for htter consumption are excluded in the study (Figure 1).

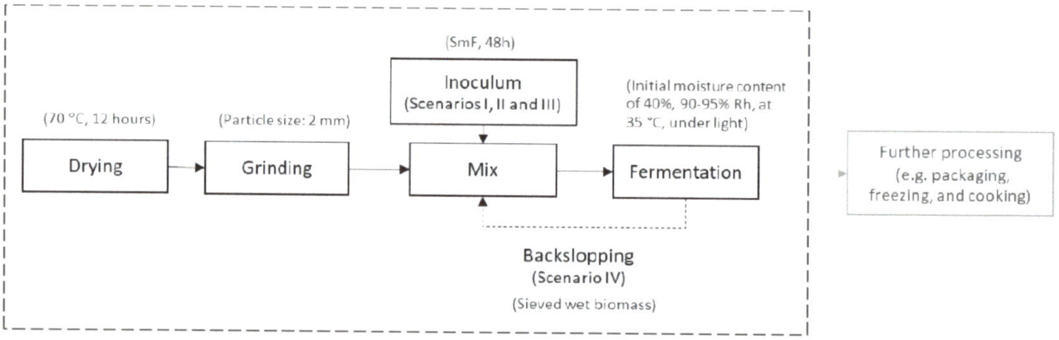

Figure 1. Basic process flow chart. The conditions of each process are described in brackets. The dotted box represents the system boundaries of the study. The further processing of the fungal biomaas after fermenfation is not included in the system boundaries.

2.1.1. Substrate and Fungus

Steinbrenner & Nyberg (Molndal, Sweden) kindly provided unsold fresh sourdough bread that showed no signs of mould contamination. The breael was prepared according to ehe process described in Gmoser et al [25]. Samples were augoclaved and stored in airtight containers at room temperature prior to use. T^{g}e edible food grade filamentous fungus *Neurospora intermedia* CBS 131.92 (Centraalbureau voor Schimmelcultures, Utrecht, The Netherlands) was used. The fungus was maintained on potato dextrose agar (PDA) plates containing 20 g/L glucose 20, 15 g/L agar and 4 g/L potato extract. The PDA plates were prepared by incubation for 5 days at 30°C followed by storage at 4°C. The spore solution was prepared by flooding each plate with 20 mL of distilled water, using a disposable plastic spreader to gently release the spores and transferring them to a sterile slant tube. The spore solution (2.5 x 10^6 spores/mL) was used to prepare the four different inoculations (one for each of the scenarios). A detailed description of the conditions of fermentation is given in Section 2.1.3.

2.1.2. Design of the Inoculation

The choice of the nutrient source was based on existing technologies. The inoculum for Scenario I is produced in a similar way to the inoculum production for tempeh as described in Wiloso et al. [28], but instead of using white rice or other grains, it is based on stale bread. Production of the inoculum for Scenario II, which is based on a similar process to cultivate yeast [29], used a synthetic medium based on molasses and ammonium chloride. Scenario III is based on glucose, and it was adapted from Jungbluth et al. [30], which describes the production of mycoprotein for Quorn™ (Table 1).

Table 1. Inventory analysis for the different scenarios for the production of the fermented fungal product per functional unit. The indented substances in italic refer to the inputs related to the production of the fungal biomass (inoculum) used for the inoculation.

	Unit	Scenario I	Scenario II	Scenario. III	Scenario IV
Inputs					
Bread	kg	2.0	2.0	2.0	2.0
Inoculum	g	10	10	10	29*
Water	l	2.63	1.81	1.92	
Glucose	g	-	-	38.50	
NH_4Cl	g	-	16.28	14.50	
NaOH	g	-	-	0.62	
Molasses	g	-	0.36	-	
Bread	g	52.60	-	-	
Electricity	kWh	0.065	0.033	0.033	0.033
Water for mixing	l	1.34	1.34	1.34	1.34
Electricity for drying	kWh	0.53	0.53	0.53	0.53
Electricity for grinding	kWh	0.024	0.024	0.024	0.024
Electricity for SSF	kWh	1.55	1.55	1.55	1.55
Outputs					
Fermented fungal product*	kg	1	1	1	1

** Calculation based on the mass balance revealed by Wang et al. [31] and the starch, protein and moisture content in the samples.*

The three submerged fermentation (SmF) processes (Scenarios I, II and III) were inoculated with 10 mL/L of the spore solution, cultivated aerobically for 48 h in cotton plugged 250 mL Erlenmeyer flasks containing 100 mL of medium autoclaved at 121°C for 20 min. The initial pH of the medium was adjusted to 5.5 with 2 M HCl or

NaOH. The different medium compositions are summarized in Table 1. A water bath (Grant OLS-Aqua pro, Cambridge, UK) was used to maintain the temperature with orbital shaking at 125 rpm (radius of 9 mm) for 48 h. At the end of the fermentation, the wet biomass from each shake flask was harvested by pouring the cultivation medium through a 1 mm^2 pore area fine mesh and washed with distilled water. The liquid fraction was collected separately and frozen for further analysis. The moisture content of the fungal biomass was measured gravimetrically and sieved wet biomass was used directly as inoculum for the solid-state fermentation step.

Scenario IV relies on the recirculation (backslopping) of part of the fermented product from a previous batch as the inoculum. The solid-state fermentation (SSF) on stale bread used as inoculum was prepared according to the SSF process below, except that 0.5 mL of spore solution was used as inoculum instead. Based on an assumption that the fungal fermented final product contained 35% fungal biomass, with a moisture content of 60%, in the substrate-fungal matrix after 5 days of solid-state fermentation [25], the amount needed for backslopping was set to 0.01% dry weight of the substrate.

2.1.3. Solid State Fermentation

Cultivation of *N. intermedia* was conducted in sterile petri dishes (100 mm x 20 mm). A total dry weight of 15 g breadcrumbs was inoculated with *N. intermedia* separately for each petri dish. The initial moisture content was adjusted to 40% (on a dry basis (*w/w*, db) with distilled water, after which each sample was mixed evenly, and a lid was placed on the petri dish. Solid-state fermentation was conducted batch wise for 2-10 days in a climatic test cabinet (NUVE test cabinet TK 120, Ankara, Turkey) at 90%-95% relative humidity (Rh) ± 1%, at 35°C under light. The light source was applied to increase pigment production in the fungal biomass in order to form a visually appealing product with improved nutritional value [6,25]. The petri dishes were flipped every second day. At the end of fermentation, samples were dried at 45°C. Effects of days of fermentation on the final weight and protein content in the fungal-bread product were investigated as well as a visual observation on morphology after the SmF step.

2.2. Analyses

The fungal spore concentration was determined using a Burker counting chamber. The final moisture content, the amount of fungal biomass harvested after SmF, and fungal- bread product used as inoculum in Sc. IV were determined gravimetrically. Preparation of samples and measurement of the protein content by Kjeldahl digestion using a 2020 Kjeltec Digestor and a 2400 Kjeltec Analyser unit (FOSS Analytical A/S, Hillerod, Denmark) was performed following the same procedure as in Gmoser et al. [25]. Total starch analysis was accomplished by a-amylase/ amylogluco-sidase/hexokinase method following the procedure provided in the

Total Starch HK Assay Kit (Megazyme International Ltd., Bray, Ireland).

2.3. Statistical Analysis

Experiments on the SSF were performed in duplicate (n = 2) and two analytical replicates were performed on each batch (a = 2). Values present the mean value of the measurement ± SD. Raw data were statistically analysed using the software package MINITAB®® (version 17.1.0, Minitab Inc., State College, PA, USA). Analysis of variance (ANOVA) was determined using general linear model with a 95% confidence level followed by Tukey's multiple comparison test. A linear factorial regression model of the data was also conducted with the four inoculation methods as factor levels and days of SSF (2-8 days) as continuous control with protein concentration as output.

2.4. Life Cycle Assessment

The functional unit for the environmental assessment is 1 kg of fermented fungal product after the fermentation step. The selection of a mass unit was connected with the ability to compare results with those available in the literature. The system boundaries are cradle-to-gate as described in Section 2.1. The geographical scope is Sweden. The characterization method ReCiPe (ReCiPe 2016 v1.1 midpoint and endpoint methods, Hier3archist version) [32], was used due to its broad set of midpoint impact categories and the possibility of using both midpoint and endpoint methods [32].

The inventory was built using primary data flows from the laboratory experiments, which were initially based on data from the literature [28-30,33,34] and further developed based on the results from the trial experiments. The energy consumption of the equipment was estimated based on the methodology described in Piccinno et al. [35] for scaling up laboratory data. Capital goods are not included in the inventory, as all scenarios are expected to use similar equipment. The ecoinvent database version 3.6 was used for the background processes. Inventory data for the scenarios is available in Table 1.

3. RESULTS

This section describes the experimental results of the solid-state fermentation, in relation to the increase of the protein content over time and the morphology of the inoculum. Moreover, the result for the life cycle assessment is presented.

3.1. Experimental Results

There is a significant difference in protein content of the final fungal product after SSF for the different inoculation methods (*p* value 0.00). The inoculation scenario using molasses (Sc. II) resulted in a higher protein content after SSF compared to glucose (Sc. III), with no significant difference between Scenarios I and IV. The inoculum scenario based on glucose medium (Sc. III) also resulted in a longer lag phase and slower initial exponential growth, possibly due to change in substrate during SSF, while *N. intermedia* grown on bread (Sc. I) and molasses (Sc. III) was in a more active phase, adapted to bread as substrate and hypothetically, already producing enzymes for complex carbohydrate assimilation, when applied to the SSF step. However, considering the natural variation in biological systems, the relatively small differences between the inoculation methods do not contribute with sufficient support for choosing one scenario over the other.

As expected, the protein content increased over time for all four groups, with a significant increase over every second day (p value = 0.00) (Figure 2). The highest increase in protein content happened between day four and six for all inoculation methods based on SmF, varying between 38%, 49% and 51%, for Scenarios I, II and III respectively. Inoculum based on backslopping resulted in a slightly higher increase in protein content between day two and four (39% increase) compared to the other days. The linear regression equations of the four inoculation methods revealed that SmF based on molasses gave the highest average protein increase per day of 4.0, compared to SmF based on bread, backslopping and SmF based on glucose of 3.6%, 3.4% and 3.2% increase per day respectively (consider a 6.6% lack of fit of the model). By comparing the linear regression equations, it can be seen that the daily increase based on glucose medium is only slightly offset in x-axis (timescale), whereas the protein value reaches approximately the same amount as the other inoculation methods after 8 days SSF.

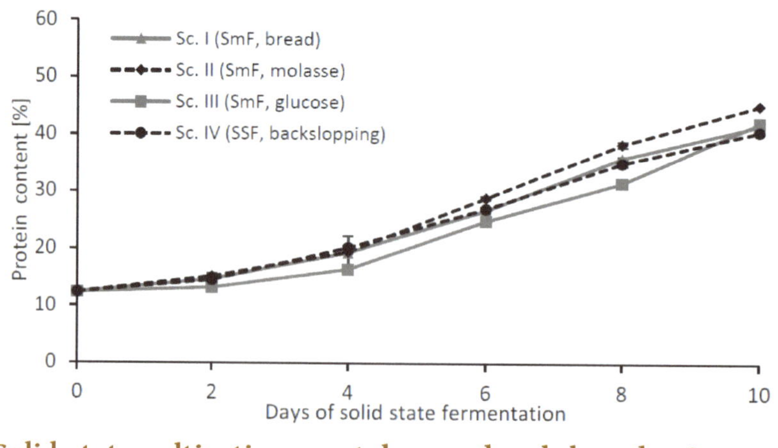

Figure 2. Solid-state cultivation on stale sourdough bread using N. intermedia after 0-10 days cul- tivationunder light a 135°C , 90% Rh, and 40% initial moisture content using different inoculation methods: backslopping (black circle, dotted line), submerged fermentation with synthetic medium based on glucose (grey square, solid line), submerged fermentation based on molasses (black diamonds, dotted line) and submerged fermentation based on breadcrumbs (grey triangles, solid line). Results are expressed as the mean value ± 95% confidence level.

The fungal morphology in the SmF biased on molasses medium (Sc. II) resulted in fine, dense pellets, whereas the other SmF cultures resulded in loose filamentous mycelium. SSF based on inoculum drom molasses med ium gave the highest mean percentage of protein.

3.2. Life Cycle Assessment

The relative results for five selected impact assessment categories for the scenarios and separately for the inoculum production process are presented in Figures 3 and 4, respectively. The impact categories described in Figures 3 and 4 are global warming, terrestrial acidification, freshwater eutrophication, marine ecotoxicity and fossil resource scarcity. The results for the remaining thirteen impact categories available in the ReCiPe characterization method [32] is available in Table 2.

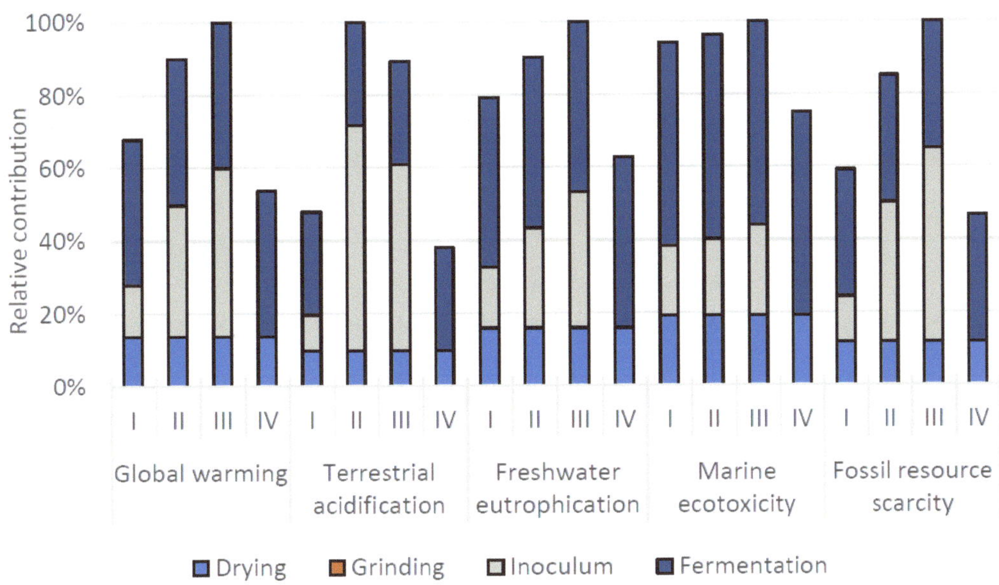

Figure 3. Seleoaed LCIA results for rhe scenarios for fee fermented fungal product production. Sc. I uses bread as fee carbon source. Sc. II uses mnlasses. Sc. III is based on glucose and Sc. IV relies on backslopping, as described in Table 1.

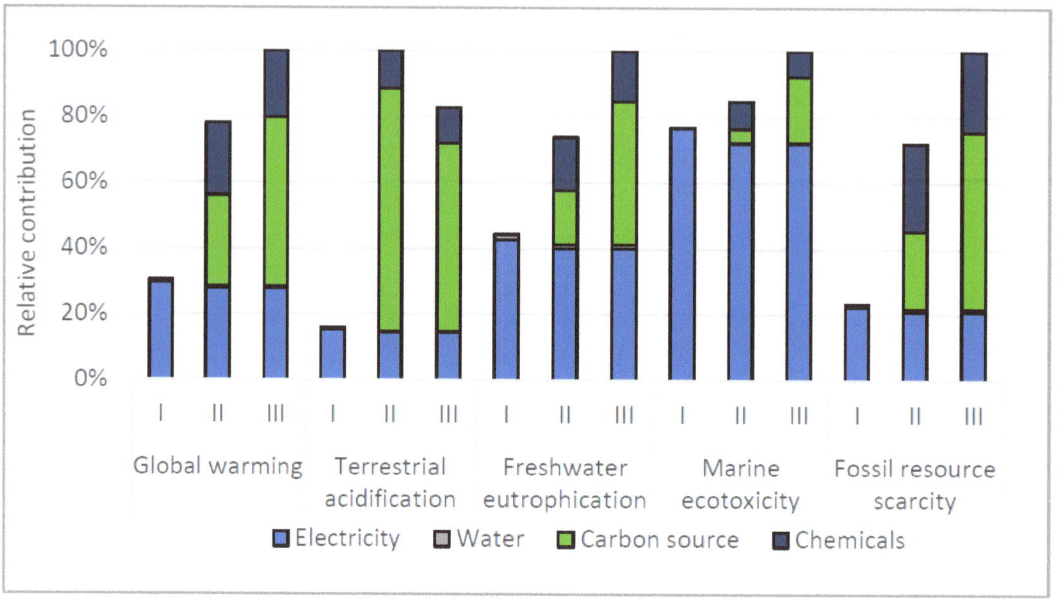

Figure 4. Selected LCIA results for the different inoculation production processes. Sc. I uses breadcrumbs as the carbon source. Sc. II uses molasses. Sc. III is based on glucose.

Table 2. Impact assessment results for Scenarios I, II, III and IV based on the ReCiPe midpoint impact categories.

Impact Category	Unit	Sc. I	Sc. II	Sc. III	Sc. IV
Global warming	kg CO_2 eq	1.4×10^{-1}	1.9×10^{-1}	2.1×10^{-1}	1.1×10^{-1}
Stratospheric ozone depletion	kg CFC-11 eq	3.0×10^{-7}	6.2×10^{-7}	5.2×10^{-7}	2.4×10^{-7}
Ionizing radiation	kBq Co-60 eq	8.0×10^{-1}	7.9×10^{-1}	7.9×10^{-1}	6.4×10^{-1}
Ozone formation, Human health	kg NOx eq	3.5×10^{-4}	4.8×10^{-4}	5.1×10^{-4}	2.8×10^{-4}
Fine particulate matter formation	kg PM2.5 eq	2.3×10^{-4}	3.7×10^{-4}	3.8×10^{-4}	1.9×10^{-4}
Ozone formation, TE	kg NOx eq	3.6×10^{-4}	4.9×10^{-4}	5.2×10^{-4}	2.8×10^{-4}
Terrestrial acidification	kg SO2 eq	5.8×10^{-4}	1.2×10^{-3}	1.1×10^{-3}	4.6×10^{-4s}
Freshwater eutrophication	^{k}g p ^{e}q	9.4×10^{-5}	1.1×10^{-4}	1.2×10^{-4}	7.54×10^{-5}
Marine eutrophication	kg n eq	1.8×10^{-5}	1.0×10^{-4}	8.5×10^{-5}	1.5×10^{-5}
Terrestrial ecotoxicity	kg 1,4-DCB	2.2×10^{0}	2.4×10^{0}	2.7×10^{0}	1.8×10^{0}
Freshwater ecotoxicity	kg 1,4-DCB	9.4×10^{-2}	9.6×10^{-2}	9.9×10^{-2}	7.5×10^{-2}
Marine ecotoxicity	kg 1,4-DCB	1.1×10^{-1}	1.2×10^{-1}	1.2×10^{-1}	9.1×10^{-2}

Impact Category	Unit	Sc. I	Sc. II	Sc. III	Sc. IV
Human carcinogenic toxicity	kg 1,4-DCB	1.9 x 10^{-2}	2.0 x 10^{-2}	2.1 x 10^{-2}	1.5 x 10^{-2}
Human non-carcinogenic toxicity	kg 1,4-DCB	6.0 x 10^{-1}	6.4 x 10^{-1}	6.5 x 10^{-1}	4.8 x 10^{-1}
Land use	2 m^2 a crop eq	6.1 x 10^{-2}	9.2 x 10^{-2}	8.7 x 10^{-2}	4.8 x 10^{-2}
Mineral resource scarcity	kg Cu eq	2.0 x 10^{-3}	2.4 x 10^{-3}	2.5 x 10^{-3}	1.6 x 10^{-3}
Fossil resource scarcity	kg oil eq	2.6 x 10^{-2}	3.8 x 10^{-2}	4.5 x 10^{-2}	2.1 x 10^{-2}
Water consumption	3 m^3	2.1 x 10^{-2}	2.3 x 10^{-2}	2.2 x 10^{-2}	1.5 x 10^{-2}

Figure 3 indicates that the fermentation step and the choice of inoculum production has a large influence on the environmental burdens associated with the fermented fungal product.

The production of inoculum had a contribution higher than 40% in 9 out of 18 impact categories assessed in both Sc. II and Sc. III. More specifically, the carbon source used for the inoculum production for these scenarios is a hotspot, as shown in more detail in Figure 4. In Scenario I the inoculation has a relatively lower contribution, ranging from 20% to 29%. The reason is that it uses stale bread as the carbon source, which is modelled as waste and therefore burden free.

The results indicate that Scenario IV, which is based on backslopping, has the lowest environmental impacts in all of the eighteen categories assessed, followed by Scenario I. Both scenarios substituted glucose or molasses for stale bread, significantly reducing the environmental impacts on the inoculum production. Scenarios II and III had the worst performance in the categories assessed. Scenario II had the worst performance in five impact categories, while Scenario III had the highest impact in 12 environmental impact categories (Table 2).

The solid-state fermentation is also a hotspot, and it contributes significantly to the majority of the impact categories. These impacts are related to the electricity consumption during the process, as shown in Table 1. The impact is the same for all scenarios, but it has a different relative contribution as shown in Figure 3. Overall, the fermentation step, drying of bread and the inoculation were identified as hotspots in the production of the fungal product (Figure 3).

Figure 4 shows the detailed impact assessment for the different inoculation methods. Sc. IV has no contribution from the inoculum production since it is obtained from back- slopping. This assumes that the contribution of the initial inoculum required to make the first batch of fungi is insignificant.

The results indicate that the carbon source used in the inoculum production is a hotspot in the majority of impacts categories in both Scenario II, which uses molasses and Scenario III, based on glucose (Figure 4). The carbon source is responsible for 36% and 51% of the impacts on the global warming impact category for Sc. II and Sc. III, respectively. In terrestrial acidification, the contribution is higher, 74% and 69% for Sc. II and Sc. III, respectively. Scenario I, conversely, uses stale bread as the carbon source, which is modelled as waste and is, therefore, burden-free.

Electricity consumption has a significant contribution for both the inoculum production and the overall process. The geographical scope is Sweden, which has a relatively low impact energy system. Therefore, the results of the study are sensitive to the energy mix used.

4. DISCUSSION

The environmental impacts of a product are often determined by the decisions made at the early stages of development. For this reason, the development of the product should integrate environmental considerations to minimize the environmental impacts throughout its life cycle without compromising essential characteristics such as function, cost and quality. The estimation of environmental impact at this early stage is challenging due to the limited amount of information available, for example, the energy consumption in the processes. In contrast, the later stages in the product development provide more quantitative data but fewer opportunities to change the system.

Table 3 summarizes and ranks the main technical and environmental differences between the four scenarios studied here, namely the environmental impacts, protein content, lag phase and morphology. These aspects are critical to support the decision of the more suitable method for fungal biomass production using bread as substrate. The environmental impact ranking in Table 3 is based on the single score from ReCiPe endpoint (H) as available in the SimaPro software version 9.1.

Table 3. Summary of technical and environmental indicators for the scenarios assessed. The protein content refers to SSF after 6 days.

Scenarios	Environmental Impact (mPt)	Protein Content	Lag Phase	Morphology
Scenario I	9.7	27%	Short	Dispersed mycelium
Scenario II	12.5	29%	Short	Pellets
Scenario III	13.0	25%	Long	Dispersed mycelium
Scenario IV	7.6	27%	Short	Dispersed mycelium

Filamentous fungi grow in different morphological forms in submerged cultures, including freely suspended filamentous mycelia and pellets depending on the genotype of the strain and culture conditions [36]. The growth mode affects the rheological properties of the cultivation medium, and consequently, the overall process performance and final product yields. Generally, the excessive growth of free filamentous mycelia is connected to practical and technical difficulties, such as lower oxygen diffusion, laborious harvesting, high medium viscosity and a relative lower yield of products [37]. These problems associated with the filamentous morphology can be overcome when the fungi grow in the form of pellets, which also have a higher potential for cell reuse and higher productivity due to the possibility of using high-density cultivations [38]. However, including in the pellet form, critical characteristics such as size and compactness versus fluffiness can influence oxygen and substrate transfer rates. The trend in research works has been towards the production of small fluffy pellets. Among the scenarios assessed, only Scenario II grew in the pellet morphology. The slightly shorter lag phase in Scenario II compared to Scenario III further promote this inoculation method since it promotes a faster initiation of the subsequent SSF step. Nevertheless, the environmental impact of Sc. II is significantly higher in comparison with Sc. I and IV (Table 3).

Scenario IV had the lowest environmental impact, and the protein content was sufficient for the intended use. However, one potential drawback with this method is the higher risk of contamination since it is more problematic to sustain a sterile condition at a larger scale when backslopping part of the final product as inoculum compared to inoculum using pure fungal biomass from SmF. Thus, microbiology analysis of the final product and industrial trials need to be assessed before this method can be applied.

Considering the weighted results at Table 3, Scenario I had an environmental impact that is 28% higher in comparison with Scenario IV, but the results indicate a lower impact in comparison with Scenarios II and III. Moreover, the main advantage of this scenario in comparison to Scenario IV is that it is less sensitive to contamination. Scenario I also yields filamentous mycelia morphology. However, further development of this scenario by changes in cultivation conditions such as pH and aeration can potentially alter the morphology.

The fermentation time is another important aspect influencing both the protein content and the environmental impacts of the final product. The results showed that the protein content for all scenarios reached similar values after 8 days of SSF. However, based on previous studies [6] on the visual appearance and smell of the final product, the SSF step should not be longer than a maximum of eight days and preferably six days if the protein content is sufficient for the intended purpose as food. A direct comparison of the environmental performance of this fungal product with established food products based on mycoprotein requires special attention due to scale-up issues and technology uncertainties inherent in the majority LCAs of products in an early stage of development [18]. Moreover, it is expected that the technology under assessment will need further development in relation to technical aspects and consumer's acceptance.

Jungbluth et al. [30] reported an electricity consumption of 1.17 kWh and a heat consumption of 13 MJ for the industrial production of 1 kg of mycoprotein at plant. Smetana et al. [9] reported 6 kWh/kg of mycoprotein ready to eat. The electricity consumption in this study ranged from 2.61 to 2.64 kWh/functional unit for the different scenarios. Nevertheless, it is important to note that a direct comparison is not possible, due to differences in the system boundaries of each study. The system boundaries in this study and Jungbluth et al. [30] end at the plant gate, while Smetana et al. [9] consider the system boundaries until the consumer's plate. There are also discrepancies in the technology readiness level between this study and Jungbluth et al. [30] in which inventory is based on an operational plant.

A huge advantage is that we now have an LCA model that can be used to get updated environmental impacts when the scenarios are adjusted to improve factors such as protein content, lag time and morphology. This allows for environmental impacts to be assessed throughout the development process.

5. CONCLUSIONS

The current analysis, which included different production pathways for the valorisation of stale bread into fungal biomass according to the environmental impacts, protein content, lag phase and morphology, indicated the potential use of the fungal biomass as a substitute for traditional food products. The technology proposed uses an abundant waste flow and has the potential to be implemented in supermarkets, small and medium-sized bakeries and industrial bakeries on its returned flows.

Technical and environmental analyses were performed to determine potential tradeoffs of four proposed scenarios to valorise stale bread into a fungal based food product rich in protein. The results from the hotspot analysis indicate that the choice of the medium used in the inoculum production has a large influence on the environmental and technical performance. Inoculum production using molasses or glucose as medium, Sc. II and III respectively, are not preferred from an environmental perspective. Sc. I and Sc. IV, which uses surplus bread and backslopping, showed the lowest environmental impacts in all categories studied here, while maintaining important characteristics such as high protein content. Therefore, it is recommended that such alternatives should be further investigated for the technical feasibility of the process. Moreover, the hotspot analysis indicated that for all scenarios assessed, the fermentation step and drying of bread have a significant impact on the environmental burdens. Therefore, improvements aimed at environmental gains in the fungal biomass production should focus on alternatives for decreasing the fermentation time and consequently the electricity consumption, and on the development of alternative processes in which the drying of bread is not required.

None of the scenarios performed best in all of the parameters analysed. Therefore, it is not possible to draw a simple conclusion regarding the preferred scenario. The decision will ultimately be made according to the technical, environmental or economic agenda of the decision-makers. The contribution of this study is to highlight the trade-offs inherently involved in the decision process within the product development and to provide guidance for further development of the process.

Bread waste has the potential to be retained in the food chain by applying a SSF process using the edible fungus *N. intermedia*. The study has integrated environmental considerations into the early stages of the development of a fungal food product, showing which scenario has the best environmental performance and highlighting trade-offs and the parts of the process that are hotspots and should, thereby, be in focus when optimizing the process. This approach is suggested to contribute to a sustainable way to handle otherwise wasted bread, consistent with a circular economy, and it provides a broader base for the developers of the technology to make sustainable decisions during process optimisation.

REFERENCES

1. FAO. *The State of Food and Agriculture 2019: Moving Forward on Food Loss and Waste Reduction;* Food and Agriculture Organization of the United Nations: Rome, Italy, 2019.
2. Hartmann, C.; Siegrist, M. Consumer perception and behaviour regarding sustainable protein consumption: A systematic review. *Trends Food Sci. Technol.* **2017**, 61,11-25. [CrossRef]
3. Godfray, H.C.J.; Aveyard, P.; Garnett, T.; Hall, J.W.; Key, T.J.; Lorimer, J.; Pierrehumbert, R.T.; Scarborough, P.; Springmann, M.; Jebb, S.A. Meat consumption, health, and the environment. *Science* **2018**, 361, eaam5324. [CrossRef] [PubMed]
4. Filho, P.F.S.; Nair, R.B.; Andersson, D.; Lennartsson, P.R.; Taherzadeh, M.J. Vegan-mycoprotein concentrate from pea-processing industry byproduct using edible filamentous fungi. *Fungal Biol. Biotechnol.* **2018**, 5, 5. [CrossRef]
5. Moore, D.; Chiu, S.W. Fungal products as food. *Fungal Divers. Res. Ser.* **2001**, 6, 223-251.

6. Gmoser, R.; Fristedt, R.; Larsson, K.; Undeland, I.; Taherzadeh, M.J.; Lennartsson, P.R. From stale bread and brewers spent grain to a new food source using edible filamentous fungi. *Bioengineered* **2020**,11, 582-598. [CrossRef] [PubMed]
7. Shurtleff, W.; Aoyagi, A. *The Book of Tempeh;* Harper & Row, Publishers: New York, NY, USA, 1979.
8. Wiebe, M. Myco-protein from Fusarium venenatum: A well-established product for human consumption. *Appl. Microbiol. Biotechnol.* **2002**, 58, 421-427. [CrossRef] [PubMed]
9. Smetana, S.; Mathys, A.; Knoch, A.; Heinz, V. Meat alternatives: Life cycle assessment of most known meat substitutes. *Int. J. Life Cycle Assess.* **2015**,20,1254-1267. [CrossRef]
10. Melikoglu, M.; Webb, C. Chapter 4—Use of Waste Bread to Produce Fermentation Products. In *Food Industry Wastes;* Kosseva, M.R., Webb, C., Eds.; Academic Press: San Diego, CA, USA, 2013; pp. 63-76.
11. WRAP, H. Link Consumer Strategies, Campden BRI. In *Reducing Household Bakery Waste;* WRAP, Waste and Resource Action Programme: Banbury, UK, 2011.
12. Brancoli, P.; Rousta, K.; Bolton, K. Life cycle assessment of supermarket food waste. *Resour. Conserv. Recycl.* **2017**, 118, 39-46. [CrossRef]
13. Brancoli, P.; Lundin, M.; Bolton, K.; Eriksson, M. Bread loss rates at the supplier-retailer interface—Analysis of risk factors to support waste prevention measures. *Resour. Conserv. Recycl.* **2019**,147,128-136. [CrossRef]
14. Ghosh, R.; Eriksson, M. Food waste due to retail power in supply chains: Evidence from Sweden. *Glob. Food Sec.* **2019**, 20, 1-8. [CrossRef]
15. Eriksson, M.; Ghosh, R.; Mattsson, L.; Ismatov, A. Take-back agreements in the perspective of food waste generation at the supplier-retailer interface. *Resour. Conserv. Recycl.* **2017**,*122*, 83-93. [CrossRef]
16. Papargyropoulou, E.; Lozano, R.; Steinberger, J.K.; Wright, N.; bin Ujang, Z. The food waste hierarchy as a framework for the management of food surplus and food waste. *J. Clean. Prod.* **2014**, 76,106-115. [CrossRef]
17. Brancoli, P.; Bolton, K.; Eriksson, M. Environmental impacts of waste management and valorisation pathways for surplus bread in Sweden. *Waste Manag.* **2020**,117,136-145. [CrossRef]
18. Cucurachi, S.; van der Giesen, C.; Guinee, J. Ex-ante LCA of Emerging Technologies. *Procedia CIRP* **2018**, 69,463-468. [CrossRef]
19. Jeswiet, J.; Hauschild, M. EcoDesign and future environmental impacts. *Mater. Des.* **2005**, 26, 629-634. [CrossRef]

20. McAloone, T.C.; Bey, N. *Environmental Improvement through Product Development: A Guide;* Danish Environmental Protection Agency: Copenhagen, Denmark, 2009.

21. European Commission. *Innovating for Sustainable Growth: A Bioeconomy for Europe;* Publications Office of the European Union: Brussels, Belgium, 2012.

22. Eriksson, M.; Strid, I.; Hansson, P.A. Carbon footprint of food waste management options in the waste hierarchy—A Swedish case study. *J. Clean. Prod.* **2015**, *93*,115-125. [CrossRef]

23. Vandermeersch, T.; Alvarenga, R.A.F.; Ragaert, P.; Dewulf, J. Environmental sustainability assessment of food waste valorization options. *Resour. Conserv. Recycl.* **2014**, 87, 57-64. [CrossRef]

24. Lam, C.-M.; Yu, I.K.M.; Hsu, S.-C.; Tsang, D.C.W. Life-cycle assessment on food waste valorisation to value-added products. *J. Clean. Prod.* **2018**,199, 840-848. [CrossRef]

25. Gmoser, R.; Sintca, C.; Taherzadeh, M.J.; Lennartsson, P.R. Combining submerged and solid state fermentation to convert waste bread into protein and pigment using the edible filamentous fungus N. intermedia. *Waste Manag.* **2019**, 97, 63-70. [CrossRef]

26. Melikoglu, M.; Lin, C.S.K.; Webb, C. Solid state fermentation of waste bread pieces by Aspergillus awamori: Analysing the effects of airflow rate on enzyme production in packed bed bioreactors. *Food Bioprod. Process.* **2015**, 95, 63-75. [CrossRef]

27. Buyle, M.; Audenaert, A.; Billen, P.; Boonen, K.; Van Passel, S. The future of ex-ante LCA? Lessons learned and practical recommendations. *Sustainability* **2019**,11, 5456. [CrossRef]

28. Wiloso, E.I.; Sinke, P.; Setiawan, A.A.R.; Sari, A.A.; Waluyo, J.; Putri, A.M.H.; Guinee, J. Hotspot identification in the Indonesian tempeh supply chain using life cycle assessment. *Int. J. Life Cycle Assess.* **2019**, 24,1948-1961. [CrossRef]

29. Dunn, J.B.; Mueller, S.; Wang, M.; Han, J. Energy consumption and greenhouse gas emissions from enzyme and yeast manufacture for corn and cellulosic ethanol production. *Biotechnol. Lett.* **2012**, 34, 2259-2263. [CrossRef]

30. Jungbluth, N.; Nowack, K.; Eggenberger, S.; Konig, A.; Keller, R. *Untersuchungen zur umweltfreundlichen Eiweissversorgung- Pilotstudie;* ESU-services GmbH fur das Bundesamt fur Umwelt (BAFU): Bern, Switzerland, 2016.

31. Wang, R.; Gmoser, R.; Taherzadeh, M.J.; Lennartsson, P.R. Solid-state fermentation of stale bread by an edible fungus in a semi-continuous plug-flow bioreactor. *Biochem. Eng. J.* **2021**,169,107959. [CrossRef]

32. Huijbregts, M.A.J.; Steinmann, Z.J.N.; Elshout, P.M.F.; Stam, G.; Verones, F.; Vieira, M.D.M.; Hollander, A.; Zijp, M.; Van Zelm, R. ReCiPe 2016: A harmonized life cycle impact assessment method at midpoint and endpoint level Report I: Characterization. *Int. J. Life Cycle Assess.* **2017**,22,138-147. [CrossRef]

33. Gmoser, R.; Lennartsson, P.R.; Taherzadeh, M.J. From surplus bread to burger using filamentous fungi at bakeries: Techno- economical evaluation. *Clean. Environ. Syst.* **2021**, 2,100020. [CrossRef]

34. Ingledew, W.M.; Austin, G.D.; Kraus, J.K. Commercial yeast production for the fuel ethanol and distilled beverage industries. In *The Alcohol Textbook,* 5th ed.; Nottingham University Press: Nottingham, UK, 2009; pp. 127-144.

35. Piccinno, F.; Hischier, R.; Seeger, S.; Som, C. From laboratory to industrial scale: A scale-up framework for chemical processes in life cycle assessment studies. *J. Clean. Prod.* **2016**,135,1085-1097. [CrossRef]

36. Ward, O.P. Production of recombinant proteins by filamentous fungi. *Biotechnol. Adv.* **2012**, 30,1119-1139. [CrossRef]

37. Nair, R.B. Integration of first and second generation bioethanol processes. In *Swedish Centre for Resource Recovery;* University of Boras: Boras, Sweden, 2017.

38. Ferreira, J.A.; Lennartsson, P.R.; Edebo, L.; Taherzadeh, M.J. Zygomycetes-based biorefinery: Present status and future prospects. *Bioresour. Technol.* **2013**,135, 523-532. [CrossRef] [PubMed]

CHAPTER 7

Aquafeed Production from Fermented Fish Waste and Lemon Peel

In order to obtain a high-protein-content supplement for aquaculture feeds, rich in healthy microorganisms, in this study, *Saccharomyces cerevisiae* American Type Culture Collection (ATCC) 4126 and *Lactobacillus reuteri* ATCC 53608 strains were used as starters for fermenting fish waste supplemented with lemon peel as a prebiotic source and filler. Fermentation tests were carried out for 120 h until no further growth of the selected microorganisms was observed and the pH value became stable. All the samples were tested for proteins, crude lipids, and ash determination, and submitted for fatty acid analysis. Moreover, microbiological analyses for coliform bacteria identification were carried out. At the end of the fermentation period, the substrate reached a concentration in protein and in crude lipids of 48.55 ± 1.15% and 15.25 ± 0.80%, respectively, representing adequate levels for the resulting aquafeed, whereas the ash percentage was 0.66 ± 0.03. The main fatty acids detected were palmitic, oleic, and linoleic acids. Saturated fatty acids concentration was not affected by the fermentation process, whereas monounsaturated and polyunsaturated ones showed an opposite trend, increasing and decreasing, respectively, during the process. Coliform bacteria were not detected in the media at the end of the fermentation, whereas the amount of *S. cerevisiae* and *L. reuteri* were around 10^{11} and 10^{12} cells per g, respectively.

1. INTRODUCTION

World capture fishery production reached a peak of approximately 96 million tons in 2018; the most recent estimates suggest that 52% of marine stocks are fully exploited, 17% are overexploited, and 7% are totally depleted, while human population and the demand for marine and other aquatic resources continue to increase [1]. Global aquaculture would make a considerable contribution toward bridging the gap between supply and demand. Unfortunately, its development is hampered by an inadequate supply of feed, particularly fishmeal, which is scarce and expensive [2]. This has stimulated the evaluation of a variety of alternative dietary protein sources with the objective of partially, or totally, replacing fishmeal protein in aquafeeds [3]. The use of food industry waste as animal feed is an alternative of high interest because it stands to bring both environmental and public benefits, besides reducing the costs of animal production [4-6]. In particular, with reference to the fishing industry, the cost of waste management for aquaculture is typically in the range of USD 0.05 to USD 0.065 per pound of fish produced, representing concerns from both an economic and an environmental point of view. Waste management, in fact, contributes to the overall costs of production and reduces farmers' net income. Moreover, the improper management of fish wastes could have a negative environmental impact, such as eutrophication effects, on natural aquatic ecosystems [7].

By-products coming from the fishing industry, such as viscera, skin, scales and bones, representing up to 30-80% of the fish body weight, are discarded as solid wastes by industrial fish-processing operations [8] but, due to their composition, they have great potential to be used as protein supplements in aquaculture feeds [9,10]. Their conversion to aquafeed is also encouraged by the significant advantage that they do not require any thermal-chemical and/or enzymatic hydrolysis pretreatment steps. Since the pretreatment step is neither economically favorable nor environment-friendly, its elimination from the process makes the utilization of fish waste economic and more environmentally friendly [7].

Biotechnological methods like fermentation with microbe cultures are gaining more popularity for the treatment of waste [11,12].

Among the different microbes used, especially for the fermentation of animal/fish processing wastes, lactic bacteria have advantages over other microbes, as they are generally recognized as safe (GRAS) [13]. In addition, the products obtained upon fermentation with *Lactobacillus* are also reported to have additional beneficial effects on various aquatic animal intestines (anti-microbial properties, antioxidative properties), making them suitable for food/feed applications. In fact, they easily adapt to the intestinal environment of both aquatic and domestic animals, making them favorable for use in probiotic aquaculture feeds [14-18].

Among the microorganisms applied, yeasts have also been used as inoculum, along with lactic bacteria, to ferment fish waste [19] for converting it to a useful product that can be used as an ingredient to balance the food rations of animals. Yeast has many different immunostimulatory compounds, e.g., nucleic acid, b-glucans, and mannan oligosaccharides [20,21]. These compounds may enhance the growth of different fish species and therefore can be considered as the best health promoters for fish culture [22].

Feed nutritional composition is important; the major growth-promoting factors are proteins and lipids, since they are known to influence the growth and the body composition of fish [23,24].

Fermented fish waste is a liquid product, obtained by the liquefaction of tissues carried out by the enzymes already present in the fish and accelerated by an acid pH [10]. Natural fillers, such as agricultural by-products, can also be added to the substrate [25].

Citrus peel can be used as filler [26] during fermentation, playing at the same time an important role as a prebiotic source [27-29]. Among its beneficial effects, it has been reported that prebiotics can elevate fish resistance to pathogens and improve growth performance, feed utilization and lipid metabolism, as well as stimulating the immune response through the modulation of intestinal microbiota [18,30-32].

The aim of this research was to process non-sterilized fish wastes, supplemented with lemon peel as a filler and prebiotic source, by biological fermentation using combined starter cultures of *Saccharomyces cerevisiae* American Type Culture Collection (ATCC)4126 and *Lactobacillus reuteri* ATCC 53608 for bio-transforming these by-products into a high protein content supplement, rich in healthy microorganisms, for aquaculture feeds.

For this purpose, and to verify the optimum nutritional composition of aquafeed, proteins, crude lipids, ash and lipid content percentages were monitored throughout the process. The influence of the fermentation process on fatty acid concentrations was also evaluated. Finally, microbiological analyses of the starters and total and fecal coliform bacteria quantification were carried out, to evaluate the healthiness of the final product.

2. MATERIALS AND METHODS

2.1. Substrate

Fish by-products (non-edible parts) of *Dicentrarchus labrax,* represented by the head, viscera, skin and bones, were provided by Acqua Azzurra S.p.a. (Pachino, Italy). Samples were collected directly at the farm, forwarded to the laboratory under refrigerated conditions, and stored at -20°C until tests were performed. Lemon peel was provided by Simone Gatto S.r.l. (San Pier Niceto, Italy) and stored at -20°C until use.

2.2. Microorganisms

Saccharomyces cerevisiae ATCC 4126 was maintained on yeast medium (YM) agar (yeast extract 3 g/L, malt extract 3 g/L, peptone 5 g/L, glucose 10 g/L, agar 20 g/L; Oxoid, Basingstoke, UK) and *Lactobacillus reuteri* ATCC 53608, then maintained on MRS (de Man, Rogosa, Sharpe) agar (peptone 10 g/L, "Lab-Lemco" powder 8 g/L, yeast extract 4 g/L, sorbitan mono-oleate 1 mL, di-potassium hydrogen phosphate 2 g/L, sodium acetate tri-hydrate 5 g/L, tri-ammonium citrate 2 g/L, magnesium sulfate heptahydrate 0.2 g/L, manganese sulfate tetrahydrate 0.05 g/L, agar 10 g/L;

Oxoid, Basingstoke, UK) at 4°C. To carry out the tests, *S. cerevisiae* and *L. reuteri* were cultured overnight at 35°C and 37°C, respectively, on a rotary shaker (INNOVA 44, Incubator Shaker Series, New Brunswick Scientific, Edison, NJ, USA) at 200 rpm, in tubes containing 20 mL YM medium for the yeast and MRS broth for the bacteria.

2.3. Experimental Set-Up

Fermentation tests were carried out in a 5 L batch fermenter (Biostat Biotech B, Sartorius Stedim Biotech, Goettingen, Germany. The fermenter was equipped with one four-bladed Rushton turbine and the usual control systems: temperature, pH, pO_2 and a foam detector. Fish waste and lemon peel (2:1 *w/w*) were homogenized in a blender for 5 min.

The resulting homogenate, with a dry matter content of 40% (*w/w*) was supplemented with 20 mL of *S. cerevisiae* (10^8 cells per mL) and 20 mL of *L. reuteri* culture (10^8 cells per mL), simultaneously. No sterilization procedures were adopted.

Fermentation parameters were 35°C, with pH 5.0 and constant stirring at 200 rpm, and a final working volume of 3.5 L.

All fermentations were carried out for 120 h until no further growth of the selected microorganisms was observed, and the pH value became stable. The pH was not controlled by alkali addition during cultivation.

Medium samples were withdrawn daily from the reaction vessel using a sterile 20 mL syringe and immediately frozen at -20°C until analysis.

2.4. Yeast Cell, Lactic Acid Bacteria and Coliform Bacteria Numbers

Suspended yeast cells and lactic acid bacteria were counted via the dilution plating method, whereas coliform bacteria were counted via the MPN (most probable number) method.

The CFU (colony-forming unit) of suspended yeast was counted by culturing at 35°C in yeast media agar (pH 5.8) containing 100 mg/mL chloramphenicol (Oxoid, Basingstoke, UK). Lactic acid bacterial colonies were counted by assessing acid formation at 37°C in the MRS agar (pH 6.8) containing 50 mg/mL bromocresol purple (Oxoid, Basingstoke, UK). Coliform bacteria grown at 37°C in lauryl tryptose broth (Oxoid, Basingstoke, UK) (pH 6.8) and the gas-forming bacteria were confirmed on green bile medium (Oxoid, Basingstoke, UK) (pH 7.4).

2.5. Protein, Moisture, and Ash Determination

Representative samples were drained off daily for protein content testing, using the method suggested by the AOAC (Association of Official Agricultural Chemists, Rockville, MD, USA) [33]. The protein percentage was calculated considering a conversion factor of 6.25. The dry weights, both of the fresh waste and fermentation samples, were calculated as steady weights after 2 h at 110°C, using a Mettler PM 200 equipped with a Mettler LP16 IR balance (Mettler-Toledo GmbH, Laboratory & Weighing Technologies, Greifensee, Switzerland). Ash determination was carried out according to the AOAC method [33]. All samples were analyzed in triplicate.

2.6. Crude Fat and Fatty Acid Determination

Samples were extracted with a mixture of chloroform and methanol (2:1). The mixture was allowed to stand overnight and the lower lipid layer, transferred into a pretreated and weighed flask, was dried off. The difference in the two weights established the weight of the fat [24].

The fatty acid analysis was performed by gas chromatography after transmethylation with 2% H2SO4 in methanol at 80 °C. The separation and quantification of fatty acid methyl esters were conducted with a Dani Master GC 1000, equipped with an FID detector (Dani Instruments, Milan, Italy) and a capillary column Supelco SLB-IL100 60 m x 0.25 mm, film 0.20 gm(Merck KGaA, Darmstadt, Germany), using the following experimental conditions: injector temperature 220 °C; oven temperature from 120 °C to 200 °C (10 min hold) at a rate of 1 °C/min; detector temperature 240 °C; carrier gas He at a constant velocity rate of 34 cm/sec; and a split ratio of 1:50.

Fatty acids were identified by comparing the samples with reference standards, using Supelco 37 component Fatty Acid Methyl Esters (FAME) mix in methylene chloride. All samples were analyzed in triplicate.

All chemicals were provided by Merk Life Science (Merck KGaA, Darmstadt, Germany).

2.7. Statistical Analysis

The studies of significant differences were carried out via Kruskal-Wallis tests, using the SPSS 13.0 software package for Windows (SPSS Inc., Chicago, IL, USA). The H statistic and asymptotic significance are the Kruskal-Wallis test output values that allow the significance evaluation of differences when more than two groups are considered.

3. RESULTS AND DISCUSSION

3.1. Substrate Fermentation

The time course of fermentation by yeast and lactobacilli is shown in Figure 1, as well as the pH trend.

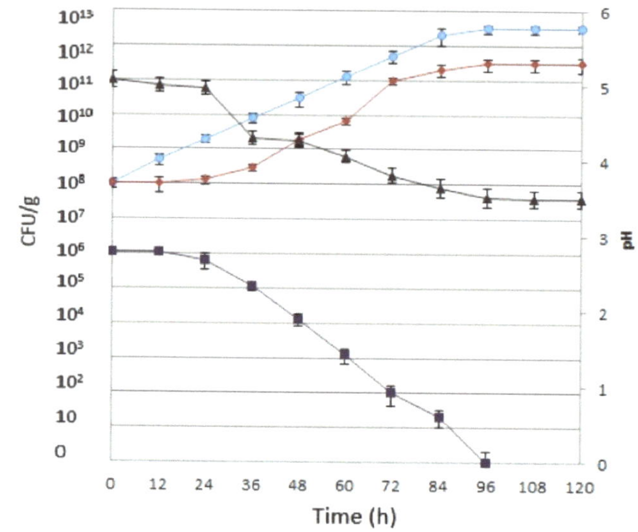

Figure 1. *Lactobacillus reuteri (circle), Saccharomyces cerevisiae (diamond), and coliform (square) concentrations, reported as colony-forming unit (CFU) per g, and pH (triangle) values recorded during the fermentration.*

The growth of[2] *S. cerevisiae* was slow during the first 24 h of fermentation, maintaining a concentration of 10^8 CFU/g. The amount of *S. cerevisiae* reached a concentration of 10^{11} CFU/g after 72 h, remaining stable until the end of the process.

L. reuteri increased constantly from the beginning of the fermentation until after 96 h, rising from 10^8 CFU/g up to 10^{12} CFU/g, reaching a steady state until the end of the process. According to Giraffa et al. [17] and Hoseinifar et al. [18], this represents an added value for the resulting aquafeed.

The reduction in pH was slow during the first 24 hours of fermentation because of microorganism adaptations at the beginning of the process [19,34]. In the presence of acid lactic bacteria and yeast, after 24 h the pH of the mixture became stable at 3.5 after 96 h. The decrease in pH in the substrate offers evidence of good acidification through lactic fermentation by the starter cultures and represents the most important factor to control in biotransformation. Acidification must be achieved as quickly as possible, in order to inhibit the growth of pathogenic and spoilage microorganisms in the substrate, increasing the shelf life of the resulting fermented substrate [10,35,36]. Moreover, considering that no sterilization procedures were carried out, the quick drop in pH was found to be necessary for maintaining microbial hygiene, along with retaining the quality of the product as an aquaculture feed [37].

In fact, the amount of initial substrate total coliforms was 10^6 CFU/g, whereas no fecal coliforms were detected. The microbiological analysis for total coliform determination showed a net decrease during the fermentation, to reach a complete absence after 96 h (Figure 1).

The reduction in coliform numbers could be due to some inhibitory compounds (bacteriocins) formed by the microorganisms employed during lactic acid fermentation and/or to the acidification of the medium [19]. Moreover, the decrease in coliforms may ensure good biopreservation against undesirable and/or hazardous microorganisms.

The final fermented products were low in spoilage microorganisms and rich in healthy microorganisms, representing a healthy final substrate enriched by added value.

The starter cultures' capability of growing at low pH can be ascribed to the lemon peel supplementation since polysaccharides, such as pectins, show a protective effect on lactic acid bacteria LAB against low pH [38,39]. Their ability to achieve this on fermenting fish waste supplemented by lemon peel was confirmed by the protein level's increasing during the process, up to 48.55%, making these wastes an excellent raw material for aquafeed production with *Lactobacillus reuteri* and *Saccharomyces cerevisiae.*

3.2. Substrate Protein, Ash and Crude Lipids Concentration

In Table 1, the protein, crude lipid, and ash percentages at different fermentation times are reported. All statistical evaluations were performed at a = 0.05.

Table 1. *Protein, crude lipid, and ash percentages, after different fermentation times.*

		Protein %	Crude Lipid %	Ash %
H statistic		16.290	7.058	13.386
Asymp. Sign.		**0.006**	0.0216	**0.020**
0h	Mean ± S.D.	11.68 ± 0.48 (A)	13.74 ± 0.72 (A)	0.83 ± 0.04 (A)
24 h	Mean ± S.D.	15.46 ± 0.40 (a)	14.04 ± 0.74 (a)	0.82 ± 0.04 (a)
48 h	Mean ± S.D.	20.01 ± 0.26 (a)	14.65 ± 0.77 (a)	0.78 ± 0.04 (AB)
72 h	Mean ± S.D.	32.09 ± 0.77 (B)	14.85 ± 0.78 (a)	0.71 ± 0.04 (AB)
96 h	Mean ± S.D.	48.55 ± 1.15(b)	15.05 ± 0.79 (a)	0.66 ± 0.03 (B)
120 h	Mean ± S.D.	48.55 ± 1.15(b)	15.25 ± 0.80 (a)	0.66 ± 0.03 (b)

Bold values are significant at $p < 0.05$. A and B indicate homogeneous groups at a = 0.05: fermentation times that do not differ from each other are designated by the same letter.

The substrate's initial protein content was 11.68 ± 0.48%. It increased slowly by 72 h, reaching up to 32.09 ± 0.77%. The highest protein percentage in the substrate, 48.55 ± 1.15%, was reached after 96 h. This value remained stable until the end of the fermentative process, allowing to obtain a substrate rich in protein, achieving a suitable percentage for aquafeed formulation according to Nasseri et al. [34]. The protein content against the CFU of the yeast and LAB is reported in Figure 2.

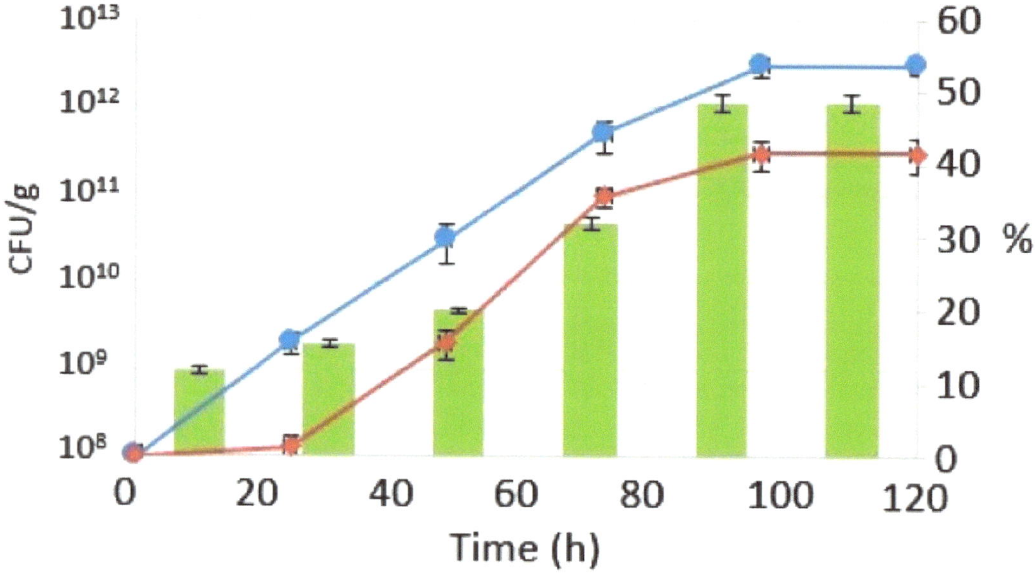

Figure 2. Lactobacillus reuteri (circle), Saccharomyces cerevisiae (diamond), reported as colony-forming unit (CFU) per g, and protein (bars) levels increasing during the fermentation process, reported as percentages.

During all fermentations, ash concentrations dpcreased significantly, from 0.83 ± 0.04% to 0.66 ± 0.03%. This could ᴾe due to partial ash utilization by the yeast as a source of minerals [40].

The crude lipid content calculated on the initial substrate was 13.74 ± 0.72%. Throughout the process, this value did not increase sigmficantly; at the end of the fermentation period, it reached just 15.25 ± 0.80%.

According to the; literature [85,35,41], final protein and lipid contents had reached adequate levels in the resulting aquafeed, ofering a way to amyliorate the problem of a lack of protein sources in aquaculture by encouraging the conversion of fish waste into feed, using a low-cost process such as lactic fermentation.

3.3. Fatty Acid Composition

Fatty acid contents at different fermentation times are shown in Table 2. At the end of the fermentation period, as confirmed by statistical tests with significances higher than 95%, MUFAs (monounsaturated fatty acids) increased significantly (+23.5%) and PUFAs (polyunsaturated fatty acids) decreased (-22.7%), whereas the SFA (saturated fatty acids) content was not affected by the fermentation process. Changes in fatty acid composition during the fermentation process are shown in Table 3. The main fatty acids detected throughout the process were C18:1 n-9 cis (33.02%), C18:2 n-6 cis (20.49%) and C16:0 (17.53%). C14:1, C17:0, C17:1, C20:0, and C21:0 were detected, starting after 72 h of the process. A significant increase in concentration for C20:1 n-9, C 20:2 n-6 and C 24:1 n-9 was observed, whereas C18:2 n-6 cis, C20:3 n-6, C20:4 n-6, C20:5 n-3, C22:2, C23:0 and C24:0 showed a significant opposite trend. According to Nadege et al. [23] and Babalola and Apata [42], this fatty-acid composition is suitable for aquafeed formulations in which the shelf life could be extended because of the decrease in polyunsaturated fatty acids.

Table 2. Fatty acid contents at different fermentation times.

		SFA (%)	MUFA (%)	PUFA (%)
H statistic		5.398	13.070	13.819
Asymp. Sign.		0.369	**0.023**	**0.017**
0h	Mean ± S.D.	28.92 ± 1.51	39.02 ± 2.04 (A)	31.73 ± 1.66 (A)
24 h	Mean ± S.D.	28.45 ± 1.49	39.68 ± 2.08 (a)	29.44 ± 1.54 (A)
48 h	Mean ± S.D.	27.68 ± 1.45	42.20 ± 2.21 (a)	28.08 ± 1.47 (a)
72 h	Mean ± S.D.	27.53 ± 1.44	44.90 ± 2.35 (a)	26.43 ± 1.38 (B)
96 h	Mean ± S.D.	26.96 ± 1.41	47.48 ± 2.49 (B)	24.63 ± 1.29 (b
120 h	Mean ± S.D.	26.49 ± 1.39	48.19 ± 2.52 (B	24.53 ± 1.29 (B

Values in bold are significant at p < 0.05. A and B indicate homogeneous groups at a = 0.05; fermentation times that do not differ from each other are designated by the same letter.

Table 3. Fatty acid contents (%) at different fermentation times.

		C14:0	C15:0	C16:0	C17:0	C18:0	C20:0	C21:0	C22:0	C23:0	C24:0	
H statistic		13.757	11.294	4.273	14.537	10.584	14.868	14.434	7.602	15.690	15.409	
Asymp. Sign.		0.017	0.031	0.511	0.013	0.060	0.011	0.013	0.207	**0.008**	**0.009**	
0h	Mean	2.38	0.24	17.53	0.005 *	3.72	0.005 *	0.005 *	0.55	3.85 (B)	0.65 (B)	
	S.D.	0.12	0.01	0.92	0.00	0.19	0.00	0.00	0.03	0.20	0.03	
24 h	Mean	2.57	0.22	17.69	0.005 *	3.92	0.005 *	0.005 *	0.60	3.20 (AB)	0.26 (AB)	
	S.D.	0.13	0.01	0.93	0.00	0.21	0.00	0.00	0.03	0.17	0.01	
48 h	Mean	3.09	0.26	17.21	0.005 *	3.63	0.005 *	0.005 *	0.58	2.75 (AB)	0.17 (A)	
	S.D.	0.16	0.01	0.90	0.00	0.19	0.00	0.00	0.03	0.14	0.01	
72 h	Mean	2.31	0.22	17.62	0.20	3.21	0.23	0.06	0.52	3.00 (AB)	0.15 (A)	
	S.D.	0.12	0.01	0.92	0.01	0.17	0.01	0.00	0.03	0.16	0.01	
96 h	Mean	2.62	0.25	17.16	0.15	3.65	0.10	0.05	0.57	2.25 (A)	0.16 (A)	
	S.D.	0.14	0.01	0.90	0.01	0.19	0.01	0.00	0.03	0.12	0.01	
120 h	Mean	2.62	0.25	17.16	0.15	3.65	0.10	0.05	0.57	2.25 (A)	0.16 (A)	
	S.D.	0.14	0.01	0.90	0.01	0.19	0.01	0.00	0.03	0.12	0.01	
		C14: 1	C16: 1	C17: 1	C18: 1n9t	C18: 1n9c	C20: 1n9	C22: 1n9	C24: 1n9			
H statistic		14.434	12.251	14.110	15.058	10.708	15.199	10.081	16.579			
Asymp. Sign.		0.013	0.031	0.015	0.011	0.057	**0.010**	0.073	**0.005**			

		C14:0	C15:0	C16:0	C17:0	C18:0	C20:0	C21:0	C22:0	C23:0	C24:0	
0h	Mean	0.005 *	2.77	0.005*	0.52	33.02	1.58 (A)	0.95	0.19(A)			
	S.D.	0.00	0.14	0.00	0.03	1.73	0.08	0.05	0.01			
24 h	Mean	0.005 *	3.23	0.005*	0.63	33.33	1.47 (A)	0.85	0.17(A)			
	S.D.	0.00	0.17	0.00	0.03	1.75	0.08	0.04	0.01			
48 h	Mean	0.005 *	3.67	0.005*	0.87	34.95	1.63 (AB)	0.96	0.13(A)			
	S.D.	0.00	0.19	0.00	0.05	1.83	0.09	0.05	0.01			
72 h	Mean	0.06	3.58	0.16	0.74	36.06	2.95 (AB)	1.03	0.33 (AB)			
	S.D.	0.00	0.19	0.01	0.04	1.89	0.15	0.05	0.02			
96 h	Mean	0.06	3.33	0.14	0.65	38.12	3.77(B)	1.01	0.40 (B)			
	S.D.	0.00	0.17	0.01	0.03	2.00	0.20	0.05	0.02			
120 h	Mean	0.07	3.38	0.15	0.69	38.39	3.94 (B)	1.02	0.55 (B)			
	S.D.	0.00	0.18	0.01	0.04	2.01	0.21	0.05	0.03			
		C18: 2n6t	C18: 2n6c	C18: 3n6	C18: 3n3	C20: 2n6	C20: 3n6	C20: 3n3	C20: 4n6	C20: 5n3	C22: 2	C22: 6 n3
H statistic		14.427	14.184	16.064	11.854	15.503	15.592	14.868	16.155	16.251	15.827	15.082
Asymp. Sign.		0.013	**0.004**	0.017	0.037	**0.008**	**0.008**	0.011	**0.006**	**0.006**	**0.007**	0.011
0h	Mean	0.27	20.49 (A)	0.29	4.71	1.44 (AB)	0.26 (B)	0.18	0.60 (B)	0.16 (B)	0.39 (B)	2.93
	S.D.	0.01	1.07	0.02	0.25	0.08	0.01	0.01	0.03	0.01	0.02	0.15

		C14:0	C15:0	C16:0	C17:0	C18:0	C20:0	C21:0	C22:0	C23:0	C24:0	
24 h	Mean	0.26	20.50 (A)	0.22	3.70	1.12 (A)	0.10 (A)	0.28	0.75 (B)	0.15 (B)	0.38 (B)	1.97
	S.D.	0.01	1.07	0.01	0.19	0.06	0.01	0.01	0.04	0.01	0.02	0.10
48 h	Mean	0.21	19.07 (A)	0.20	4.25	1.96 (B)	0.09 (A)	0.23	0.09 (A)	0.04 (AB)	0.06 (A)	1.87
	S.D.	0.01	1.00	0.01	0.22	0.10	0.00	0.01	0.00	0.00	0.00	0.10
72 h	Mean	0.29	16.98 (AB)	0.33	4.36	1.83 (B)	0.08 (A)	0.12	0.10(A)	0.02 (A)	0.05 (A)	2.26
	S.D.	0.02	0.89	0.02	0.23	0.10	0.00	0.01	0.01	0.00	0.00	0.12
96 h	Mean	0.25	15.47 (B)	0.25	4.06	1.73 (AB)	0.10 (A)	0.12	0.08 (A)	0.05 (AB)	0.06 (A)	2.45
	S.D.	0.01	0.81	0.01	0.21	0.09	0.01	0.01	0.00	0.00	0.00	0.13
120 h	Mean	0.22	15.59 (B)	0.26	4.07	1.52 (AB)	0.11 (A)	0.12	0.10(A)	0.03 (AB)	0.07 (A)	2.43
	S.D.	0.01	0.82	0.01	0.21	0.08	0.01	0.01	0.01	0.00	0.00	0.13

** Values corresponding to the half of the determination limit; values in bold are significant at $p < 0.01$. A and B indicate homogeneous groups at $a = 0.01$: fermentation times that do not differ from each other are designated by the same letter.*

Previous studies carried out by Fickers et al. [43] and Yano et al. [44] reported similar fatty acid behavior. This trend can be ascribed to the yeast and lactic bacteria activities that degrade fats for single-cell protein production [44] and to obtain the energy necessary for metabolic activities during the fermentation process [45]. The fatty acids resulting from the degradation of lipids are subsequently degraded through the |3 -oxidation system in the yeast cells [44], resulting in a reduction in polyunsaturated fatty acids.

4. CONCLUSIONS

This study demonstrated an effective approach to utilizing the demonstrated fermented substrate for aquafeed, starting with fish and lemon peel wastes, allowing the conversion of both animal and vegetable food wastes into an added-value product.

The final fermented product is low in spoilage microorganisms and rich in healthy microorganisms, representing a healthy final substrate enriched by added value.

The microorganisms' ability to feed on fermenting fish waste that is supplemented by lemon peel was confirmed by the protein level's increasing during the process, up to 48.55%, making these wastes an excellent raw material for aquafeed production via *Lactobacillus reuteri* and *Saccharomyces cerevisiae.* In fact, the final protein and lipid contents represent adequate levels in the resulting aquafeed, reducing the problem of a lack of protein sources for aquaculture by encouraging the conversion of fish waste and lemon peel into feed.

This study pointed out the possibility of setting up a fermentation process based on the simultaneous addition of two different microorganisms, reaching a plateau after 96 h. Further studies are in progress for converting the resulting fermentation product into pellets and for testing the effect of the final product on the growth and immune response of fish from aquaculture and, consequently, in human consumers.

Finally, additional work will be needed to further optimize production to facilitate future larger-scale production, also evaluating it from an economic point of view.

REFERENCES

1. Agriculture Organization of the United Nations. *The State of World Fisheries and Aquaculture;* FAO: Rome, Italy, 2020; ISSN 2410-5902.
2. Tacon, A.G.J.; Metian, M. Global overview on the use of fishmeal and fish oil in industrially compounded aquafeeds: Trends and future prospects. *Aquaculture* **2008**, 285,146-158. [CrossRef]
3. Nwanna, L.C. Nutritional Value and Digestibility of Fermented Shrimp Head Waste Meal by African Catfish *Clariasgariepinus. Pak. J. Nutr.* **2003**, 2, 339-345. [CrossRef]
4. Samuels, W.A.; Fontenot, J.P.; Allen, V.G.; Abazinge, M.D. Seafood processing wastes ensiled with straw: Utilization and intake by sheep. *J. Anim. Sci.* **1991**, 69, 4983-4992. [CrossRef]
5. Westendorf, M.L.; Dong, Z.C.; Schoknecht, P.A. Recycled cafeteria food waste as a feed for swine: Nutrient content digestibility, growth, and meat quality. *J. Anim. Sci.* **1998**, 76, 2976-2983. [CrossRef] [PubMed]

6. Westendorf, M.L. Food Waste as Animal Feed: An Introduction. In *Food Waste to Animal Feed*; Michael, L., Ed.; Iowa State University Press: Ames, IA, USA, 2000; pp. 3-16, 69-90.
7. Suan, S.; Jing, L.; Wenjian, G.; David, B. Nutrient value of fish manure waste on lactic acid fermentation by *Lactobacillus pentosus. R. Soc. Chem.* **2018**, 8, 31267-31274. [CrossRef]
8. Zhiwen, Z.; Baiyu, Z.; Qinhong, C.; Jingjing, L.; Kenneth, L.; Bing, C. Fish Waste Based Lipopeptide Production and the Potential Application as a Bio-Dispersant for Oil Spill Control. *Front. Bioeng. Biotechnol.* **2020**, 8, 734. [CrossRef]
9. Hua, K.; Cobcroft, J.M.; Andrew, C.; Condon, K.; Jerry, D.R.; Mangott, A.; Praeger, C.; Vucko, M.J.; Zeng, C.; Zenger, K.; et al. The Future of Aquatic Protein: Implications for Protein Sources in Aquaculture Diets. *One Earth* **2019**,1, 316-329. [CrossRef]
10. Arvanitoyannis, I.S.; Kassaveti, A. Fish industry waste: Treatments, environmental impacts, current and potential uses. *Int. J. Food Sci. Technol.* **2008**, 43, 726-745. [CrossRef]
11. Coello, N.; Montiel, E.; Concepcion, M.; Christen, P. Optimization of a culture medium containing fish silage for L-lysine production by *Corynebacterium glutamicum. Bioresour. Technol.* **2002**, 85,207-211. [CrossRef]
12. Tropea, A.; Wilson, D.; La Torre, G.L.; Lo Curto, R.B.; Saugman, P.; Troy-Davies, P.; Dugo, G.; Waldron, K.W. Bioethanol Production From Pineapple Wastes. *J. Food Res.* **2014**, 3, 60-70. [CrossRef]
13. Amit, K.R.; Swapna, H.C.; Bhaskar, N.; Halami, P.M.; Sachindra, N.M. Effect of fermentation ensilaging on recovery of oil from fresh water fish viscera. *Enzym. Microb. Technol.* **2010**, *46*, 9-13. [CrossRef]
14. Bernardeau, M.; Guguen, M.; Vernoux, J.P. Beneficial lactobacilli in food and feed: Long-term use biodiversity and proposal for specific and realistic safety assessments. *FEMS Microbiol.* **2006**, 30, 487-513. [CrossRef] [PubMed]
15. Bucio, A.; Hartemink, R.; Schrama, J.W.; Verreth, J.; Rombouts, F.M. Presence of lactobacilli in the intestinal content of freshwater fish from a river and from a farm with a recirculation system. *Food Microbiol.* **2006**, 23, 476-482. [CrossRef]
16. Balcazar, J.L.; Venderll, D.; de Blas, I.; Ruiz-Zarzuela, I.; Muzquiz, J.L.; Girones, O. Characterization of probiotic properties of lactic acid bacteria isolated from intestinal microbiota of fish. *Aquaculture* **2008**, 278,188-191. [CrossRef]
17. Giraffa, G.; Chanishvili, N.; Widyastuti, Y. Importance of lactobacilli in food and feed biotechnology. *Res. Microbiol.* **2010**,161, 480-487. [CrossRef]

18. Hoseinifar, S.H.; Sharifian, M.; Vesaghi, M.J.; Khalili, M.; Esteban, M.A. The effects of dietary xylooligosaccharide on mucosal parameters, intestinal microbiota and morphology and growth performance of Caspian white fish *(Rutilus frisiikutum)* fry. *Fish Shellfish. Immunol.* **2014**, 39, 231-236. [CrossRef] [PubMed]

19. Ennouali, M.; Elmoualdi, L.; Labioui, H.; Ouhsine, M.; Elyachioui, M. Biotransformation of the fish waste by fermentation. *Afr. J. Biotechnol.* **2006,** 5,1733-1737. [CrossRef]

20. White, L.A.; Newman, M.C.; Cromwell, G.L.; Lindemann, M.D. Brewers dried yeast as a source of mannan oligosaccharides for weanling pigs. *J. Anim. Sci.* **2002**, 80, 2619-2628. [CrossRef]

21. Tropea, A.; Wilson, D.; Cicero, N.; Potorti, A.G.; La Torre, G.L.; Dugo, G.; Richardson, D.; Waldron, K.W. Development of minimal fermentation media supplementation for ethanol production using two *Saccharomyces cerevisiae* strains. *Nat. Prod. Res.* **2016,** 30, 1009-1016. [CrossRef]

22. Lara-Flores, M.; Olvera-Novoa, M.A.; Guzma'n-Me'ndez, B.E.; Lo'-pez-Madrid, W. Use of the bacteria *Streptococcus faecium* and *Lactobacillus acidophilus,* and the yeast *Saccharomyces cerevisiae* as growth promoters in Nile tilapia *(Oreochromis niloticus). Aquaculture* **2002,** 216,193-201. [CrossRef]

23. Nadege, R.; Mourente, G.; Sadasivam, K.; Corraze, G. Replacement of a large portion of fish oil by vegetable oils does not affect lipogenesis, lipid transport and tissue lipid uptake in European seabass *(Dicentrarchus labrax* L.). *Aquaculture* **2006,** 261,1077-1087. [CrossRef]

24. Mahmud, N.A.; Robiul Hasan, M.D.; Hossain, M.B.; Minar, M.H. Proximate Composition of Fish Feed Ingredients Available in Lakshmipur Region, Bangladesh. *Am.-Eurasian J. Agric. Environ. Sci.* **2012**,12, 556-560.

25. Soltan, M.A.; El-Laithy, S.M. Evaluation of fermented silage made from fish, tomato and potato by-products as a feed ingredient for *Nile tilapia. Oreochromis niloticus. Egypt. J. Aquat. BioiFish* **2008**,12, 25-41. [CrossRef]

26. Soltan, M.A.; Hanafy, M.A.; Wafa, M.I.A. An evaluation of fermented silage made from fish by-products as a feed ingredient for african catfish (*Clariasgariepinus*). *Glob. Vet.* **2008,** *2*, 80-86.

27. Manderson, K.; Pinart, M.; Tuohy, K.M.; Grace, W.E.; Hotchkiss, A.T.; Widmer, W.; Yadhav, M.P.; Gibson, G.R.; Rastall, R.A. In Vitro Determination of Prebiotic Properties of Oligosaccharides Derived from an Orange Juice Manufacturing By-Product Stream. *Appl. Environ. Microbiol.* **2005**, 71, 8383-8389. [CrossRef] [PubMed]

28. Roberfroid, M. Prebiotics: The concept revisited. *J. Nutr.* **2007**,137, 830-837. [CrossRef]

29. Gomez, B.; Gullon, B.; Remoroza, C.; Schols, H.A.; Parajo, J.C.; Alonso, J.L. Purification, Characterization, and Prebiotic Properties of Pectic Oligosaccharides from Orange Peel Wastes. *J. Agric. Food Chem.* **2014**, 62, 9769-9782. [CrossRef]

30. Gibson, G.R. Fibre and effects on probiotics (the prebiotic concept). *Clin. Nutr. Suppl.* **2004**,1, 25-31. [CrossRef]

31. Ringo, E.; Hoseinifar, S.H.; Ghosh, K.; Doan, H.V.; Beck, B.R.; Song, S.K. Lactic acid bacteria in finfish—An update. *Front. Microbiol.* **2018**, 9,1818. [CrossRef]

32. Ringo, E.; Dimitroglou, A.; Hoseinifar, S.H.; Davies, S.J. Prebiotics in finfish: An update. In *Aquaculture Nutrition: Gut Health, Probiotics and Prebiotics;* Ringo, E., Merrifield, D.L., Eds.; Wiley-Blackwell Scientific Publication: London, UK, 2014. [CrossRef]

33. Latimer, G.W. *Official Methods of Analysis of AOAC International;* AOAC International: Gaithersburg, MD, USA, 2016.

34. Nasseri, A.T.; Rasoul-Amini, S.; Morowvat, M.H.; Ghasemi, Y. Single Cell Protein: Production and Process. *Am. J. Food Technol.* **2011**, 6,103-116. [CrossRef]

35. Shigeaki, I.; Kyoko, S.U.; Yukako, K.; Akemi, K.; Isao, Y.; Koudai, T.; Ayumi, M.; Kunimasa, K. Fermentation of non-sterilized fish biomass with a mixed culture of film-forming yeasts and lactobacilli and its effect on innate and adaptive immunity in mice. *J. Biosci. Bioeng.* **2013**,116, 682-687. [CrossRef]

36. Garcia-Diez, J.; Saraiva, C. Use of Starter Cultures in Foods from Animal Origin to Improve Their Safety. *Int. J. Environ. Res. Public Health* **2021**,18, 2544. [CrossRef] [PubMed]

37. Ayan, S. A review of fish meal replacement with fermented biodegradable organic wastes in aquaculture. *Int. J. Fish. Aquat. Stud.* **2018**, 6, 203-208.

38. Nadja, L.; Thiago, B.C.; Susana, M.I.S.; Andreas, B.; Lene, J. The effect of pectins on survival of probiotic *Lactobacillus spp.* in gastrointestinal juices is related to their structure and physical properties. *Food Microbiol.* **2018**, 74,11-20. [CrossRef]

39. Huan, L.; Mingyong, X.; Shaoping, N. Recent trends and applications of polysaccharides for microencapsulation of probiotics. *Food Front.* **2020**,1, 45-59. [CrossRef]

40. Araya-Cloutier, C.; Rojas-Garbanzo, C.; Velazquez-Carrillo, C. Effetct of initial sugar concentration on the production of L(+) lactic acid by simultaneous enzymatic hydrolysis and fermentation of an agro-industrial waste product of

pineapple (*Ananas comosus*) using *Lactobacillus casei* subspeies *rhamnosus*. *Int. J. Wellness Ind.* **2012**,*1*, 91-100. [CrossRef]

41. Craig, S.; Helfrich, L.A. Understanding Fish Nutrition, Feeds and Feeding. *Va. Coop. Ext.* **2002**. Available online: https: //vtechworks.lib.vt.edu/bitstream/handle/10919/80712/FST-269.pdf (accessed on 19 November 2021).
42. Babalola, T.O.O.; Apata, D.F. Chemical and quality evaluation of some alternative lipid sources for aqua feed production. *Agric. Biol. J. N. Am.* **2011**, 2, 935-943. [CrossRef]
43. Fickers, P.; Benetti, P.H.; Wache, Y.; Marty, A.; Mauersberger, S.; Smit, M.S.; Nicaud, J.M. Hydrophobic substrate utilization by the yeast *Yarrowia lipolytica*, and its potential applications. *FEMS Yeast Res.* **2005**, 5, 527-543. [CrossRef]
44. Yano, Y.; Oikawa, H.; Satomi, M. Reduction of lipids in fishmeal prepared from fish waste by a yeast *Yarrowia lipolytica*. *Int. J. Food Microbiol.* **2007**,121, 302-307. [CrossRef] [PubMed]
45. Oseni, O.A.; Akindahunsi, A.A. Some phytochemical properties and effect of fermentation on the seed of *Jatropha curcas* L. *J. Food Technol.* **2011**, 6,158-165. [CrossRef]

CHAPTER 8

Strategies to Increase the Value of Pomaces with Fermentation

1. INTRODUCTION

Pomace is the main residue (a humid, solid material) generated from the pressing of fruits and olives to obtain juices and olive oil, respectively. This residue is heterogeneous and may contain seeds, pulp, stems, and peels, depending on the source [1,2]. In terms of the global production of juices and olive oil, the amount of pomace produced every year achieves several millions of tons [3,4]. Its high organic matter, nutrients, and moisture content favor the growth of microorganisms to decompose this residue (the generation of greenhouse gases, unpleasant odors, and contamination of groundwater) and can attract pests, which ultimately leads to an important environmental impact [5]. Additionally, the consumption of juices [6,7] and olive oil [8-10] is expected to increase in the upcoming years. In this sense, the residues from these two sectors of the food industry are expected to increase.

Another important aspect related to pomaces is the presence of bioactive compounds that are lost when these residues are discarded. One of the most studied classes of phytochemicals are polyphenols. This class of compounds is characterized by the antioxidant [11-13], antimicrobial [3,12], anti-inflammatory [13], and anti-diabetic [14] activities tested in vitro and in vivo. This scenario can be seen as a relevant opportunity to explore strategies to improve the management of pomace and reduce its environmental impact.

In this sense, the concept of a circular economy is favored to improve the sustainability in this sector of the food industry, i.e., transforming residues into raw materials with high- added value and connecting them with other chains of food processing [15]. Moreover, a circular economy is on of the principles if the Europian Green Deal that aims to improve the efficiency of resource use and to cut potlufion, for instance [16,1n]. recent publications support the potential utilization of this strategy [18-24]. It is also important to mention that the reutilization of residues of the food industry and the consequent development of food products are concepts supported and well-accepted by consumers [25,26].

Among the possible solutions So manage pomaces, fermentation has been suggested to obtain high-added value products and compounds. Moreover, fermentation can be seen as an important and more sustainable strategy to treat food industry residues [2,27,28]. Thus, this review aims to provide an overview of the utilization of fermentation (mainly involving lactic acid bacteria and yeasts) and biotransformation (biotransformation) of pomace in the production of silage and supplement feed for animal, enzymes, polyphenols, bioactive compounds (release of bound polyphenols and the synthesis of fatty acids and carotenoids), odor-active volatile compounds, and organic acid production.

2. UTILIZATION IN SILAGE OR AS A FEED SUPPLEMENT FOR ANIMAL PRODUCTION

The feeding of animals reared for food Prodaction is one of the Possicle applications of fermented pomaces (Figure 1), for which there are two main strategies: adding the pomace in silage production or fermenting/biotransforming; the pomace and using it as feed supplement (Tattle 1). Regarding the first strategy, the production of silage consisos of preserving pasture grass for further use (especially during dry periods). The process occurs mainly by fermenting pasture with bacteria that are naturally or strategically added to acidify the material and delay microbial and biochemical spoilage [29]. Considering the importance of silage and the fermentation process, many studies have explored the effect of pomace in the characteristics of silage and its effect in animal health and performance.

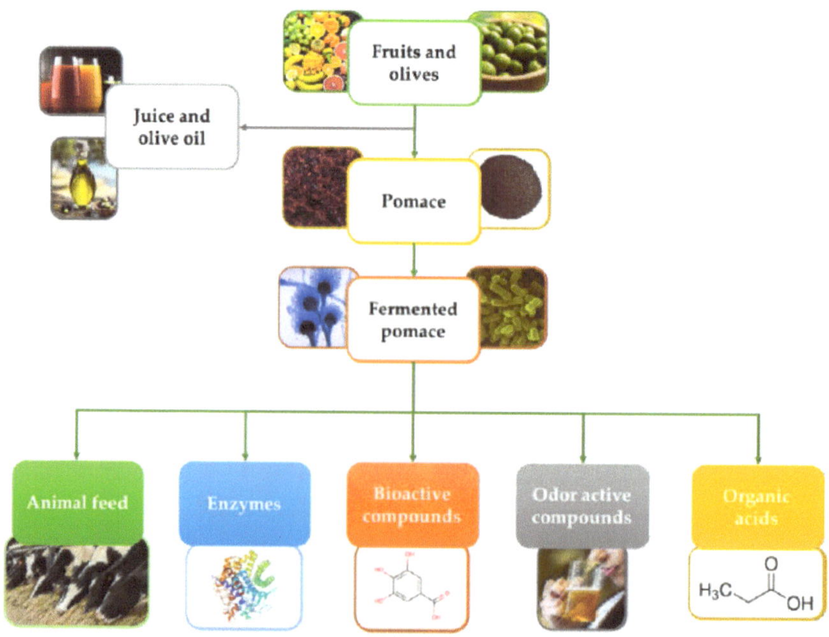

Figure 1. Schematic representation of strategies to valorize pomaces with fermentation.

Table 1. Effect of fermentation in the characteristics of silage produced from apple, white mulberry, and grape pomace.

Source	Experimental Conditions	Effect in Silage Characteristics	Ref.
Apple pomace	Alfalfa hay, timothy hay, soybean meal, and vitamin and mineral supplement, pomace (0, 5,10, and 20%), 3.4-22.4°C, and 60 days	Increased pH, ethanol, acetic acid, and ammonia nitrogen levels; reduced lactic acid content	[30]
Apple pomace	Maize, wheat bran, soybean meal, timothy hay, alfalfa hay and vitamin-mineral supplement, pomace (20% in silage), 4.4-25.8°C, and 60 days	Increased pH and ethanol content; reduced lactic and acetic acid, and ammonia nitrogen contents	[31]
Grape pomace	*Calotropis procera*, pomace (0,10, 20, 40% in silage), and 90 days	Increased ethanol, acetic, propionic and butyric acid contents, effluent loss and gas loss; reduced soluble carbohydrate and lactic acid content and digestibility; no effect in pH	[32]
Grape pomace	Sweet sorghum silage, pomace (0, 5,10, and 15% in silage), and 90 days	Increased acetic acid (only 10%) and total polyphenol content; reduced water-soluble carbohydrates, lactic acid (only 15% treatment), and butyric acid contents; no effect in dry matter and protein neutral and detergent fiber contents, pH, and ammonia nitrogen level	[33]
White mulberry pomace	Meadow grass, pomace (0, 25, 50, 75, and 100% in silage), and 60 days	Increased gas production, organic matter digestibility, and metabolizable energy	[34]

Using pomace as a raw material for silage production may shift the characteristics of silage and change its content and composition of organic acids, digestible matter, and pH. These results were reported in studies with apple pomace that also indicated a reduction in the production of lactic acid [30,31]. Along with the increase in ethanol content in silage, the pH was increased, and the accumulation of lactic acid was reduced in relation to silage without pomace. However, these studies also indicated an unclear effect in the accumulation of ammonia nitrogen.

In the case of grape pomace, the depletion in lactic acid content and the increase in the production of other organic acids, polyphenol content, effluent and gas loss were also reported in two recent studies [32,33]. Both studies did not indicate significant differences in the pH of silage. It is relevant to mention that the study carried out by Li et al. [33] also evaluated the combination of grape pomace with the starter culture composed of *Lactobacillus plantarum* and *Lactobacillus buchneri.* These microorganisms led to a better control of fermentation and quality of silage by favoring the accumulation of lactic and acetic acid, water soluble carbohydrates, and crude protein. Moreover, ammonia nitrogen levels were reduced and no effect in the neutral detergent fiber content and the pH of silage were reported. A related experiment evaluated the production of silage with white mulberry pomace with meadow grass [34]. In this case, significant effects in organic matter digestibility and metabolizable energy, as well as in gas production, were reported.

Since silage is an important component for animal production in periods and regions of reduced feed availability, some studies reported the effect of silage with pomace and fermented pomace in animal nutrition, health and the composition and characteristics of foods obtained from animals in these experimental diets (Table 2). For instance, recent experiments reported the effect of silage added with apple pomace in the diet of Suffolk wethers [30,31]. In both cases, significant reductions in digestibility and nitrogen retention, in relation to the control diet, were reported. No effect in feed intake between control and experimental diets were indicated in these studies.

Table 2. Effect of fermented apple, grape, pomegranate, olive, and tomato pomaces in animal production and foods obtained from animals fed with these fermented pomaces.

Source	Experimental Conditions	Animals and Study Characteristics	Effect in Animal Production and Related Food	Ref.
Apple pomace	Silage: alfalfa hay, timothy hay, soybean meal, and vitamin and mineral supplement, pomace (0, 5,10, and 20%), 3.4-22.4°C, and 60 days	Suffolk wethers (4 animals), initial weight 50.3 kg, and 21 days of experiment	Reduced digestibility, gross energy, and nitrogen retention; no effect in feed intake and fiber content	[30]
Apple pomace	Silage: maize, wheat bran, soyabean meal, timothy hay, alfalfa hay and vitamin-mineral supplement, pomace (20% in silage), 4.4-25.8°C, and 60 days	Suffolk wethers (4 animals), initial weight 65.3 kg, and 21 days of experiment	Increased organic acid content (except propionic and butyric); reduced digestibility and nitrogen retention; no effect in feed intake	[31]

Source	Experimental Conditions	Animals and Study Characteristics	Effect in Animal Production and Related Food	Ref.
Apple pomace	Silage: control feed, pomace (14.8% in silage), 9.7-20.1°C, and 21 days	Male Yorkshire x Duroc x Landrace pigs (10 animals), initial weight 70 kg, and 53 days of experiment	Animals: increased feed efficiency; reduced average daily feed intake; no effect in finished body weight, average, daily gain, carcass weight, back fat thickness or dressing ratio Back fat: increased moisture, linoleic acid (C18:2n6), linolenic acid (C18:3) and arachidic acid; reduced water holding capacity, palmitic acid (C16:0), palmitoleic acid (C16:l) and heptadecenoic acid (C17:l) proportion	[35]
Apple pomace	Silage: minced sardine, pomace (15%), *Lactobacillus plantarum* (starter culture), 35°C for 7 days	Juvenile European sea bass fish (240 animals), initial weight 15 g, and 9 weeks of experiment	Increased feed conversion ratio, relative average daily feed intake, leukocyte count, and carcass composition (moisture, lipid and ash contents); reduced final body weight, weight gain, specific growth rate, protein efficiency, apparent net protein utilization, and microvilli density	[36]
Grape pomace	Silage: sorghum, pomace (0,10, 20, and 30%), and 7 months	Male mixed breed lambs (24 animals), initial weight 21.5 kg, and 35 days of experiment	No effect in performance, carcass composition, and meat quality	[37]
Grape pomace	Silage: corn, water, starter culture, and pomace (43.6 g/kg feed)	Landrace x Large White — Duroc — Pietrain piglets (24 animals), and 15 days of experiment	Animals: increased antioxidant defense system response, average daily gain, growth of facultative probiotic bacteria, and LAB; reduced oxidative stress and pathogen Meat: increased omega-3 fatty acids content; reduced n-6 /n-3 ratio	[38]

Source	Experimental Conditions	Animals and Study Characteristics	Effect in Animal Production and Related Food	Ref.
Grape pomace	SSF: 1 kg substrate, *Aspergillus niger,* 30°C, and 7 days; pomace (15 g/kg feed)	Male Ross 308 broiler chicks (140 animals), and 42 days of experiment	Animals: increased body weight and serum CAT level; reduced *Clostridium perfringens* count in cecum; no effect in feed intake, feed conversion ratio, serum GPx and SOD, other microorganism in cecum, and intestinal morphology Liver: no effect in pH and color	[39]
Olive pomace	SSF: Two-step fermentation: *Bacillus subtilis* var. natto N21, 37 °C, 2 days; *Lactobacillus casei,* 25-35 °C, and 5 days; pomace (7.5,15, and 30%)	Male Ross 308 broiler chicks (1400 animals), initial weight 44-47 g, and 42 days of experiment	Animals: increased feed conversion ratio, antioxidant status and defense system response; reduced body weight gain, protein efficiency ratio, nutrient digestibility, serum triglycerides and total cholesterol; no effect in feed intake, serum LDL cholesterol, ALT and AST Breast meat: increased GPx and SOD; reduced fat and cholesterol content, and lipid oxidation status; no effect in moisture and protein	[40]
Tomato pomace	SSF: pomace (10% in silage), *Lactobacillus plantarum* (starter culture), and 30 days	Pregnant Holstein dairy cows (50 animals), initial weight 710-715 kg, 7 days of experiment	Animals: increased feed intake and digestibility, blood cholesterol and HDL, IgA, IgG, IgM, and antioxidant defense system response; no effect in feed intake, digestibility, milk yield and composition Milk: increased vitamin A, C, and E contents; no effect milk yield and composition	[41]

Source	Experimental Conditions	Animals and Study Characteristics	Effect in Animal Production and Related Food	Ref.
Pomegranate pomace	SSF: 100 g substrate, *Aspergillus niger,* 30 °C, and 7 days; pomace (5 and 10 g/kg feed)	Male Ross 308 broiler chicks (175 animals), initial weight 39 g, and 42 days of experiment	Animals: increased crypt depth; reduced lipid oxidation, *Clostridium perfringens* in cecum, and villus height; no effect in body weight, feed intake and conversion ratio, carcass characteristics, antioxidant defense system response, and muscularis mucosa thickness Meat and liver: no effect in color and pH	[42]

A related experiment with pigs fed with silage containing apple pomace indicated minimal or non-significant effects in the growth performance, except for a reduction in daily feed intake and an increase in feed efficiency in animals fed with apple pomace silage [35]. Additionally, this study also indicated a significant increase in the content of some individual polyunsaturated fatty acids in back fat, whereas the content of few saturated and monounsaturated fatty acids in back fat were reduced. This effect was attributed to the dietary fiber found in apple pomace that favored the growth of probiotic microorganisms in pig intestine and led to the potential changes in back fat fatty acid composition.

Another interesting strategy to use silage with apple pomace was reported for the production of fish. Davies et al. [36] studied the effect of a silage produced with apple pomace, minced sardine, and *Lactobacillus plantarum* as a starter culture in the production of juvenile European sea bass. In these animals, the silage with apple pomace improved the health status of fish, whereas growth performance indicators were reduced in relation to the control diet (without apple pomace).

The effect of feeding animals with silage containing grape pomace was also reported in recent studies but contrasting results have been reported. In the experiment carried out by Massaro Junior et al. [37], increasing levels of silage with grape pomace (up to 30% in feed) did not cause significant changes in indicators of growth performance (initial and final body weight, average daily gain and feed conversion ratio), carcass characteristics (hot and cold carcass yield, for instance), and meat quality (such as pH, shear force, lipid oxidation, and color) in lambs. Conversely, the use of silage produced with grape pomace in piglets induced the antioxidant defense system, reduced the indicators of oxidative stress, and the counts of pathogenic microorganisms *(Campylobacter jejuni,* for instance) in fecal samples

[38]. Additionally, the meat produced from animals fed with the experimental diet had more omega-3 fatty acids in comparison to the meat from animals fed with the control diet.

Fermentation in a solid state has also been explored to obtain potential feed additives for animal production. In the case of broiler chicks, the incorporation of fermented grape pomace in animal diets produced heavier animals with increased serum levels of catalase (a component of the antioxidant defense system) [39]. Additionally, no significant reductions in other components of the antioxidant defense system, intestinal morphology, and the pH or color of liver in the animals fed with silage containing grape pomace were reported in this study.

Olive pomace has been indicated as an interesting component to improve the diet of chicks [40]. Adding fermented olive pomace in animal feed enhanced the antioxidant status and the antioxidant defense system as well as reduced serum triglycerides and total cholesterol. Conversely, body weight gain was affected and no major effects in liver enzymes were indicated by the authors. The effects on animal health were also observed in meat in terms of reduced fat, cholesterol contents and lipid oxidation levels in breast meat. Another study indicated a favorable effect of solid-state fermented pomace in animal health [41]. In this case, the consumption of fermented tomato pomace improved health indicators (serum lipids and immune and antioxidant defense systems) in Holstein cows. However, the authors indicated no effects in terms of feed intake and milk production and composition (except for vitamins A, C, and E).

The effect of silage produced with pomegranate pomace in broiler chicks was evaluated by Gungor et al. [42]. The oxidative status was improved and some effects in the internal morphology were reported in animals consuming the experimental silage. No significant effects were reported for carcass characteristics, the antioxidant defense system, and meat and liver characteristics (pH and color).

From these experiments, it seems reasonable to consider that mixing pomace with other components for silage production modifies the microbial activity as well as the characteristics of silage. These effects can be attributed to the composition and content of nutrients (such as water-soluble carbohydrates). It is important to mention that the effect is dependent on the extract composition (apple vs. grape pomace, for instance).

Additionally, the shift in the fermentation process by using pomace as a raw material in silage production (especially for the production of lactic acid to ethanol) may be reduced from the addition of starter cultures. In terms of animal production, the main benefit seems to be related to animal health and the quality of foods obtained from these animals (chicks, cows, fish, lambs, and pigs), regardless of pomace source. In terms of animal production, the use of either pomace as silage raw material or fermented feed supplement seems to have a negative impact, such as in growth performance. It is worth mentioning that the modification of foods obtained from animals fed with fermented pomace fits in the strategy to naturally enrich foods with nutrients and functional compounds [43]. This strategy is supported by studies with apple [35], grape [38,39], olive and tomato pomaces [40,41]. However, additional studies are still necessary to identify relevant sources due to the controversial results such as those reported for pomegranate pomace in chicken meat [42].

3. ENZYME PRODUCTION AND POTENTIAL APPLICATIONS

The use of pomace for the production of enzymes obtained from the agro-industrial processing of foods is an interesting strategy for producing high-added value products (Table 3). One of the main pomaces explored in the production of enzymes is obtained from apple processing. Recent studies point out that apple pomace can be used to obtain different enzymes without an additional carbohydrate source [44-48]. For instance, the production of lignin peroxidase and manganese peroxidase were reported from the fermentation of apple pomace with *Phanerochaete chrysosporium* BKM-F-1767 [48]. In this study, apple pomace was indicated as the most versatile residue to produce these enzymes in comparison to brewery residue, pulp and paper residue, and fishery waste.

The production of amylase, cellulose, pectinase, and xylanase was reported for fermentation with *Rhizopus delemar* F_2 [44]. Similarly, the production of pectinase was reported in another study carried out with *Aspergillus parvisclerotigenus* KX928754 where the fermentation was optimized in terms of pH, temperature, and the period of fermentation [45]. Similarly, the combination of two *Bacillus* strains, *Bacillus subtilis* and *Bacillus pumilus,* was indicated as a relevant strategy to produce pectinase from apple pomace [46]. In this study, the authors optimized the fermentation by exploring the effect of solid content and the ratio between *B. subtilis* and *B. pumilus* in the production of this enzyme.

Table 3. Production of enzymes from the fermentation of apple, grape, olive, tomato, orange, pea, and carrot pomaces.

Source	Microorganism	Fermentation Conditions	Enzyme and Enzymatic Activity	Ref.
Apple pomace	*Phanerochaete chrysosporium* BKM-F-1767	40 g substrate, 60% moisture, pH 4.5, 37°C, and 14 days	Lignin peroxidase: 141.4 U/gds Manganese peroxidase: 631.2 U/gds Laccase: 719.9 U/gds	[48]

Source	Microorganism	Fermentation Conditions	Enzyme and Enzymatic Activity	Ref.
Apple pomace	*Rhizopus delemar* F_2	5 g apple pomace, 10 mL moistening agent (6.0 g Na_2HPO4, 3.0 g KH_2PO4, 0.5 g NaCl, 1.0 g NH_2C1, 2 mL of 1 M MgSO4 and 0.1 mL of 1 M $CaCl_2$)	Amylase: 21.0 U/g Cellulase: 18.2 U/g Pectinase: 61.5 U/g Xylanase: 158.3 U/g	[44]
Apple pomace	*Aspergillus parvisclerotigenus* KX928754	5 g substrate, pH 7.0,168 h, 30°C, 2% sucrose and 3% peptone, 30°C, and 168 h	Pectinase: 1366.3 U/mL	[45]
Apple pomace	*Bacillus subtilis* and *Bacillus pumilus* (20 and 80% in inoculum, respectively)	15 g substrate/L, 0.2 g/L pectin, 0.2 g/L MgSC>4 $7H_2O$, and 0.2 g/L K_2HPO4, pH 9.0, 130 rpm, 30°C, and 24 h	Pectinase: 11.25 U/mL	[46]
Apple pomace and dahlia tubers	*Mucor circinelloides*	10 g substrate, apple pomace: dahlia tubers (9:1), 83.5% moisture, 0.3% $NH_4H_2PO_{4z}$ 0.2% KH_2PO4 and 0.1% KC1, pH 6.4, 30°C, 5.8 days	Inulinase: 411.3U/gds	[49]
Apple pomace	*Cellulosimicrobium* sp. CKMX1 (wild) and its mutant E5	10 g substrate, 20 mL basal salt medium, pH 8.0, 35°C, and 72 h	Xylanase: 418 (wild) and 568 (mutant E5) U/g	[47]
Grape pomace	*Aspergillus niger* NRRE3	100 mL modified Czapex minimal medium with grape pomace, 4% tannic acid, pH 5.50, 120 rpm, 30°C	Tannase: 3.0-4.5 U/mL	[50]
Grape pomace	*Bacillus subtilis* natto DSM 17766	15 g/100 mL, 3% H_2SO4, pH 6.0, and 7 days	Cellulase: 0.2 U/mL	[51]
Grape pomace	*Pleurotus ostreatus* and *Pleurotus pulmonarius*	4 g, 26°C, 140 rpm, and 15 days	Laccase: 26.2 and 15,273.0 U/g for *Pleurotus ostreatus* and *Pleurotus pulmonarius*, respectively	[52]

Source	Microorganism	Fermentation Conditions	Enzyme and Enzymatic Activity	Ref.
Grape pomace and wheat bran	*Aspergillus niger* 3T5B8	Grape pomace: wheat bran (50 and 50%), 60% moisture, 0.91% ammonium sulfate solution, 37°C, and 96 h	Tannase: 0.30 U/g	[53]
White grape pomace, olive mill wastewater, red grape pomace and wheat bran	*Aspergillus niger* B60	50 g substrate, white grape pomace and olive mill wastewater, red grape pomace, and wheat bran (15,15 and 70% of total substrate, respectively), 30°C, and 120 h	CMCase: 668 U/g Polygal acturonase: 3151 U/g Amylase: 1099 U/g Xylanase: 579 U/g Protease: 204 U/g	[54]
Olive pomace	*Kluyveromyces marxianus*	5 g substrate, 45°C, and 48 h	Tannase: 42.4 U/mg	[55]
Exhausted olive pomace	*Aspergillus niger* CECT 2915	10 g substrate, 30°C, and 6 days	Xylanase: 28 U/g Cellulase: 38 U/g	[56]
Olive pomace and wheat bran	*Aspergillus ibericus* MUM 03.49, *Aspergillus niger* MUM 03.58, and *Aspergillus tubingensis* MUM 06.152	30 g olive pomace: wheat bran (50 and 50%), 75% moisture, 30°C, and 7 days	Lipase: 223, 53.6 and 7.6 U/g for *A. ibericus, A. niger* and *A. tubingensis*, respectively	[57]
Exhausted olive pomace	*Cryptliecodinium cohnii* ATCC 30772	5 and 8 g substrate/E, 27°C, 160 rpm, and 7 days	Pectinase: 37 and 33 U/mL for 8 and 5 g/L olive pomace, respectively	[58]
Tomato pomace	*Aspergillus oryzae* NRRL 2220 in SSF or SmF	10 g substrate, 19.8 g/E casein, 0.92 g/L NaCl, 30°C, and 72 h	Protease: 21,309 and 2343.5 U/g for SSF and SmF, respectively	[59]
Tomato pomace	*Aspergillus oryzae* 2220	20 g, 50% initial moisture content, pH 6 and 1 mL of 5-day-old inoculum, 30°C, and 72 h	Protease: 12 U/gds after 42 h	[60]
Tomato pomace	*Aspergillus oryzae* 2220 (static bioreactor)	5 kg, 10 cm bed, 30°C, and 44 h	Protease: 13.6 U/gds	[60]

Source	Microorganism	Fermentation Conditions	Enzyme and Enzymatic Activity	Ref.
Tomato pomace and sorghum stalks	*Pleurotus ostreatus* and *Trametes versicolor*	500 g tomato pomace, 100 g sorghum stalks, and 28°C	Laccase: 15 and 35 U/g for *P. ostreatus* (4 days) and *T. versicolor* (18 days), respectively Protease: 13,000 and 34,000 U/g for *P. ostreatus* (4 days) and *T. versicolor* (13 days), respectively Xylanase: 9 and 50 U/g for *P. ostreatus* (4 days) and *T. versicolor* (13 days), respectively	[61]
Tomato pomace, wheat bran, and canola meal	*Bacillus subtilis* T4b	Wheat bran 30 g/L, canola meal 40 g/L, and tomato pomace 15 g/L, 180 rpm, 28°C, and 48 h	Xylanase: 315 U/mL	[62]
Orange pomace	*Aspergillus niger*	5 g substrate, 30°C, and 96 h	Pectinase: around 17 U/g (endo+exo enzyme activities)	[63]
Orange pomace	*Aspergillus niger* (tray bioreactor)	285 g substrate/tray, 30°C, and 96 h	Pectinase: around 60 U/g (endo+exo enzyme activities)	[64]
Orange pomace with sugarcane bagasse	*Aspergillus niger* (tray bioreactor)	285 g substrate/tray, 30°C, and 96 h	Pectinase: around 75 U/g (endo+exo enzyme activities)	[64]

Source	Microorganism	Fermentation Conditions	Enzyme and Enzymatic Activity	Ref.
Orange pomace	*Aspergillus niger* (rotating-drum bioreactor)	285 g substrate/batch, 30°C, and 96 h	Pectinase: around 40 U/g (endo+exo enzyme activities)	[64]
Carrot pomace	*Penicillium oxalicum* BGPUP-4	10 g substrate, 90% moisture, 0.5% inulin, 0.2% $NaNO_{3z}$ 0.2 g/mL KH_2PO_{4z} 0.1% KC1, 0.05% $MgSO_4$-$7H_2O$, 0.001% $FeSO_4$ $7H_2O$ and 0.2% $NH_4H_2PO_{4z}$ pH 7.0, 30°C, and 4 days	Inulinase: 322.10 U/g	[65]

CMCase: carboxymethyl cellulase, SmF: submerged fermentation, and SSF: solid-state fermentation.

Combining apple pomace with other sources of nutrients can improve enzyme production yields. This factor was considered in the experiment carried out by Singh et al. [49], who used dahlia tuber powder (source of inulin) to produce inulinase with apple pomace. These authors optimized the fermentation in terms of moisture, fermentation period, and pH. Another interesting strategy to obtain extracts rich in enzymes from apple pomace consist in generating mutant strains such as those indicated by Guleria et al. [47]. In this case, the new mutant of *Cellulosimicrobium* sp. CKMX1 E_5 increased the production of xylanase in relation to its parent strain.

Grape is another relevant substrate for the production of enzymes. In this case, the production of tannase was obtained from the fermentation with *Aspergillus niger* NRRL3 [50]. Similarly, the production of cellulose using *Bacillus subtilis* was also obtained from the fermentation of grape pomace [51]. Another recent experiment indicated that the production of laccase from grape pomace was dependent on the starter culture [52]. In this case, *Pleurotus pulmonarius* was more efficient for producing this enzyme than *Pleurotus ostreatus.* Moreover, the authors also indicated that solid-state fermentation was more appropriate than semiliquid and submerged fermentations.

The effect of adding wheat bran in grape pomace for the production of different enzymes was studied in a recent experiment [53]. The fermentation with *Aspergillus niger* successfully produced more tannase by combining wheat bran with grape pomace than using only wheat bran. However, the presence of grape pomace limited the production of xylanase and β-glucosidase and slowed the production of polygalacturonase. Additionally, the authors also reported a dependency on time for the production of polygalacturonase and carboxymethyl cellulase (higher enzymatic yields were obtained after 96 h of fermentation). Additionally, Papadaki et al. [54] reported the production of amylase, carboxymethyl cellulase, polygalacturonase, protease, and xylanase from a substrate composed of white grape pomace, olive mill wastewater, red grape pomace and wheat bran. *Aspergillus niger* was used to obtain these enzymes.

Olive processing for oil extraction also generates a valuable substrate for microbial enzyme production. For instance, a recent experiment with olive pomace indicated that tannase could be obtained from the fermentation with *Kluyveromyces marxianus* [55]. Another relevant example that supports the use of this pomace in the production of enzymes is the study carried out by Leite et al. [56]. In this case, the authors fermented the exhausted olive pomace with *Aspergillus niger* and reported the production of xylanase and cellulose. In the case of lipase production from grape pomace, the effect of *Aspergillus* species was evaluated in a recent study [57]. *Aspergillus ibericus* was a more efficient species in relation to *Aspergillus niger* and *Aspergillus tubingensis.* Interestingly, a related experiment with exhausted olive pomace reported the production of pectinase from the growth of the microalgae *Crypthecodinium cohnii* [58]. Additionally, no significant differences in terms of substrate concentration (5 vs. 8 g/L) in the production of this enzyme were reported.

Tomato is another relevant source of pomace that can be utilized in the production of enzymes. Proteases could be obtained from tomato pomace using *Aspergillus oryzae* according to recent studies [59,60]. Moreover, the study carried out by Belmessikh et al. [59] indicated that the production of protease from tomato pomace was more efficient in solidstate rather than submerged fermentation. The optimization also indicated that casein and NaCl levels are significant factors in improving the production of protease.

The combination of tomato pomace with other sources of nutrients for enzymatic production has also been explored [61]. Particularly, for the combination with sorghum stalks, the production in a laccase, protease, and xylanase were dependent on the starter culture [61]. In this case, *Pleurotus ostreatus* was associated with a faster but less intense production of these enzymes. Conversely, *Trametes versicolor* had higher production yields but after longer fermentation periods. Another more recent experiment with tomato pomace, wheat bran, and canola meal indicated that the fermentation with *Bacillus subtilis* was associated with high xylanase content [62].

Another relevant pomace for the production of enzymes is obtained from orange processing. In this case, recent experiments explored the generation of pectinase from the fermentation with *Aspergillus niger* [63,64]. It is also relevant to comment that a recent experiment indicated that the use of sugarcane bagasse is a relevant strategy to reduce moisture loss during fermentation and improve the production yield of pectinase from orange pomace [64]. In a similar way, carrot pomace was indicated as an interesting substrate for fermentation, which can be utilized in the production of inulinase [65]. The production of inulinase was affected by moisture content, fermentation period, and pH.

The production of enzymes from pomaces can also be improved by the use of emerging technologies such as microwave heating and ultrasound. This aspect was reported in the production of carbohydrases from apple pomace by Pathania et al. [44]. According to these authors, the intensity of microwaves (as a pre-treatment) had a significant effect on the production yield. The maximum values for amylase, pectinase, and xylanase were reported for the 450 W treatment. Additional power (up to 600 W) caused a reduction in the production of enzymes. In the same line of thought, the use of ultrasound can improve the production of enzymes. Leite et al. [56] indicated that using 750 W and 20 kHz and optimizing the time and liquid/solid ratio (12.4 min and 7.3) maximized the production yields of xylanase (75 U/g) and cellulase (35 U/g).

It is also important to highlight that some experiments to scale up the production of enzymes from pomaces have been carried out in the last decade. One relevant example that explored this aspect was performed by Boukhalfa-lezzar et al. [60] with tomato pomace fermented with *Aspergillus oryzae.* In this study, similar production yields were reported between lab scale and a bioreactor for protease production (12 U/gds after 42 h with 20 g of substrate vs. 13.6 U/gds after 44 h with 5 kg of substrate). Another relevant experiment supporting the increase in the production scale of enzymes was carried out with orange pomace in a tray reactor and a rotating-drum reactor [64]. In this case, differences in production yield were reported between these two reactors wherein the bioreactor with trays had the highest yield. Moreover, both reactors increased the production of pectinase in relation to a previous experiment from the same research group [63].

The purification of enzymes obtained from fermentation is another relevant aspect considered in recent studies. In order to explore potential solutions to improve the separation of enzymes, an experiment with lignin peroxidase and manganese peroxidase explored the use of centrifugation and filtration after the fermentation of apple pomace [48]. The results revealed that centrifugation was more efficient for separating both enzymes than filtration. A recent study compared the use of fractionation with ammonium sulfate and chromatography filtration in the purification of tannase from fermented olive pomace [55]. Both methods led to extracts with increased enzymatic activity wherein the chromatography filtration was more efficient than fractionation with ammonium sulfate (1026.1 vs. 664 U/mg, respectively). A similar outcome was obtained in another study with pectinase from apple pomace (1081.7 vs. 860.6 U/mg for chromatography filtration and ammonium sulfate fractionation, respectively) [45].

Potential applications can also be considered in the context of enzyme production. Since pomace is a by-product from food processing, the use of these enzymes in food production can be suggested. A relevant example is the experiment carried out by Mahmoodi et al. [63]. In this study, cubic pieces of fresh apple were treated with pectinase to produce apple juice. The main effects were the reduction in juice viscosity and increased juice yield, soluble sugar content, and pectate content. A similar experiment with polygalacturonase obtained from apple pomace was efficient for clarifying apple juice [66].

An interesting application for enzymes obtained from pomace fermentation is the detoxification of food. This approach was evaluated by Cuprys et al. [67] who applied laccase from apple fermentation with *Trametes versicolor* to decompose ciprofloxacin (an antibiotic). However, the presence of a reducing agent (syringaldehyde in this study) was necessary to favor the enzymatic degradation of this antibiotic in water. Although the scientific information about the application of microbial enzymes from pomace fermentation

in food processing is limited, the use of these enzymes could be considered to improve the processing of beer, bread, cheese, syrup, and wine [68] in further experiments. Moreover, potential applications in other research areas are in pharmaceutical, chemical, fuel, and paper production [69].

The use of pomace from different sources can be seen as a relevant strategy to favor the production of enzymes. Current scientific evidence indicates that the production of enzymes can be improved by adding complementary sources of nutrients (such as ingredients rich in carbohydrates for pomaces with reduced levels of this nutrient), applying emerging technologies to favor the exposure of substrates, and increasing the scale of production (minimal effect in production yield, to some extent). The purification with different techniques can also be applied and support the progression towards application in other industrial processes.

4. RELEASE AND PRODUCTION OF BIOACTIVE COMPOUNDS

Improving the biological activity of pomace from food processing is one of the potential and emerging applications of fermentation. This strategy has been applied to obtain carotenoids, fatty acids, y-linolenic acid, and polyphenols (Table 4). Polyphenols are an important class of bioactive compounds that are found in pomaces. From a broad perspective, polyphenols can be found either in free or bound forms. Polyphenols in free form are those present in the cytosol of vegetable cells, whereas the bound polyphenols are those bound to cell wall constituents [70]. For bound polyphenols in particular, their extraction is complex and conventional extraction methods have low efficiency to separate these compounds from structural components of food. In this context, the use of fermentation (by means of the action of microbial enzymes) has been indicated as a relevant strategy to recovery this compound [70,71].

Table 4. Bioactive compounds obtained from pomace fermentation.

Source	Fermentation Conditions	Bioactive Compounds	Outcome	Ref.
Grape pomace	2 g substrate, *Rhizomucor miehei* NRRL 5282, 37 ° C, and 18 days	Polyphenols	Oven dried: reduction in TPC and FRAP, no effect in DPPH Lyophilized: maximum TPC and FRAP values at day 7, no effect in DPPH	[72]
Grape pomace	10 g substrate (grape pomace:wheatbran; 1:1), *Aspergillus niger* 3T5B8, 37 °C, and 96 h	Polyphenols	Increased TPC, ABTS, and ORAC	[53]
Grape pomace	50 g substrate, *Trametes versicolor* TV-6, 5 mycelial plugs, 27 °C, and 15 days	Polyphenols	Reduced 5-lipoxygenase and hyaluronidase activities (up to 4 days of fermentation), and polyphenol content throughout fermentation period	[74]
Grape pomace	60 g substrate, *Actinomucor elegans* ATCC- 22963 or *Umbelopsis isabellina* ATCC-36671, 30 ° C, and 12 days	y-Linolenic acid and carotenoids	y-Linolenic acid: maximum at 4 days for *Umbelopsis isabellina* and 6 days for *Actinomucor elegans* Carotenoids: carotene increased throughout fermentation and maximum at 8 days for lutein	[73]
Apple pomace	2 g substrate, *Rhizomucor miehei* NRRL 5282, 37 ° C, and 18 days	Polyphenols	Oven dried: reduced TPC, maximum FRAP value at day 3, no effect in DPPH Lyophilized: slight increase in TPC and DPPH up to day 10, maximum FRAP value at day 10	[72]
Apple pomace	12.5 g, natural fermentation, 30 °C, and 72 h	Polyphenols	Reduced throughout the fermentation period	[75]

Source	Fermentation Conditions	Bioactive Compounds	Outcome	Ref.
Apple pomace	250 g substrate, *Saccharomyces cerevisiae* ref: 32, *Saccharomycodes bayanus* ref: C6, and *Hanseniaspora uvarum* ref: 62, 25 °C, and 7 days	Fatty acids and polyphenols	Increased fatty acids Slight reduction in polyphenols	[76]
Apple pomace	40, 60 and 80 g substrate/L, *Yarrowia lipolytica*, 28 °C, and 6 days	Fatty acids	Maximum production after day 3	[77]
Elderberry and dwarf elderberry pomace	50 g substrate, *Aspergillus niger* ATCC-6275, 30 °C, and 6 days	Polyphenols and fatty acids	TPC: maximum release up to 3-4 days of fermentation DPPH: maximum after 3-4 days of fermentation Lipids: slight increase in linoleic and oleic acids up to 4 days of fermentation	[78]
Olive pomace	5 g substrate, *Kluyveromyces marxianus* NRRL Y-8281, 45 °C, and 48 h	Tannic and gallic acids	Reduced tannic acid and increased gallic acid content	[79]
Exhausted olive pomace	5 and 8 g substrate/L, *Crypthecodinium cohnii* ATCC 30772, 27 °C, 160 rpm, and 7 days	Fatty acids	Increased total lipid and DHA content in dry cells	[58]
Exhausted olive pomace	25 g substrate/L, *Crypthecodinium cohnii* ATCC 30772, 27 °C, 160 rpm, and 5 days	Fatty acids	High production yield; negative effect of detoxification prior to fermentation	[80]
Chokeberry pomace	40 g substrate, *Aspergillus niger* ATCC-6275 or *Rhizopus oligosporus* ATCC-22959, 30 °C, and 12 days	Polyphenols	TPC: maximum at 6 days for or *Rhizopus oligosporus* and 9 days for *Aspergillus niger;* DPPH and TEAC: maximum at 6 days for *Aspergillus niger* and 9 days for *Rhizopus oligosporus*	[81]

Source	Fermentation Conditions	Bioactive Compounds	Outcome	Ref.
Plum pomace	15 g substrate, *Aspergillus niger* ATCC-6275 or *Rhizopus oligosporus* ATCC-22959, 30 °C, and 14 days	Polyphenols	TPC: maximum after 9 days of fermentation; DPPH: maximum at 6 days of fermentation	[82]
Apricot pomace	15 g substrate, *Aspergillus niger* ATCC-6275 or *Rhizopus oligosporus* ATCC-22959, 30 °C, and 14 days	Polyphenols	TPC: maximum at 9 days for *Rhizopus oligosporus*; reduced after 6 days for *Aspergillus niger*; DPPH: maximum at 2 days for both	[83]
Pitahaya pomace	2 g substrate, *Rhizomucor miehei* NRRL 5282, 37 ° C, and 18 days	Polyphenols	Oven dried: slight decrease in TPC, decreased FRAP, and no effect in DPPH Lyophilized: slight increase in TPC, maximum FRAP value at day 10 and DPPH value at day 15	[72]
Red bayberry pomace	0.02% live yeast, 25 ° C, 16 h followed by 0.1% probiotic mix, 28 °C, 24 h, and let for up to 7 days	Polyphenols	Increased TPC and TFC values; reduced DPPH value	[84]

DHA: Docosahexanoic acid, DPPH: (2,2-diphenyl-1-picrylhydrazyl) free radical, FRAP: ferric reducing antioxidant power, ORAC: oxygen radical absorbance capacity, TEAC: trolox equivalent antioxidant capacity, TFC: total flavonoid content, and TPC: total polyphenol content.

For instance, studies carried out with grape pomace indicate that polyphenols [53,72], y-linolenic acid and carotenoids [73] can be obtained from fermentation. In addition to the characterization of the content of these bioactive compounds, these studies also revealed aspects related to the preparation of samples, fermentation period, and the effect of the starter culture.

Regarding the effects of sample preparation and fermentation period in the release of polyphenols, a recent experiment indicated that lyophilization is a better pre-treatment than oven-drying to improve the extraction of polyphenols from grape pomace [72]. Moreover, this study also indicated that long fermentation periods do not favor the accumulation of polyphenols. Additionally, this effect could be explained by the instability of free polyphenols during fermentation. The gradual decomposition of free polyphenols can occur, which may be compensated by the release of bound polyphenols from microbial activity. Another related study with pomace supports this consideration and the necessity to define the optimum fermentation period. The high polyphenol and bioactivity in the beginning of the fermentation period were followed by the reduction in both indicators (polyphenol content and biological activity) as fermentation progressed up to 15 days [74]. Additionally, Teles et al. [53] reported increasing polyphenol content and antioxidant activity during the fermentation of grape pomace with *Aspergillus niger* during a shorter period (96 h) in relation to these aforementioned studies. This study also indicated that polyphenol content was positively correlated with antioxidant potential.

The production of y-linolenic acid and carotenoids by solid-state fermentation also displayed the same dependency on fermentation time, wherein maximum yields were obtained after 6 days of fermentation [73]. In the case of carotenoids, the synthesis of lutein had a maximum yield after 8 days, whereas the production of carotene increased throughout the fermentation period (18 days).

Apple pomace has also been explored as a relevant source of polyphenols and fatty acids. For instance, the effect of pre-treatment and fermentation on polyphenol accumulation during fermentation was studied by Zambrano et al. [72]. The maximum polyphenol content was not affected by the pretreatment (lyophilization vs. over-drying), but significant changes were reported during the fermentation period. The maximum polyphenol yield and antioxidant potential were obtained at day 10. Conversely, Lohani and Muthuku- marappan [75] reported a gradual reduction in the polyphenol content of naturally fermented apple pomace. Madrera et al. [76] reported a slight reduction in the polyphenol content of fermented apple pomace with different yeasts. Additionally, this study also indicated that the production of fatty acids can be obtained from the fermentation of apple pomace with yeasts. An interesting experiment with apple pomace explored the production of fatty acids in a 5 L bioreactor [77]. In this case, different concentrations of apple pomace were used as a carbon source for lipid biosynthesis. A concentration-dependent effect (40, 60 and 80 g substrate/L) in the production of fatty acids was reported. Moreover, the maximum yield for each tested apple pomace concentration was achieved in a short period (3 days).

In the case of olive pomace, the fermentation with *Kluyveromyces marxianus* led to a reduction in tannic acid content and an increase in the concentration of its depolymerized form, gallic acid [79]. Another interesting application of exhausted olive pomace (residue obtained after the removal of residual oil from olive pomace) is the production of microbial fatty acids, especially docosahexaenoic acid (DHA). A recent experiment indicated that the concentration of exhausted olive pomace had a concentration-dependent effect in the production of DHA by the microalgae *Crypthecodinium cohnii* [58]. Interestingly, another study with the same microalga revealed that detoxification with activated carbon reduced the production of fatty acids [80].

The simultaneous production of polyphenols and fatty acids from fruit pomace was also explored in a recent study with two *Sambucus* species [78]. In these fruits, optimum polyphenol production yield and antioxidant activity were obtained at day 3 and 4 (regardless of species), respectively. A similar effect was observed for the accumulation of linoleic and oleic fatty acids, which had maximum values at day 4. Similarly, the accumulation of polyphenols and antioxidant activity during the fermentation of chokeberry pomace were dependent on the time and starter culture [81]. Maximum values for total phenolic content were obtained between day 6 and 9 of fermentation for *Rhizopus oligosporus* and 9 days for *Aspergillus niger.*

Studies carried out with plum [82] and apricot [83] pomaces indicated that optimum fermentation periods for polyphenol accumulation and antioxidant activity from *Aspergillus niger* fermentation were 9 and 6 days, respectively. Another recent experiment indicated that the accumulation of polyphenols in pitahaya pomace from the activity of *Rhizomucor miehei* was improved by lyophilizing samples before fermentation [72]. A related experiment evaluated the accumulation of polyphenols and antioxidant activity in red bayberry pomace during 7 days during the sequential fermentation with *Saccharomyces cerevisiae* and a mix of lactic acid bacteria (*Lactobacillus bulgaricus, Bifidobacterium lactis,* and other lactic acid bacteria) [84]. A gradual increase in the polyphenol content was reported throughout the 7 days of fermentation. Moreover, the antioxidant activity of fermented pomace after this period was improved in relation to non-fermented pomace.

Since the fermentation of pomaces can lead to high polyphenol content and antioxidant activity (Table 4), the biological response to the consumption of fermented pomace was also explored in recent studies. Improvements in the antioxidant defense system and a reduction in the oxidative status of liver and ilium in mice fed with fermented blueberry pomace were reported [85]. The intestine inflammatory response (tumor necrosis factor-alpha and interleukin-10) was also improved in animals that consumed the diet supplemented with fermented blueberry pomace. Concentration-dependent effects were observed in the antioxidant and anti-inflammatory activities. Moreover, these effects ameliorated the modifications induced by a high-fat diet in terms of antioxidant and anti-inflammatory responses.

A further experiment carried out by the same research group explored the functional effect of fermented blueberry pomace in indicators of gut health of mice [86]. The consumption of supplemented diets improved the gut immunological response (secretory immunoglobulin A), affected the gut microbiota and also favored the production of butyric acid (a short fatty acid associated with health benefits).

Again, the supplemented diet ameliorated the modifications induced by a high-fat diet in the gut immunological response and gut health. Another experiment in vivo that supports the health benefits associated with the fermentation of pomaces was carried out by Yan et al. [87]. In this case, the consumption of fermented blueberry pomace (rich in polyphenols) improved the resistance to fatigue in relation to control animals that ingested sterile water.

Along with the production of fermented pomaces with increased biological activity, it is also important to develop strategies to isolate active components from the bulk of fermented pomace. This aspect was recently explored by Espinosa-Pardo et al. [88] who optimized the extraction of polyphenols with super-critical CO2 and co-solvents. The authors indicated that the extraction with CO_2 (25 MPa at 60°C) and 90% ethanol as co-solvent was the most efficient extraction condition to obtain the highest polyphenol content and antioxidant activity. Another important aspect to consider is the effect of digestion in the stability of active compounds. Yan et al. [87] evaluated the impact of simulated digestion and indicated significant reduction in the polyphenol content and antioxidant activity of blueberry pomace fermented by *Lactobacillus rhamnosus* GG and *Lactobacillus plantarum-1* (1:1).

The fermentation of pomaces can be seen as a relevant strategy to produce functional supplements with interesting biological effects, especially from berries. However, additional advances, especially in the application of extraction technologies and the characterization of biological effects in vivo, are still necessary.

5. PRODUCTION OF ORGANIC ACIDS

Organic acids are multipurpose compounds that have been applied in animal production [89], food processing [90], cosmetic preservation [91] and battery recycling [92], for instance. Due to their importance and the current trends to improve sustainability within the organic acid production sector, several studies have been carried out to explore the use of pomaces in the production of high-added value compounds (Table 5). One relevant example of this strategy is the study performed by Vashisht et al. [93] who evaluated the production of acetic acid using *Acetobacter pasteurianus* SKYAA25 from apple pomace. These authors indicated that the production of acetic acid was affected by the temperature (37°C), concentration of bioethanol (8%, produced from the same strain), and apple pomace (2%) in fermentation media. Similarly, the production of acetic acid from the fermentation of apple pomace was reported in another study using *Acetobacter aceti* [94].

Table 5. Organic acids produced from pomace fermentation.

Source	Fermentation Conditions	Organic Acid and Yield	Ref.
Apple pomace	120 g substrate/L, *Acetobacter pasteurianus*, 37°C, 180 rpm, and 24 h	Acetic acid: 52.4 g/100 g DM	[93]
Apple pomace	1.5 L of substrate, *Acetobacter aceti*, pH 7.0, 28°C, and 7 days	Acetic acid: 61.4 g/100 g DM	[94]
Apple pomace	14 g substrate/100 g, *Propionibacterium freudenreichii*, 37°C, and 120 h	Propionic acid: 38 g/100 g DM	[95]
Apple pomace	250 mL substrate, *Propionibacterium freudenreichii*, 37°C, and 120 h	Acetic acid: 5.01 g/L Propionic acid: 14.54 g/L	[96]
Apple pomace	25 g substrate, *Aspergillus ornatus* and *Alternaria alternate*, pH 5.0, 30°C, and 48 h	Citric acid: 0.5 g/L	[97]
Apple pomace	25 g substrate, *Rhizopus oryzae*, 30°C, and 14 days	Fumaric acid: 52 g/kg	[98]
Apple pomace	3-4 L working volume, 50% moisture, *Rhizopus oryzae*, 1.97 atm, and 14 days	Fumaric acid: 138 g/kg	[99]

Piwowarek et al. [95] studied the optimization of the production of propionic acid from apple pomace fermentation with *Propionibacterium freudenreichii* T82. According to these authors, the accumulation of propionic acid was increased due to a better control of the fermentation process, i.e., adding biotin to fermentation media, carrying out the pH control at 24 and 48 h of fermentation, and increasing the nitrogen level (supplementing the apple pomace with peptone). However, no significant effects were obtained for the variations in temperatures from 30 to 37°C. In another study from the same research group with apple pomace, the effect of supplementation (potato wastewater, yeast extract, and peptone) to increase the production yield of propionic acid was evaluated [96]. The use of yeast extract and peptone in apple pomace in the fermentation medium improved the propionic acid yield to a maximum of 14.54 g/L after 120 h of fermentation with *Propionibacteriumfreudenreichii.* Additionally, the production of acetic acid was also evaluated in this study. A continuous increase in the accumulation of this acid was reported until the end of the fermentation period (120 h) and the most efficient supplement for apple pomace was potato wastewater (maximum yield of 5.01 g/L).

Apple pomace can also be fermented to produce citric acid [97]. In this case, a recent experiment explored the effect of temperature, pH, and substrate amount in the fermentation batch with the combination of *Aspergillus ornatus* and *Alternaria alternate.* The pH and temperature had optimum values of 5 and 30°C, respectively. Increasing the substrate caused a significant increase in the production of citric acid, which led to choosing the maximum substrate amount tested in this study (25 g). Additionally, supplementing the apple pomace with arginine favored the production of citric acid (maximum yield of 2.7 g/L).

Another relevant acid produced from pomaces is the fumaric acid. The production of this acid with *Rhizopus oryzae* was dependent on the fermentation time [98]. The maximum yield was reported after 14 days (52 g/kg) and no additional increase was observed at up to 21 days of fermentation. The production of fumaric acid using the same microorganism and pomace was also explored in a bench scale fermenter [99]. The system comprised by a rotary drum increased the production of fumaric acid to 138 g/kg within the same fermentation period (14 days).

6. PRODUCTION OF BIOFLAVORS

The production of high-added value compounds from pomaces has also been shown to produce bioflavors. The production of aromas from apple pomace fermentation was explored in a recent experiment with yeasts *(Hanseniaspora uvarum, Hanseniaspora valbyensis,* and *Saccharomyces cerevisiae)* [100]. This study indicated a strain-dependent effect in the formation of volatile compounds wherein the use of *Saccharomyces cerevisiae* led to a bigger accumulation of volatile fatty acids and their respective ethyl esters, whereas the fermentation with *Hanseniaspora* strains favored the generation of volatile acetic acid esters. A related study evaluated the effect of fermented pomace in a volatile composition of beer [101]. In this case, apple pomace was fermented with lactic acid bacteria *(Lactobacillus rhamnosus* 1473 and 1019, and *Lactobacillus casei* 2246) and significant differences were reported among volatile compositions of apple pomace. However, the fermented pomace *(Lactobacillus rhamnosus* 1473) led to slight modifications in the volatile composition (particularly for ketones and alcohols) of beer. The production of bioflavors was also explored using *Lacticaseibacillus rhamnosus* to ferment orange pomace [102]. This study revealed that fermented pomace had floral (citronellyl formate, 1-nonanol, and β-linalool), citrus (citral and limonene), fruity (p-cyclocitral and benzaldehyde), herbaceous (1-hexanol), bready and caramelly (furfural), and spice (eugenol and carveol) notes.

Finally, another aspect to be considered in the context of the utilization of high-added value compounds obtained from the fermentation/biotransformation of pomaces is their safety. Mycotoxins and pesticides are relevant contaminants in the peels of fruits that may persist in pomaces [103-105]. The effect of fermentation to decontaminate fruits and pomaces is still poorly studied.

7. CONCLUSIONS

The use of fermentation/biotransformation to obtain high-added value compounds is a valuable solution to improve the reutilization of pomaces from the food industry. The advances in incorporating and optimizing the use of pomaces in animal feed by generating silages and feeds that improve animal health is a relevant alternative to using fermented pomaces. Growth performance can be affected, whereas animal health status can be improved. The absence of negative effects and the improvement in the nutritional quality of the foods obtained from animals fed with fermented pomaces is another favorable characteristic to support this strategy.

In terms of industrial processes, the production of high-added value products (especially from grape, apple, and olive) such as enzymes and organic acids for application in food processing as well as in other areas is a relevant application. The release of bound phenolics for the development of functional foods (supported by studies in vitro and in vivo), the synthesis of carotenes and fatty acids, and the production of volatile compounds to improve the aroma of food products are potential applications.

One of the main limitations in terms of industrial application consists of its current poor incorporation into other processing chains. Extraction and purification technologies can be seen as current bottlenecks to strengthening the connections between the pomace generation in food industries and their incorporation into other productions chains. In this sense, further studies could aim to explore strategies to improve the isolation of high-added value compounds. Additional studies are still necessary to define strategies to apply the high-added value compounds obtained from pomaces from fer- mentation/biotransformation in the development of food products. Studies about the detoxification and reduction of potential health risks associated with mycotoxins and pesticides are necessary.

REFERENCES

1. Bordiga, M.; Travaglia, F.; Locatelli, M. Valorisation of grape pomace: An approach that is increasingly reaching its maturity—A review. Int. *J. Food Sci. Technol.* **2019**, *54*, 933-942. [CrossRef]
2. Gassara, F.; Brar, S.K.; Pelletier, F.; Verma, M.; Godbout, S.; Tyagi, R.D. Pomace waste management scenarios in Quebec-Impact on greenhouse gas emissions. *J. Hazard. Mater.* **2011**,192,1178-1185. [CrossRef]
3. Difonzo, G.; Troilo, M.; Squeo, G.; Pasqualone, A.; Caponio, F. Functional compounds from olive pomace to obtain high-added value foods—A review. *J. Sci. Food Agric.* **2021**,101,15-26. [CrossRef] [PubMed]
4. Perussello, C.A.; Zhang, Z.; Marzocchella, A.; Tiwari, B.K. Valorization of Apple Pomace by Extraction of Valuable Compounds. *Compr. Rev. Food Sci. Food Saf.* **2017**,16, 776-796. [CrossRef]

5. Jin, Q.; Wang, Z.; Feng, Y.; Kim, Y.T.; Stewart, A.C.; O'Keefe, S.F.; Neilson, A.P.; He, Z.; Huang, H. Grape Pomace And Its Secondary Waste Management: Biochar Production for a Broad Range of Lead (Pb) Removal From Water. *Environ. Res.* **2020**,186, 109442. [CrossRef]
6. IMARC Fruit Juice Market Size, Share, Trends & Forecast 2021-2026. Available online: https://www.imarcgroup.com/fruit-juice- manufacturing-plant (accessed on 13 October 2021).
7. Statista Inc. Juices—Worldwide. Available online: https://www.statista.com/outlook/cmo/non-alcoholic-drinks/juices/ worldwide (accessed on 13 October 2021).
8. European Commission EU Agricultural Outlook 2019-30: Olive Oil Consumption to Rise in Non-Producing EU Countries. Available online: https://ec.europa.eu/info/news/eu-agricultural-outlook-2019-2030-olive-oil-consumption-rise-non-producing- eu-countries-2019-dec-10_en (accessed on 13 October 2021).
9. Gavahian, M.; Mousavi Khaneghah, A.; Lorenzo, J.M.; Munekata, P.E.S.; Garcia-Mantrana, I.; Collado, M.C.; Melendez-Martinez, A.J.; Barba, F.J. Health benefits of olive oil and its components: Impacts on gut microbiota antioxidant activities, and prevention of noncommunicable diseases. *Trends Food Sci. Technol.* **2019**, 88, 220-227. [CrossRef]
10. Jimenez-Lopez, C.; Carpena, M.; Louren^o-Lopes, C.; Gallardo-Gomez, M.; Lorenzo, J.M.; Barba, F.J.; Prieto, M.A.; Simal-Gandara, J. Bioactive Compounds and Quality of Extra Virgin Olive Oil. *Foods* **2020**, *9*,1014. [CrossRef]
11. Skinner, R.C.; Gigliotti, J.C.; Ku, K.M.; Tou, J.C. A comprehensive analysis of the composition, health benefits, and safety of apple pomace. *Nutr. Rev.* **2018**, 76, 893-909. [CrossRef]
12. Bobinaite, R.; Grootaert, C.; Van Camp, J.; Sarkinas, A.; Liaudanskas, M.; Zvikas, V.; Viskelis, P.; Rimantas Venskutonis, P. Chemical composition, antioxidant, antimicrobial and antiproliferative activities of the extracts isolated from the pomace of rowanberry (*Sorbus aucuparia* L.). *Food Res. Int.* **2020**,136,109310. [CrossRef] [PubMed]
13. Abbasi-Parizad, P.; De Nisi, P.; Adani, F.; Sciarria, T.P.; Squillace, P.; Scarafoni, A.; Iametti, S.; Scaglia, B. Antioxidant and anti-inflammatory activities of the crude extracts of raw and fermented tomato pomace and their correlations with aglycate- polyphenols. *Antioxidants* **2020**, 9,179. [CrossRef]
14. Costabile, G.; Vitale, M.; Luongo, D.; Naviglio, D.; Vetrani, C.; Ciciola, P.; Tura, A.; Castello, F.; Mena, P.; Del Rio, D.; et al. Grape pomace polyphenols improve insulin response to a standard meal in healthy individuals: A pilot study. *Clin. Nutr.* **2019**, 38, 2727-2734. [CrossRef] [PubMed]

15. Campos, D.A.; Gomez-Garcia, R.; Vilas-Boas, A.A.; Madureira, A.R.; Pintado, M.M. Management of Fruit Industrial By-Products— A case Study on Circular Economy Approach. *Molecules* **2020**,25, 320. [CrossRef] [PubMed]

16. European Union The EU Green Deal—A Roadmap to Sustainable Economies. Available online: https://bit.ly/39ydb5i (accessed on 1 December 2021).

17. Circular Economy Action Plan. Available online: https://ec.europa.eu/environment/strategy/circular-economy-action-plan_pt (accessed on 1 December 2021).

18. Munekata, P.E.S.; Nieto, G.; Pateiro, M.; Lorenzo, J.M. Phenolic compounds obtained from Olea europaea by-products and their use to improve the quality and shelf life of meat and meat products—A review. *Antioxidants* **2020**, 9,1061. [CrossRef] [PubMed]

19. Das, A.K.; Nanda, P.K.; Chowdhury, N.R.; Dandapat, P.; Gagaoua, M.; Chauhan, P.; Pateiro, M.; Lorenzo, J.M. Application of Pomegranate by-Products in Muscle Foods: Oxidative Indices, Colour Stability, Shelf Life and Health Benefits. *Molecules* **2021**, 26, 467. [CrossRef] [PubMed]

20. Dominguez, R.; Gullon, P.; Pateiro, M.; Munekata, P.E.S.; Zhang, W.; Lorenzo, J.M. Tomato as potential source of natural additives for meat industry. A review. *Antioxidants* **2020**, 9, 73. [CrossRef] [PubMed]

21. Mizael, W.C.F.; Costa, R.G.; Cruz, G.R.B.; de Carvalho, F.F.R.; Ribeiro, N.L.; Lima, A.; Dominguez, R.; Lorenzo, J.M. Effect of the use of tomato pomace on feeding and performance of lactating goats. *Animals* **2020**,*10*,1574. [CrossRef] [PubMed]

22. Ianni, A.; Martino, G. Dietary grape pomace supplementation in dairy cows: Effect on nutritional quality of milk and its derived dairy products. *Foods* **2020**, 9,168. [CrossRef] [PubMed]

23. Quiles, A.; Campbell, G.M.; Struck, S.; Rohm, H.; Hernando, I. Fiber from fruit pomace: A review of applications in cereal-based products. *Food Rev. Int.* **2018**, 34,162-181. [CrossRef]

24. Gullon, P.; Gullon, B.; Romani, A.; Rocchetti, G.; Lorenzo, J.M. Smart advanced solvents for bioactive compounds recovery from agri-food by-products: A review. *Trends Food Sci. Technol.* **2020**,101,182-197. [CrossRef]

25. Cattaneo, C.; Lavelli, V.; Proserpio, C.; Laureati, M.; Pagliarini, E. Consumers' attitude towards food by-products: The influence of food technology neophobia, education and information. *Int. J. Food Sci. Technol.* **2019**, 54, 679-687. [CrossRef]

26. Perito, M.A.; Di Fonzo, A.; Sansone, M.; Russo, C. Consumer acceptance of food obtained from olive by-products: A survey of Italian consumers. *Br. Food J.* **2020**,122, 212-226. [CrossRef]

27. Sabater, C.; Ruiz, L.; Delgado, S.; Ruas-Madiedo, P.; Margolles, A. Valorization of Vegetable Food Waste and By-Products Through Fermentation Processes. *Front. Microbiol.* **2020**,11, 2604. [CrossRef] [PubMed]

28. Koutinas, A.A.; Vlysidis, A.; Pleissner, D.; Kopsahelis, N.; Lopez Garcia, I.; Kookos, I.K.; Papanikolaou, S.; Kwan, T.H.; Lin, C.S.K. Valorization of industrial waste and by-product streams via fermentation for the production of chemicals and biopolymers. *Chem. Soc. Rev* **2014**, 43, 2587-2627. [CrossRef]

29. Muck, R.E.; Nadeau, E.M.G.; McAllister, T.A.; Contreras-Govea, F.E.; Santos, M.C.; Kung, L. Silage review: Recent advances and future uses of silage additives. *J. Dairy Sci.* **2018**,101, 3980-4000. [CrossRef] [PubMed]

30. Fang, J.; Xia, G.; Cao, Y. Effects of replacing commercial material with apple pomace on the fermentation quality of total mixed ration silage and its digestibility, nitrogen balance and rumen fermentation in wethers. *Grassl. Sci.* **2020**, 66,124-131. [CrossRef]

31. Xia, G.-J.; Fang, J.-C. Intake, digestibility and rumen fermentation pattern in wethers fed total mixed ration silage containing dry or fresh apple pomace. *J. Anim. Feed Sci.* **2021**, 30, 26-32. [CrossRef]

32. Belem, C.d.S.; de Souza, A.M.; de Lima, P.R.; de Carvalho, F.A.L.; Queiroz, M.A.A.; da Costa, M.M. Digestibility, fermentation and microbiological characteristics of *Calotropis procera* silage with different quantities of grape pomace. *Cienc. Agrotecnologia* **2016**, 40, 698-705. [CrossRef]

33. Li, P.; Shen, Y.; You, M.; Zhang, Y.; Yan, J.; Li, D.; Bai, S. Effect of grape pomace on fermentation quality and aerobic stability of sweet sorghum silage. *Anim. Sci. J.* **2017**, *88*,1523-1530. [CrossRef]

34. Koksal, Y.; Bolukbas, B.; Seljuk, Z. An in vitro Evaluation of the Silages in White Mulberry Pomace/Meadow Grass mixtures containing different levels of White Mulberry Pomace. *Kocatepe Vet. J.* **2021**,14, 309-315. [CrossRef]

35. Fang, J.; Cao, Y.; Matsuzaki, M.; Suzuki, H.; Kimura, H. Effects of apple pomace-mixed silage on growth performance and meat quality in finishing pigs. *Anim. Sci. J.* **2016**, 87,1516-1521. [CrossRef] [PubMed]

36. Davies, S.J.; Guroy, D.; Hassaan, M.S.; El-Ajnaf, S.M.; El-Haroun, E. Evaluation of co-fermented apple-pomace, molasses and formic acid generated sardine based fish silages as fishmeal substitutes in diets for juvenile European sea bass *(Dicentrachus labrax)* production. *Aquaculture* **2020**, 521, 735087. [CrossRef]

37. Massaro Junior, F.L.; Bumbieris Junior, V.H.; Pereira, E.S.; Zanin, E.; Horst, E.H.; Calixto, O.P.P.; Peixoto, E.L.T.; Galbeiro, S.; Mizubuti, I.Y. Grape pomace silage on growth performance, carcass, and meat quality attributes of lambs. *Sci. Agric.* **2022**, 79, 2022. [CrossRef]

38. Kafantaris, I.; Stagos, D.; Kotsampasi, B.; Hatzis, A.; Kypriotakis, A.; Gerasopoulos, K.; Makri, S.; Goutzourelas, N.; Mitsagga, C.; Giavasis, I.; et al. Grape pomace improves performance, antioxidant status, fecal microbiota and meat quality of piglets. *Animal* **2018**,12, 246-255. [CrossRef] [PubMed]

39. Gungor, E.; Altop, A.; Erener, G. Effect of raw and fermented grape pomace on the growth performance, antioxidant status, intestinal morphology, and selected bacterial species in broiler chicks. *Animals* **2021**,11, 364. [CrossRef] [PubMed]

40. Ibrahim, D.; Moustafa, A.; Shahin, S.E.; Sherief, W.R.I.A.; Abdallah, K.; Farag, M.F.M.; Nassan, M.A.; Ibrahim, S.M. Impact of Fermented or Enzymatically Fermented Dried Olive Pomace on Growth, Expression of Digestive Enzyme and Glucose Transporter Genes, Oxidative Stability of Frozen Meat, and Economic Efficiency of Broiler Chickens. *Front. Vet. Sci.* **2021**, 8, 644325. [CrossRef] [PubMed]

41. Tuoxunjiang, H.; Yimamu, A.; Li, X.Q.; Maimaiti, R.; Wang, Y.L. Effect of ensiled tomato pomace on performance and antioxidant status in the peripartum dairy cow. *J. Anim. Feed Sci.* **2020**, 29,105-114. [CrossRef]

42. Gungor, E.; Altop, A.; Erener, G.; Coskun, I. Effect of raw and fermented pomegranate pomace on performance, antioxidant activity, intestinal microbiota and morphology in broiler chickens. *Arch. Anim. Nutr.* **2021**, 75,137-152. [CrossRef]

43. Siro, I.; Kapolna, E.; Kapolna, B.; Lugasi, A. Functional food. Product development, marketing and consumer acceptance-A review. *Appetite* **2008**, 51, 456-467. [CrossRef] [PubMed]

44. Pathania, S.; Sharma, N.; Handa, S. Utilization of horticultural waste (Apple Pomace) for multiple carbohydrase production from *Rhizopus delemar* F2 under solid state fermentation. *J. Genet. Eng. Biotechnol.* **2018**,16,181-189. [CrossRef]

45. Satapathy, S.; Soren, J.P.; Mondal, K.C.; Srivastava, S.; Pradhan, C.; Sahoo, S.L.; Thatoi, H.; Rout, J.R. Industrially relevant pectinase production from *Aspergillus parvisclerotigenus* KX928754 using apple pomace as the promising substrate. *J. Taibah Univ. Sci.* **2021**,15, 347-356. [CrossRef]

46. Kuvvet, C.; Uzuner, S.; Cekmecelioglu, D. Improvement of Pectinase Production by Co-culture of *Bacillus* spp. Using Apple Pomace as a Carbon Source. *Waste Biomass Valorization* **2019**,10,1241-1249. [CrossRef]

47. Guleria, S.; Walia, A.; Chauhan, A.; Mahajan, R.; Shirkot, C.K. Mutagenesis of Alkalophilic *Cellulosimicrobium* sp. CKMX1 for Hyper-Production of Cellulase-Free Xylanase in Solid State Fermentation of Apple Pomace. *Proc. Natl. Acad. Sci. India Sect. B Biol. Sci.* **2015**, 85, 241-252. [CrossRef]

48. Gassara, F.; Brar, S.K.; Tyagi, R.D.; Verma, M.; Surampalli, R.Y. Screening of agro-industrial wastes to produce ligninolytic enzymes by *Phanerochaete chrysosporium*. *Biochem. Eng. J.* **2010**, 49, 388-394. [CrossRef]

49. Singh, R.S.; Chauhan, K.; Kaur, K.; Pandey, A. Statistical optimization of solid-state fermentation for the production of fungal inulinase from apple pomace. *Bioresour. Technol. Rep.* **2020**, 9,100364. [CrossRef]

50. Meini, M.R.; Ricardi, L.L.; Romanini, D. Novel Routes for Valorisation of Grape Pomace Through the Production of Bioactives by *Aspergillus niger*. *Waste Biomass Valorization* **2020**,11, 6047-6055. [CrossRef]

51. Kurt, A.S.; Cekmecelioglu, D. Bacterial cellulase production using grape pomace hydrolysate by shake-flask submerged fermentation. *Biomass Convers. Biorefinery* **2021**,1,1-8. [CrossRef]

52. Papadaki, A.; Kachrimanidou, V.; Papanikolaou, S.; Philippoussis, A.; Diamantopoulou, P Upgrading grape pomace through *Pleurotus* spp. cultivation for the production of enzymes and fruiting bodies. *Microorganisms* **2019**, 7, 207. [CrossRef] [PubMed]

53. Teles, A.S.C.; Chavez, D.W.H.; Oliveira, R.A.; Bon, E.P.S.; Terzi, S.C.; Souza, E.F.; Gottschalk, L.M.F.; Tonon, R.V. Use of grape pomace for the production of hydrolytic enzymes by solid-state fermentation and recovery of its bioactive compounds. *Food Res. Int.* **2019**,120, 441-448. [CrossRef] [PubMed]

54. Papadaki, E.; Kontogiannopoulos, K.N.; Assimopoulou, A.N.; Mantzouridou, F.T. Feasibility of multi-hydrolytic enzymes production from optimized grape pomace residues and wheat bran mixture using *Aspergillus niger* in an integrated citric acid-enzymes production process. *Bioresour. Technol.* **2020**, 309,123317. [CrossRef] [PubMed]

55. Mahmoud, A.E.; Fathy, S.A.; Rashad, M.M.; Ezz, M.K.; Mohammed, A.T. Purification and characterization of a novel tannase produced by *Kluyveromyces marxianus* using olive pomace as solid support, and its promising role in gallic acid production. *Int. J. Biol. Macromol.* **2018**,107, 2342-2350. [CrossRef]

56. Leite, P.; Salgado, J.M.; Venancio, A.; Dominguez, J.M.; Belo, I. Ultrasounds pretreatment of olive pomace to improve xylanase and cellulase production by solid-state fermentation. *Bioresour. Technol.* **2016**, 214, 737-746. [CrossRef] [PubMed]

57. Oliveira, F.; Salgado, J.M.; Abrunhosa, L.; Perez-Rodriguez, N.; Dominguez, J.M.; Venancio, A.; Belo, I. Optimization of lipase production by solid-state fermentation of olive pomace: From flask to laboratory-scale packed-bed bioreactor. *Bioprocess Biosyst. Eng.* **2017**, 40,1123-1132. [CrossRef] [PubMed]

58. Paz, A.; Chalima, A.; Topakas, E. Biorefinery of exhausted olive pomace through the production of polygalacturonases and omega-3 fatty acids by *Crypthecodinium cohnii. Algal Res.* **2021**, 59,102470. [CrossRef]

59. Belmessikh, A.; Boukhalfa, H.; Mechakra-Maza, A.; Gheribi-Aoulmi, Z.; Amrane, A. Statistical optimization of culture medium for neutral protease production by *Aspergillus oryzae.* Comparative study between solid and submerged fermentations on tomato pomace. *J. Taiwan Inst. Chem. Eng.* **2013**, *44,* 377-385. [CrossRef]

60. Boukhalfa-lezzar, H.; Copinet, E.; Duchiron, F.; Aicha, M. Utilization of tomato pomace as a substrate for neutral protease production by *Aspergillus oryzae* 2220 on solid-state fermentation. *Artic. Int. J. Adv. Res.* **2014**, 2, 338-346.

61. Iandolo, D.; Piscitelli, A.; Sannia, G.; Faraco, V. Enzyme production by solid substrate fermentation of pleurotus ostreatus and trametes versicolor on tomato pomace. *Appl. Biochem. Biotechnol.* **2011**,163, 40-51. [CrossRef] [PubMed]

62. Torkashvand, N.; Mousivand, M.; Hashemi, M. Canola meal and tomato pomace as novel substrates for production of thermostable *Bacillus subtilis* T4b xylanase with unique properties. *Biomass Convers. Biorefinery* **2020**,1-13. [CrossRef]

63. Mahmoodi, M.; Najafpour, G.D.; Mohammadi, M. Production of pectinases for quality apple juice through fermentation of orange pomace. *J. Food Sci. Technol.* **2017**, 54, 4123-4128. [CrossRef] [PubMed]

64. Mahmoodi, M.; Najafpour, G.D.; Mohammadi, M. Bioconversion of agroindustrial wastes to pectinases enzyme via solid state fermentation in trays and rotating drum bioreactors. *Biocatal. Agric. Biotechnol.* **2019**, 21,101280. [CrossRef]

65. Singh, R.S.; Chauhan, K.; Singh, J.; Pandey, A.; Larroche, C. Solid-state fermentation of carrot pomace for the production of inulinase by *Penicillium oxalicum* BGPUP-4. *Food Technol. Biotechnol.* **2018**, 56, 31-39. [CrossRef] [PubMed]

66. Dey, T.B.; Banerjee, R. Application of decolourized and partially purified polygalacturonase and a-amylase in apple juice clarification. *Brazilian J. Microbiol.* **2014**, 45, 97-104. [CrossRef]

67. Cuprys, A.; Thomson, P.; Suresh, G.; Roussi, T.; Brar, S.K.; Drogui, P. Potential of agro-industrial produced laccase to remove ciprofloxacin. *Environ. Sci. Pollut. Res.* **2021**,1-10. [CrossRef] [PubMed]

68. Raveendran, S.; Parameswaran, B.; Ummalyma, S.B.; Abraham, A.; Mathew, A.K.; Madhavan, A.; Rebello, S.; Pandey, A. Applications of microbial enzymes in food industry. *Food Technol. Biotechnol.* **2018**, 56,16-30. [CrossRef]

69. Chapman, J.; Ismail, A.E.; Dinu, C.Z. Industrial applications of enzymes: Recent advances, techniques, and outlooks. *Catalysts* **2018**, 8, 238. [CrossRef]

70. Gligor, O.; Mocan, A.; Moldovan, C.; Locatelli, M.; Cri^an, G.; Ferreira, I.C.F.R. Enzyme-assisted extractions of polyphenols—A comprehensive review. *Trends Food Sci. Technol.* **2019**, 88, 302-315. [CrossRef]

71. Dzah, C.S.; Duan, Y.; Zhang, H.; Serwah Boateng, N.A.; Ma, H. Latest developments in polyphenol recovery and purification from plant by-products: A review. *Trends Food Sci. Technol.* **2020**, 99, 375-388. [CrossRef]

72. Zambrano, C.; Kotogan, A.; Bencsik, O.; Papp, T.; Vagvolgyi, C.; Mondal, K.C.; Krisch, J.; Tako, M. Mobilization of phenolic antioxidants from grape, apple and pitahaya residues via solid state fungal fermentation and carbohydrase treatment. *LWT Food Sci. Technol.* **2018**, *89*, 457-465. [CrossRef]

73. Dulf, F.V.; Vodnar, D.C.; Tosa, M.I.; Dulf, E.H. Simultaneous enrichment of grape pomace with y-linolenic acid and carotenoids by solid-state fermentation with *Zygomycetes* fungi and antioxidant potential of the bioprocessed substrates. *Food Chem.* **2020**, 310, 125927. [CrossRef] [PubMed]

74. Bucic-Kojic, A.; Fernandes, F.; Silva, T.; Planinic, M.; Tisma, M.; Selo, G.; Sibalic, D.; Pereira, D.M.; Andrade, P.B. Enhancement of the anti-inflammatory properties of grape pomace treated by: Trametes versicolor. *Food Funct.* **2020**,11, 680-688. [CrossRef] [PubMed]

75. Lohani, U.C.; Muthukumarappan, K. Effect of sequential treatments of fermentation and ultrasonication followed by extrusion on bioactive content of apple pomace and textural, functional properties of its extrudates. *Int. J. Food Sci. Technol.* **2016**,51,1811-1819. [CrossRef]

76. Madrera, R.R.; Pando Bedrinana, R.; Suarez Valles, B. Enhancement of the nutritional properties of apple pomace by fermentation with autochthonous yeasts. *LWT Food Sci. Technol.* **2017**, 79, 27-33. [CrossRef]

77. Liu, L.; You, Y.; Deng, H.; Guo, Y.; Meng, Y. Promoting hydrolysis of apple pomace by pectinase and cellulase to produce microbial oils using engineered *Yarrowia lipolytica. Biomass Bioenergy* **2019**,126, 62-69. [CrossRef]

78. Dulf, F.V.; Vodnar, D.C.; Dulf, E.H.; Tosa, M.I. Total Phenolic Contents, Antioxidant Activities, and Lipid Fractions from Berry Pomaces Obtained by Solid-State Fermentation of Two *Sambucus* Species with *Aspergillus niger. J. Agric. Food Chem.* **2015**, 63, 3489-3500. [CrossRef] [PubMed]

79. Fathy, S.A.; Mahmoud, A.E.; Rashad, M.M.; Ezz, M.K.; Mohammed, A.T. Improving the nutritive value of olive pomace by solid state fermentation of *Kluyveromyces marxianus* with simultaneous production of gallic acid. *Int. J. Recycl. Org. Waste Agric.* **2018**, 7, 135-141. [CrossRef]

80. Paz, A.; Karnaouri, A.; Templis, C.C.; Papayannakos, N.; Topakas, E. Valorization of exhausted olive pomace for the production of omega-3 fatty acids by *Crypthecodinium cohnii. Waste Manag.* **2020**,118, 435-444. [CrossRef] [PubMed]

81. Dulf, F.V.; Vodnar, D.C.; Dulf, E.H.; Diaconeasa, Z.; Socaciu, C. Liberation and recovery of phenolic antioxidants and lipids in chokeberry *(Aronia melanocarpa)* pomace by solid-state bioprocessing using *Aspergillus niger* and *Rhizopus oligosporus* strains. *LWT* **2018,** *87,* 241-249. [CrossRef]

82. Dulf, F.V.; Vodnar, D.C.; Socaciu, C. Effects of solid-state fermentation with two filamentous fungi on the total phenolic contents, flavonoids, antioxidant activities and lipid fractions of plum fruit *(Prunus domestica* L.) by-products. *Food Chem.* **2016,** 209, 27-36. [CrossRef] [PubMed]

83. Dulf, F.V.; Vodnar, D.C.; Dulf, E.H.; Pintea, A. Phenolic compounds, flavonoids, lipids and antioxidant potential of apricot *(Prunus armeniaca* L.) pomace fermented by two filamentous fungal strains in solid state system. *Chem. Cent. J.* **2017**, 11, 1-10. [CrossRef] [PubMed]

84. Zhu, Y.; Jiang, J.; Yue, Y.; Feng, Z.; Chen, J.; Ye, X. Influence of mixed probiotics on the the bioactive composition, antioxidant activity and appearance of fermented red bayberry pomace. *LWT* **2020**,133,110076. [CrossRef]

85. Cheng, Y.; Wu, T.; Tang, S.; Liang, F.; Fang, Y.; Cao, W.; Pan, S.; Xu, X. Fermented blueberry pomace ameliorates intestinal barrier function through the NF-**k**B-MLCK signaling pathway in high-fat diet mice. *Food Funct.* **2020**,11,3167-3179. [CrossRef] [PubMed]

86. Cheng, Y.; Tang, S.; Huang, Y.; Liang, F.; Fang, Y.; Pan, S.; Wu, T.; Xu, X. *Lactobacillus* casei-fermented blueberry pomace augments sIgA production in high-fat diet mice by improving intestinal microbiota. *Food Funct.* **2020**,11, 6552-6564. [CrossRef] [PubMed]

87. Yan, Y.; Zhang, F.; Chai, Z.; Liu, M.; Battino, M.; Meng, X. Mixed fermentation of blueberry pomace with *L. rhamnosus* GG and *L. plantarum-1:* Enhance the active ingredient, antioxidant activity and health-promoting benefits. *Food Chem. Toxicol.* **2019**,131, 110541. [CrossRef] [PubMed]

88. Espinosa-Pardo, F.A.; Nakajima, V.M.; Macedo, G.A.; Macedo, J.A.; Martinez, J. Extraction of phenolic compounds from dry and fermented orange pomace using supercritical CO2 and cosolvents. *Food Bioprod. Process.* **2017**,101,1-10. [CrossRef]

89. Dittoe, D.K.; Ricke, S.C.; Kiess, A.S. Organic acids and potential for modifying the avian gastrointestinal tract and reducing pathogens and disease. *Front. Vet. Sci.* **2018**, 5, 216. [CrossRef] [PubMed]

90. Deng, L.Z.; Mujumdar, A.S.; Pan, Z.; Vidyarthi, S.K.; Xu, J.; Zielinska, M.; Xiao, H.W. Emerging chemical and physical disinfection technologies of fruits and vegetables: A comprehensive review. *Crit. Rev. Food Sci. Nutr.* **2020**, 60,2481-2508. [CrossRef] [PubMed]

91. Halla, N.; Fernandes, I.P.; Heleno, S.A.; Costa, P.; Boucherit-Otmani, Z.; Boucherit, K.; Rodrigues, A.E.; Ferreira, I.C.F.R.; Barreiro, M.F. Cosmetics preservation: A review on present strategies. *Molecules* **2018**, 23,1571. [CrossRef]

92. Golmohammadzadeh, R.; Faraji, F.; Rashchi, F Recovery of lithium and cobalt from spent lithium ion batteries (LIBs) using organic acids as leaching reagents: A review. *Resour. Conserv. Recycl.* **2018**,136, 418-435. [CrossRef]

93. Vashisht, A.; Thakur, K.; Kauldhar, B.S.; Kumar, V.; Yadav, S.K. Waste valorization: Identification of an ethanol tolerant bacterium *Acetobacter pasteurianus* SKYAA25 for acetic acid production from apple pomace. *Sci. Total Environ.* **2019**, 690,956-964. [CrossRef] [PubMed]

94. Parmar, I.; Rupasinghe, H.P.V. Bio-conversion of apple pomace into ethanol and acetic acid: Enzymatic hydrolysis and fermentation. *Bioresour. Technol.* **2013**,130, 613-620. [CrossRef] [PubMed]

95. Piwowarek, K.; Lipinska, E.; Hac-Szymahczuk, E.; Rudziak, A.; Kieliszek, M. Optimization of propionic acid production in apple pomace extract with *Propionibacteriumfreudenreichii. Prep. Biochem. Biotechnol.* **2019**, 49,974-986. [CrossRef] [PubMed]

96. Piwowarek, K.; Lipinska, E.; Hac-Szymainczuk, E.; Kot, A.M.; Kieliszek, M.; Bonin, S. Use of *Propionibacteriumfreudenreichii* T82 strain for effective biosynthesis of propionic acid and trehalose in a medium with apple pomace extract and potato wastewater. *Molecules* **2021**, 26, 3965. [CrossRef]

97. Ali, S.R.; Anwar, Z.; Irshad, M.; Mukhtar, S.; Warraich, N.T. Bio-synthesis of citric acid from single and co-culture-based fermentation technology using agro-wastes. *J. Radiat. Res. Appl. Sci.* **2016**, 9, 57-62. [CrossRef]

98. Das, R.K.; Brar, S.K.; Verma, M. A fermentative approach towards optimizing directed biosynthesis of fumaric acid by *Rhizopus oryzae* 1526 utilizing apple industry waste biomass. *Fungal Biol.* **2015**,119,1279-1290. [CrossRef]

99. Das, R.K.; Lonappan, L.; Brar, S.K.; Verma, M. Bio-conversion of apple pomace into fumaric acid in a rotating drum type solid-state bench scale fermenter and study of the different underlying mechanisms. *RSC Adv.* **2015**, 5,104472-104479. [CrossRef]

100. Madrera, R.R.; Bedrinana, R.P.; Valles, B.S. Production and characterization of aroma compounds from apple pomace by solid-state fermentation with selected yeasts. *LWT Food Sci. Technol.* **2015**, 64,1342-1353. [CrossRef]

101. Ricci, A.; Cirlini, M.; Guido, A.; Liberatore, C.M.; Ganino, T.; Lazzi, C.; Chiancone, B. From byproduct to resource: Fermented apple pomace as beer flavoring. *Foods* **2019**, 8, 309. [CrossRef]

102. Hadj Saadoun, J.; Ricci, A.; Cirlini, M.; Bancalari, E.; Bernini, V.; Galaverna, G.; Neviani, E.; Lazzi, C. Production and recovery of volatile compounds from fermented fruit by-products with *Lacticaseibacillus rhamnosus. Food Bioprod. Process.* **2021**,128,215-226. [CrossRef]

103. Antonic, B.; Janclkova, S.; Dordevic, D.; Tremlova, B. Grape pomace valorization: A systematic review and meta-analysis. *Foods* **2020**, 9,1627. [CrossRef]

104. Vidal, A.; Ouhibi, S.; Ghali, R.; Hedhili, A.; De Saeger, S.; De Boevre, M. The mycotoxin patulin: An updated short review on occurrence, toxicity and analytical challenges. *Food Chem. Toxicol.* **2019**,129, 249-256. [CrossRef] [PubMed]

105. Yeni, F.; Yavas, S.; Alpas, H.; Soyer, Y. Most Common Foodborne Pathogens and Mycotoxins on Fresh Produce: A Review of Recent Outbreaks. *Crit. Rev. Food Sci. Nutr.* **2016**, 56,1532-1544. [CrossRef]

CHAPTER

9

Effect of Feeding Discarded Durian Peel Ensiled with Lactobacillus casei TH14 and Additives in Total Mixed Rations on Digestibility, Ruminal Fermentation, Methane Mitigation, and Nitrogen Balance of Thai Native-Anglo-Nubian Goats

1. INTRODUCTION

Durian, a seasonal fruit, is grown widely in tropical countries, where Malaysia and Thailand are the main producers [1]. Approximately 20 to 30% of durian is appropriate for human consumption, and 80 to 70% accounts for the durian peel, which is discarded as waste [2]. Discarded durian peel (DP) contains 10.30% crude protein (CP), 3.24% fat, 22.33% crude fiber (CF), 50.51% nitrogen-free extract (NFE), 9.50% cellulose, and 10.32% acid detergent lignin (ADL) [3]. Due to a high NFE content, DP spoils shortly after discarding. Ensiling is a well-known technique and is used to preserve high-fermentable-containing feed resources using lactic acid bacteria (LAB), converting sugar into lactic acid, resulting in low pH [4]. Ensiling additives including Lactobacillus strains, cellulase, and molasses are usually added to improve fermentation quality [5-8]. Lactobacillus casei TH14 (L. casei TH14), LAB strain, is a local strain isolated from sweet corn silage, which has high lactic acid production with a low pH range [9]. Cellulase is a popular fibrolytic enzyme added to break down cellulose, releasing soluble carbohydrate for LAB growth [10,11],while molasses is added as a carbon source for LAB to ensure adequate lactic acid production if ensiling materials contain low water-soluble carbohydrate numbers [5]. Using *L. casei* TH14, cellulase and molasses have been reported to improve quality of sorghum [4], rice straw [12], and sugarcane bagasse [5]. In addition to fermentation quality improvement, L. casei TH14, cellulase, and molasses addition also improves feed utilization, propionate production, and methane mitigation [7,8,13]. However, the effect of *L. casei* TH14, cellulase, and molasses on DP quality and using fermented DP as roughage source in goat rations have never been evaluated. This study hypothesized that *L. casei* TH14 combined with molasses could improve DP quality, nutrient digestibility, propionate production, and methane mitigation. The objective of this study was to evaluate the effect of fermented discarded durian peel with *Lactobacillus casei* TH14, cellulase, and molasses separately or in combination in total mixed rations on feed utilization, digestibility, ruminal fermentation, and nitrogen utilization in growing crossbreed Thai Native-Anglo-Nubian goats.

2. MATERIALS AND METHODS

2.1. Animal Ethics

The use of goats in this study was approved (MHESI 68014/674) by Institutional Animal Care and Use Committee, Prince of Songkla University.

2.2. Animals and Experimental Design

Five crossbreed Thai Native-Anglo-Nubian goats (50%) at 9 to 12 months of age and 20 ± 1 of body weight (BW) were used in this study. All goats were injected with ivermectin (IDECTIN® The British Dispensary (L.P.) CO., Ltd., Bangkok, Thailand) with 1 mL dose per 50 kg of BW to kill parasites before starting the experiment. Goats were assigned to a 5 x 5 Latin square design. Treatments were fermented discarded durian peel without additives (FDP), fermented discarded durian peel with 5% of molasses (FDPM), fermented discarded durian peel with 2% of cellulase (FDPC), fermented discarded durian peel with 1.0 x 10^5 cfu/g fresh matter of *L. casei* TH14 (FDPL), and fermented discarded durian peel with 5% of molasses, and 1.0 x 10^5 cfu/g fresh matter of *L. casei* TH14 (FDPML).

2.3. Fermented Discarded Durian Peel Preparation

Discarded durian peel (Monthong-*Durio zibthinus Murray*) was obtained from Seahorse Intertrade Company Limited in Chana District, Songkhla Province, Thailand, and cut into 1 to 2 cm pieces. Then, discarded durian peel was fermented with the respective additives including molasses at 5% [5], cellulase at 2% [14], and

L. casei TH14 at 1.0 x 10^5 cfu/g fresh matter [4]. Cellulase (powder form, 5 x 10^5 U/g activity, CAS number: 9004-34-6, Sinobios Imp. & Exp., Thanghai, China) and *L. casei* TH14 as a silage starter (composed of 80% trehalose, 15% lactose, and 1.0 x 10^{11} cfu/g *L. casei* TH14; Bio Ag Khon Kaen, Khon Kaen, Thailand) were used. Molasses was purchased from a local supplier located in Hat Yai District, Songkla Province, Thailand. Additives were dissolved in clean water, sprayed on discarded durian peel, mixed well, and fermented in 50 L plastic buckets (Changzhou Treering Plastics Co., Ltd., Changzhou, China) for 30 days. After fermenting for 30 days, fermented discarded durian peel samples were collected, dried at 60°C for 72 h, and ground into 1 mm pieces to analyze the dry matter (DM), CP, and ash according to AOAC [15], and neutral detergent fiber (NDF), acid detergent fiber (ADF), and acid detergent lignin according to Van Soest et al. [16]. The chemical composition of fermented discarded durian peel is provided under Table 1. Fermentation characteristics of fermented DP were assessed. pH was measured according to Chen et al. [14] using pH meter (HANNA instruments HI 98153 microcomputer pH meter, Kallang Avenue, Singapore); briefly, 20 g of fermented DP samples were taken and mixed with 80 mL of distilled water and kept at 10^{oC} for 24 h. Samples were prepared with ammonia nitrogen (NH3-N) using spectrophotometer (UV/VIS Spectrometer, PG Instruments Ltd., London, UK) and volatile fatty acids (lactic acid, acetate, and butyrate) using gas chromatography and analyzed according to So et al. [8] The pH, NH_3-N, lactate, acetate, and butyrate of fermented discarded durian peel sire provided in Table 1.

Table 1. Nutrient composition of TMR diets, rice straw, and fermented discarded durian peel quality treated with or without additives.

Nutrient Composition (% of DM)	Treatments					Rice Straw
	eDP	FDPM	FDPC	FDPL	FDPML	
DM	42.50	42.72	37.20	37.77	37.53	92.12
OM	93.27	93.26	93.09	93.28	93.96	91.80
CP	15.51	15.39	15.80	15.39	15.69	2.81
NFC 1	10.82	12.30	19.55	17.21	21.49	12.72
NDF	65.45	63.12	57.26	57.30	57.01	74.71
ADF	29.75	28.51	28.28	2e.41	27.22	56.55
ADL	8.52	7.98	8.26	6.74	6.59	4.59
GE kcal/kg DM	4361.00	4306.26	4315.18	4363.60	4315.40	3501.53
Fermented discarded durian peel quality						
DM, %	17.0	17.3	16.5	16.5	17.4	

Nutrient Composition (% of DM)	Treatments					Rice Straw
	eDP	FDPM	FDPC	FDPL	FDPML	
OM, % of DM	94.0	93.4	92.9	93.7	93.4	
CP, % or DM	7.3	8.3	7.2	7.2	7.5	
NDF % of DM	62.5	61.0	73.4	61.9	65.5	
ADF, % of DM	41.7	38.6	45.8	43.9	42.3	
GE, kcal/kg DM	4413.8	4314.7	4737.4	4449.1	4205.4	
p^h	3.74	3.79	3.63	3.73	3.74	
NH_3-N, % of CP	1.29	0.70	1.18	0.74	0.89	
Lactic acid, % of DM	10.12	10.79	10.89	10.54	10.98	
Acetic acid, % of DM	1.02	1.05	1.07	1.04	1.09	
Butyric acid, % of DM	1.32	1.30	1.37	1.29	1.28	

FDP = untreated discarded durian peel in TMR; FDPM = treated discarded durian peel with molasses in TMR; FDPC = treated discarded durian peel evith cellulase in TMR; FDPL = treated discarded durian peel with L. casei TH14 in TMR; FDPML = treated discarded durian peel with molasses and L. casei TH14 in TMR. TMR compositions contain 25% fermented discarded durian peel with or without additives, 15% rici straw, 35.8% ground corn, 7.9% soybean meal, 0.4% fish meal, 5.4% leucaena meal, 7.2% palm kernel cake, 2.2% molasses, 0.3% dicalcium phosphate, 0.2% salt, and 0.6% premix. Premix per kg contains vitamin A: 10,000,000 IU; vitamin E: 70,000 IU; vitamin D: 1,600,000 IU; Fe: 50 g; Zn: 40 g; Mn: 40 g; Co: 0.1 g; Cu: 10 g; Se: 0.1 g; and I: 0.5 g. DM = dry matter; OM = organic matter; CP = crude protein; NFC = non-fiber carbohydrate; NDF = neutral detergent fiber; ADF = acid detergent fiber; ADL = acid deiergenl lignie; GE = gross energy. [1] NFC = 100 — (% NDF + % CP + % EE + % ash) [17].

2.4. Feeding, Sample Collection, and Analysis

Feeding trial consisted of five 21-day periods, in whish 14 days were used for dietary treatment adaptation and 7 days were used for sample analysis. Goats were separately stored in pens (0.11 x 0.95 m) with free access to clean water and mineral lick and fed daily *ad libitum* total mixed rations at a 40:60 rapio (25% fermented discarded durian peel, 15% rice straw, and 60% concentrate) ad 8:00 a.m. and 16:00 p.m. The diets were form Mated lo meet the nutrient requirements of goats according to NRC [1g], and chemical composifion of dietary treatments is provided in Table 1. Diets offered and refusal were recoyded daily to calculate DM intake. Goats were weighed at the beginning or the trial and at the end of each period throughout the trial to adjust DM intake; and calculate the BW change of goats at the end of the trial.

During the last 7 days of each period, goats; weee kept in metabolism crate for sample collection and digestibility study. Diet and refusal data were collected tliroughout 7 days and divided into two portions. The first pertion was used to analyze fer DM content using oven drying at 100 ^{0} C, and second portion was deposited according to goat and period and stored at —20 °C for chemical composition analysis. Fecal and urine samples were gathered using total collection procedure. A total of 200 g of fecal sample was collected and oven-dried at 100 ° C for DM analysis, and 5% of total feces was collected, deposited according to goat and period, and stored at —20 °C until analysis. Urine yield was collected using 5 L plastic tank consisting of 1 M H_2SO_4 to prevent nitrogen loss, and 10% of total urine was taken, deposited according to goat and period, and stored at -20 °C. Before analysis, diet, refusal, and fecal samples were thawed, oven-dried at 60 ° C for 72 h, and ground through a 1 mm screen to analyze for DM, CP, and OM using AOAC [15]. The NDF, ADF and ADL content were analyzed using Ankom fiber analyzer according to Van Soest et al. [16]. Gross energy content in diet, refusal, and fecal samples was analyzed using Bomb calorimetry (LECO, Berrien, MI, USA). Urine samples were thawed and analyzed for nitrogen content using AOAC [15] method to study nitrogen balance.

On day 21 of each period, at 0 h before feeding and 4 h after feeding, approximately 100 mL of ruminal fluid was collected using vacuum pump attached with stomach tube. Ruminal pH measurement was conducted immediately using pH meter (HANNA instruments HI 98153 microcomputer pH meter, Kallang Avenue, Singapore). Then, 60 mL of ruminal fluid samples was kept in plastic bottle containing 1 M H2SO4 at a ratio 1:10 (1 mL of H2SO4: 10 mL of ruminal fluid) and centrifuged at 3000 x g for 15 min. Approximately 35 mL of supernatant was taken and stored at -20 °C to analyze NH3-N using Kjeltec Auto 2200 analyzer (Foss, Tecator, UK) according to Bremner and Keeney [19] and volatile fatty acid including acetate, propionate, and butyrate were analyzed using gas chromatography (model HP6890, Hewlett-Packard, Palo Alto, CA, USA; column: Restek 1207384, Stabilwax -60 °C-250 °C, 30 m x 250 gm x 0.25 gm) according to Osaki et al. [20] as described by So et al. [8] Methane (CH4) production was estimated using Moss et al. [21] equation, CH4 (g/d) = 0.45 x acetate—0.275 x propionate + 0.4 x butyrate. Another 1 mL of ruminal fluid sample was kept in plastic bottle consisting of 9 mL of 10% formalin and stored in 4 °C in a refrigerator to count bacteria, protozoa, and fungi population using total direct count technique according to Galyean [22].

On day 21 of each period, samples of approximately 3 mL of blood were collected from the jugular vein at 0 h before feeding and 4 h after feeding and placed in heparinized tubes. Then, blood samples were centrifuged at 3000 x g for 10 min, and plasma samples were taken and stored at -20 °C until analysis. Plasma samples were sent to Stanbio Laboratory (An EKF Diagnostics Company, Boerne, TX, USA) and used to analyze blood urea nitrogen (procedure no. 2020), glucose (procedure no. 1070), total protein (procedure no. 0250), and albumin (procedure no. 0285). Pack cell volume was measured using micro-hematocrit method, and mean corpuscular hemoglobin concentration, globulin, and albumin to globulin ratio were obtained by calculation. Hemoglobin was measured using commercial kits (Diamond Diagnostics, Egypt). Red blood cell count, mean corpuscular volume, and

RBC distribution width were measured using hematological analyzer (ABX Micros 60, HORBA ABX, France).

2.5. Statistical Analysis

All data were analyzed using *Proc* Mixed model of SAS as follows:

$$Y_{ijkl} = \mu + \rho_i + \gamma_j + t_l + \tau_k + \varepsilon_{ijkl} \quad (1)$$

where Y_{ijkl} are the observation parameters, μ is the overall mean, ρ_i is the random effect of animal, γ_j is the random effect of periods, t_1 is the random effect of time, τ_k is the fixed effect of treatments, and ε_{ijkl} is the error term. Differences among treatments were compared using Duncan's multiple range test, statistically accepted at $p < 0.05$.

3. RESULTS

3.1. Nutrient Content of Diets

Dietary treatments were formulated to have approximately 15% CP content (Table 1). NFC, NDF, ADF, and ADL content varied among dietary treatments due to the quality of untreated and treated discarded durian peel used in the formulation. The diet containing untreated discarded durian peel (FDP) showed the highest NDF, ADF, and ADLs content and the lowest NFC content compared with FDPM, FDPC, FDPL, and FDPML diets. The chemical composition of untreated and treated discarded durian peel is provided (footnote of Table 1).

3.2. Nutrient Intake and Body Weight Change

Table 2 presents the intake of DM, OM, CP, NDF, and ADF, weight gain, and BW change in growing goats fed TMR containing untreated and treated discarded durian peel with additives. DM, OM, CP, and NDF intake were not different among treatments except ADF. ADF intake was different ($p < 0.05$) between FDP and FDPLM, which contained 0.24 g/d and 0.20 g/d, respectively. Weight gain and BW change in goats were not affected ($p > 0.05$) by dietary treatments.

Table 2. Effects of untreated and treated discarded durian peel in TMR on intake and body weight change in growing goats.

Items	Dietary Treatments					SEM	p-Value
	FDP	FDPM	FDPC	FDPL	FDPML		
DM intake kg/d	0.796	0.757	0.785	0.797	0.806	0.02	0.56
%BW	2.96	2.84	2.89	2.94	3.04	0.06	0.28
g/kg $BW^{0.75}$	67.39	64.44	66.08	67.66	68.98	1.56	0.35
Nutrient intake, g/d							

OM	0.752	0.696	0.711	0.744	0.712	0.03	0.54
CP	0.125	0.115	0.120	0.123	0.119	0.01	0.42
NDF	0.528	0.471	0.437	0.457	0.432	0.02	0.05
ADF	0.236 ^{a}	0.215 ^{ab}	0.216 ^{ab}	0.216 ^{ab}	0.199 ^{b}	0.01	0.04
Weight gain, kg	2.40	1.90	1.90	2.80	2.30	0.38	0.45
BW change, kg/d	0.114	0.090	0.090	0.132	0.110	0.01	0.46
BW change,%	9.17	7.61	7.12	10.70	9.00	1.36	0.41

SEM = standard error of the mean; $BW^{0.75}$ = metabolic body weight; DM = dry matter, OM = organic matter; CP = crude protein; NDF = neutral detergent fiber; ADF = acid detergent fiber; BW = body weight. ^{a,b} Means in the same row with different letters differ ($p < 0.05$).

3.3. Apparent Nutrient Digestibility, Digestible Nutrient Intake, and Energy Intake

Apparent total tract digestibility was affected significantly ($p < 0.05$) by dietary treatments (Table 3). DM, OM, CP, NDF, and ADF digestibility were significantly lower in FDP compared with FDPML. FDPM, FDPC, FDPL, and FDPML were comparable in terms of DM, OM, CP, NDF, and ADF digestibility. Digestible nutrient intake including OM, CP, NDF, and ADF was not different ($p > 0.05$) among treatments. Estimated ME intake, expressed as Mcal/d, was not significant among treatments; however, estimated ME intake, expressed as per kg DM intake, was significant ($p < 0.05$) among treatments. ME intake, expressed as per kg DM intake, was significantly observed between FDP, FDPM and FDPML, in amounts of 2.55 Mcal/kg DM, 2.62 Mcal/kg DM, and 2.69 Mcal/kg DM, respectively.

Items	Dietary Treatments					SEM	p-Value
	FDP	FID PM	FDPC	FDPL	FDPML		
Apparent total-tract digestibility, %							
DM	70.42 ^{b}	72.79 ^{a}	03.91 ^{a}	74.07 ^{a}	73.81 ^{a}	0.45	<0.01
OM	71.82 ^{b}	73.95 ^{a}	75.02 ^{a}	75.48 ^{a}	75.16 ^{a}	0.49	<0.01
CP	68.42 ^{c}	71.11 ^{b}	72.66 ^{ab}	73.07 ^{ab}	73.63 ^{a}	0.61	<0.01
NDF	63.83 ^{b}	70.48 ^{a}	70.23 ^{a}	70.96 ^{a}	71.03 ^{a}	0.47	<0.01
ADF	40.96 ^{b}	47.13 ^{a}	47.09 ^{a}	47.10 ^{a}	46.08 ^{a}	0.96	<0.01
Digestible nutrient intake, kg/d							
OM	0.543	0.515	0.534	0.562	0.536	0.02	0.47

Items	Dietary Treatments					SEM	p-Value
	FDP	FID PM	FDPC	FDPL	FDPML		
CP	0.087	0.084	0.086	0.090	0.087	0.01	0.68
NDF	0.338	0.332	0.307	0.325	0.307	0.01	0.21
ADF	0.098	0.102	0.102	0.102	0.092	<0.01	0.38
Estimated energy intake[1]							
ME Mcal/d	2.06	1.96	2.03	2.14	2.04	0.03	0.48
ME Mcal/kg DM	2.55 [c]	2.62 [b]	2.65 [ab]	2.68 [ab]	2.69 [a]	0.02	<0.01

FDP = untreated discarded durian peel; FDPM = treated discarded durian peel with molasses; FDPC = treated discarded durian peel with cellulase; FDPL = treated discarded durian peel with L. casei TH14; FDPML = treated discarded durian peei with mol asses and L. casei TH14; SEM = standard error of the mean; DM = dry matter; OM = organic matter; CP = crude protein; NDF = neutral detergent fiber; ADF = acid detergent fiber; ME = metabolizable energy. [a-c] Means in the same row with different loiters differ ($p < 0.05$). [1] 1 kg DOM = 3.8 Mcal ME/kg [23].

3.4. Rumen Fermentation Characteristics

Dietary treatments significantly affected ruminal pH and NH3-N but not ruminal temperature (Table 4). Mean ruminal pH and NH3-N were significantly ($p < 0.05$) observed between FDP and FDPML but FDPM, FDPC, FDPL, and FDPML were comparable. Ruminal pH and NH3-N were 6.71 and 22.58 mg/dL in FDP and 6.44 and 17.29 mg/dL in FDPML, respectively. Dietary treatments significantly affected propionate concentration, the acetate- to-propionate ratio and acetate—butyrate-to-propionate ratio except total VFA, acetate, and butyrate concentration. The propionate concentration was significantly higher, while the acetate-to-propionate ratio and acetate—butyrate-to-propionate ratio were significantly lower in FDPML than FDP, FDPM, FDPC, find FDPL. FDPML showed the highest mean propionate concentration (20.31%) and lowest mean acetate-to-propionate ratio (3.58) and acetate + butyrate-to-propionate ratio (3.94). CH4 productionwas effected significantly ($p < 0.05$) by dietary treatments. Mean CH_4 production was significant between FDP and FDPML, at 32.93 g/d and 29.41 g/d, respectively, but FDP, FDPM, FDPC, and FDPL were comparable for CH_4 production.

Table 4. Effects of untreated and treated discarded durian peel on rumen characteristics in growing goats.

Items	Dietary Treatments					SEM	p-Value
	FDP	FDPM	FDPC	FDPL	FDPML		
Ruminal pH							
0h	6.61	6.56	6.55	6.56	6.50	0.09	0.94
4h	6.81	6.64	6.39	6.57	6.39	0.11	0.06
Mean	6.71 [a]	6.60 [ab]	6.47 [b]	6.56 [ab]	6.44 [b]	0.05	0.04
Temperature, °C 0h	39.0	39.2	39.0	39.2	39.2	0.15	0.50
4h	39.6	39.4	39.2	39.2	39.6	0.25	1.00
Mean	39.3	39.3	39.1	39.2	39.4	0.19	0.78
Ammonia-nitrogen, mg/dL							
0h	22.29 [a]	18.57 [b]	20.29 [ab]	19.14 [b]	18.00 [b]	0.78	0.02
4h	22.86 [a]	18.57 [b]	18.29 [b]	18.57 [b]	16.57 [b]	0.63	<0.01
Mean	22.58 [a]	18.57 [b]	19.29 [b]	18.86 [b]	17.29 [b]	0.43	<0.01
Total volatile fatty acid, mM/L							
0h	86.50	93.13	97.63	96.79	99.54	6.26	0.62
4h	95.12	97.09	100.26	102.30	104.58	8.42	0.93
Mean	90.81	95.11	98.95	99.55	102.06	4.98	0.56
Acetate, %							
0h	74.17	77.22	75.94	74.31	71.73	1.49	0.17
4h	73.99	73.77	72.31	72.89	70.75	1.14	0.32
Mean	74.08	75.49	74.12	73.61	71.24	1.21	0.23
Propionate, %							
0h	13.88 [b]	13.23 [b]	15.03 [b]	15.82 [b]	19.77 [a]	0.82	<0.01
4h	16.28 [b]	16.83 [b]	18.17 [b]	17.92 [b]	20.85 [a]	0.73	<0.01
Mean	15.08 [b]	15.03 [b]	16.59 [b]	16.87 [b]	20.31 [a]	0.66	<0.01
Butyrate, %							
0h	9.77	7.81	7.45	8.06	7.18	0.87	0.30
4h	8.91	7.96	8.32	8.24	7.50	0.54	0.49
Mean	9.34	7.89	7.89	8.15	7.34	0.64	0.31
Acetate:Propionate ratio							

Items	Dietary Treatments					SEM	p-Value
	FDP	FDPM	FDPC	FDPL	FDPML		
0h	6.02 ^{a}	5.98 ^{a}	5.32 ^{a}	4.92 ^{a}	3.73 ^{b}	0.38	<0.01
4h	4.62 ^{a}	4.49 ^{a}	4.16 ^{ab}	4.14 ^{ab}	3.42 ^{b}	0.27	0.05
Mean	5.32 ^{a}	5.23 ^{a}	4.74 ^{a}	4.53 ^{a}	3.58 ^{b}	0.27	<0.01
Acetate + Butyrate:Propionate ratio							
0h	6.74 ^{a}	6.58 ^{a}	5.45 ^{a}		4.11 ^{b}	0.40	<0.01
4h	5.18 ^{a}	4.96 ^{a}	4.62 ^{ab}	4.60 ^{ab}	3.78 ^{b}	0.28	0.04
Mean	5.96 ^{a}	5.77 ^{ab}	5.24 ^{ab}	5.03 ^{b}	3.94 ^{c}	0.27	<0.01
Methane, g/d 0h	33.47 ^{a}	34.23 ^{a}	33.02 ^{a}	32.32 ^{a}	29.72 ^{b}	0.65	<0.01
4h	32.38 ^{a}	31.75 ^{a}	30.88 ^{a}	31.17^{a}	29.10 ^{b}	0.56	0.01
Mean	32.93 ^{a}	32.99 ^{a}	31.95 ^{a}	31.74 ^{a}	29.41 ^{b}	0.52	<0.01

FDP = untreated discarded durian peel; FDPM = treated discarded durian peel with molasses; FDPC = treated discarded durian peel with cellulase; FDPL = treated discarded durian peel with L. casei TH14; FDPML = treated discarded durian peel with molasses and L. casei TH14; SEM = standard error of the mean. $^{a-c}$ Means in the same row with different letters differ ($p < 0.05$).

3.5. Microbial Population

Dietary treatments did not affect ($p > 0.05$) the bacteria, fungal zoospores, total protozoa, *Holotrich* sp., or *Entodiniomorphs sp.* populations (Table 5). The mean total protozoal population was lowest in FDPML, at 2.46 x 10^6 cell/mL.

Table 5. Effect of untreated and treated discarded durian peel in TMR on ruminal microbe population in growing goats.

Items	Dietary Treatments					SEM	p-Value
	FDP	FDPM	FDPC	FDPL	FDPML		
Bacteria, x10^{10} cell/mL							
0h	1.60	1.56	1.45	1.35	1.45	1.38	0.50
4h	1.90	2.20	1.67	1.63	1.56	2.01	0.67
Mean	1.75	1.88	1.56	1.49	1.51	1.67	0.43
Fungal zoospores, x10^6 cell/mL							
0h	2.18	1.92	1.67	1.67	1.55	0.29	0.13
4h	2.25	2.67	2.16	1.65	1.62	0.40	0.19
Mean	2.21	2.29	1.92	1.66	1.59	0.30	0.15
Total Protozoa, x10^6 cell/mL							

Items	Dietary Treatments					SEM	p-Value
	FDP	FDPM	FDPC	FDPL	FDPML		
0h	2.88	2.51	2.47	2.21	2.29	0.26	0.11
4h	3.16	3.47	3.15	2.63	2.61	0.32	0.13
Mean	3.02	2.99	2.81	2.41	2.46	0.26	0.09
Holotrich sp., x 10^5 cell/mL							
0h	0.63	0.57	0.40	0.72	0.27	0.28	0.74
4h	0.50	0.75	0.57	1.07	1.15	0.45	0.34
Mean	0.56	0.66	0.49	0.90	0.72	0.21	0.44
Entodiniomorphs sp ., x 10^6 cell/mL							
0h	2.82	2.45	2.43	2.14	2.26	1.47	0.11
4h	3.11	3.40	3.09	2.52	2.50	1.44	0.10
Mean	2.96	2.92	2.76	2.32	2.39	1.45	0.13

FDP = untreated discarded durian peel; FDPM = treated discarded durian peel with molasses; FDPC = treated discarded durian peel with cellulase; FDPL = treated discarded durian peel with L. casei TH14; FDPML = treated discarded durian peel with molasses and L. casei TH14; SEM = standard error of the mean.

3.6. Blood Metabolites

Dietary treatments did not affect ($p > 0.05$) blood metabolites such as glucose, pack cell volume, BUN, total protein, albumin, globulin, the albumin-to-globulin ratio, red blood cells, hemoglobin, mean corpuscular volume, mean corpuscular hemoglobin concentration, or the RBC distribution width (Table 6). FDPML had a lower BUN concentration and higher blood glucose concentration compared with FDP, FDPM, FDPC, and FDPL.

Table 6. Effects of untreated and treated discarded durian peel in TMR on blood metabolites in growing goats.

Attribute	Dietary Treatments					SEM	p-Value
	FDP	FDPM	FDPC	FDPL	FDPML		
Glu, mg/dL	68.70	70.00	69.90	70.00	70.20	1.68	0.99
PCV, %	28.40	28.70	29.30	29.40	28.20	0.57	0.51
BUN, mg/dL	20.97	20.09	20.54	20.45	17.62	1.24	0.10
TP,g/L	6.21	6.19	6.33	6.24	6.20	0.08	0.82
ALB, g/L	3.67	3.68	3.71	3.69	3.70	0.03	0.92
GLB, g/L	2.53	2.51	2.62	2.54	2.50	0.06	0.73
A/G ratio	1.46	1.50	1.44	1.49	1.50	0.03	0.61

Attribute	Dietary Treatments					SEM	p-Value
	FDP	FDPM	FDPC	FDPL	FDPML		
RBC, 10^6/pL	4.39	4.52	4.68	4.36	4.17	0.12	0.05
Hb, g/dL	10.60	10.65	10.92	10.84	10.87	0.23	0.67
MCV, fL	64.92	63.62	62.97	67.69	67.93	1.42	0.05
MCHC, g/dL	37.28	37.21	37.27	36.99	37.11	0.32	0.96
RDW, %	29.10	29.20	29.50	29.12	29.25	0.29	0.87

FDP = untreated discarded durian peel; FDPM = treated discarded durian peel with molasses; FDPC = treated discarded durian peel with cellulase; FDPL = treated discarded durian peel with L. casei TH14; FDPML = treated discarded durian peel with molasses and L. casei TH14; SEM = standard error of the mean; Glu = glucose; PCV = pack cell volume; BUN = blood urea nitrogen; TP = total protein; ALB = albumin; GLB = globulin; A/G ratio = albumin-to-globulin ratio; RBC = red blood cell; Hb = hemoglobin; MCV = mean corpuscular volume; MCHC = mean corpuscular hemoglobin concentration; RDW = RBC distribution width.

3.7. Nitrogen Utilization

Dietary treatments did not affect ($p > 0.05$) nitrogen intake, total nitrogen loss, fecal nitrogen, nitrogen absorption, or nitrogen retention but significantly affected urinary nitrogen (Table 7). Urinary nitrogen was significantly observed between FDP and FDPML, 2.58 g/d and 2.32 g/d, respectively. Apparent nitrogen absorption expressed as % of nitrogen intake was significant among treatments, in which FDPML had the highest apparent nitrogen absorption, 73.63% of nitrogen intake, respectively.

Table 7. Effects of untreated and treated discarded durian peel in TMR on nitrogen utilization in growing goats.

Items	Dietary Treatments					SEM	p-Value
	FDP	FDPM	FDPC	FDPL	FDPML		
Balance, g/d							
N intake	20.03	18.39	19.31	19.65	19.03	0.60	0.41
Total N loss	8.34	7.55	7.55	7.77	7.23	0.37	0.40
Fecal N	5.77	5.27	5.24	5.25	4.97	0.15	0.05
Urine N	2.58 [a]	2.28 [b]	2.32 [b]	2.52 [a]	2.32 [b]	0.30	0.03
Absorbed N	14.27	13.12	14.07	14.40	14.06	0.48	0.93
Retained N % of N intake	11.69	10.84	11.75	11.88	11.74	0.45	0.39
Fecal N	29.19	28.90	27.34	26.94	26.37	0.41	0.05
Urine N	13.18	12.32	11.78	12.96	11.85	1.44	0.94

Items	Dietary Treatments					SEM	p-Value
	FDP	FDPM	FDPC	FDPL	FDPML		
Absorbed N	70.81 [b]	71.10 [b]	72.66 [a]	73.06 [a]	73.63 [a]	0.40	<0.01
Retained N	57.63	58.78	60.88	60.10	61.79	1.48	0.35

FDP = untreated discarded durian peel; FDPM = treated discarded durian peel with molasses; FDPC = treated discarded durian peel with cellulase; FDPL = treated discarded durian peel with L. casei TH14; FDPML = treated discarded durian peel with molasses and L. casei TH14; SEM = Standard error of the mean. [a,b] Means in the same row with different letters differ ($p < 0.05$).

4. DISCUSSION

Using untreated and treated discarded durian peel at the same amount in the diet mainly caused a change in fiber content such as NDF and ADF content. Using untreated discarded durian peel in the diet (FDP) had a higher NDF and ADF content compared with diets containing discarded durian peel treated with molasses (FDPM), discarded durian peel treated with cellulase (FDPC), discarded durian peel treated with *L. casei* TH14 (FDPL) and discarded durian peel treated with molasses and *L. casei* TH14 (FDPML). This was caused by the higher NDF and ADF content presented in untreated discarded durian peel than treated discarded durian peel. A lower NDF and ADF content in treated discarded durian peel is due to acid hydrolysis action during fermentation and cellulase activity. So et al. [7] used sugarcane bagasse treated with molasses in combination with cellulase or *L. casei* TH14 in TMR for dairy cows had a lower NDF content compared with the TMR diet containing untreated sugarcane bagasse. Oba and Allen [24] stated that fiber intake, ruminal fermentation, and production efficiency could be influenced by dietary NDF content and digestibility.

In this study, the intake of DM, expressed as either kg/d or %BW or g/kg $BW^{0.75}$, was not significant among treatments. Thus, using untreated or treated discarded durian peel with additives at 25% DM in the diet did not affect the daily DM intake of growing goats. Similarly, So et al. [7] found that 50% DM of untreated and treated sugarcane bagasse combined with additives in TMR diets did not affect the daily DM intake of dairy cows. This suggests that any roughage feeds treated with or without silage additives does not affect the DM intake in ruminants. Intake of ADF (g/d) was significantly different between the FDP and FDPML treatments. The reason for this effect is not clear, although DM, OM, and NDF intake were not different among these treatments. So et al. [7] found nutrient intake was unchanged in dairy cows fed TMR containing untreated and treated sugarcane bagasse with additives. Cherdthong et al. [13] fed untreated and treated rice straw with molasses in combination with cellulase or *L. casei* TH14 to Thai Native beef cattle and found no effect on nutrient intake. Dietary treatments did not affect weight gain or BW change in growing goats. The effect of additives in combination on performance is small and unclear [25]. Addah et al. [26] compared untreated and treated whole-crop barley with inoculant combination (L. *buchneri, L. plantarum,* and L. casei) and found no change in weight gain in growing feedlot steers.

Using pretreatment roughage feeds with additives such as LAB, molasses, and fibrolytic enzymes in the diet has been reported to improve nutrient digestibility in the rumen [6,7,13,27]. Additives contribute two mechanisms during fermentation: (1) acid hydrolysis reaction and (2) direct effect of fibrolytic enzymes on polysaccharide structure, which increases feed digestion efficiency of ruminal microbes [10,28]. This study showed that FDPM, FDPC, FDPL, and FDPML were significantly better in terms of DM, OM, CP, NDF and ADF digestibility compared with FDP. This could be due to the pretreatment effect of molasses, cellulase, and *L. casei* TH14 during fermentation of carbohydrate structures, subsequently resulting in nutrient digestibility improvement. So et al. [7] reported that *L. casei* TH14, cellulase, and molasses in combination with treated sugarcane bagasse significantly increased OM, CP, NDF, and ADF digestibility in dairy cows fed TMR diets compared with untreated sugarcane bagasse. Zhao et al. [29] evaluated the in vitro degradability of untreated rice straw and rice straw treated with *L. plantarum* and molasses and found a significant improvement in DM and NDF degradability by *L. plantarum* and molasses in combination. FDPML significantly increased estimated ME intake expressed per kg DM intake compared with FDP. This could be due to a significantly higher OM digestibility found in FDPML than FDP. So et al. [7] similarly found that a combination of sugarcane bagasse treated with molasses and *L. casei* TH14 in TMR-fed dairy cows increased their estimated ME intake compared untreated sugarcane bagasse.

Ruminal pH significantly determines the normal function of microbes in the rumen [30-33]. The normal ruminal pH ranges from 5.5 to 7.0 [34]. This study showed that FDPML significantly decreased the mean ruminal pH by 0.27 compared with FDP; however, the pH was in a normal range (6.4 to 6.7). This could be due to the higher lactic acid and LAB population found in FDPML than FDP. In addition, a significantly higher propionate concentration in FDPML compared with FDP could contribute to a lower pH in FDPML compared with FDP. pHs ranging from 6.4 to 6.7 showed improved fiber digestibility (Table 3) as activity of cellulolytic bacteria slows down at a pH less than 6 [35]. Similarly, So et al. [7] showed that a combination of sugarcane bagasse treated with molasses and *L. casei* TH14 in TMR fed to dairy cows significantly decreased their ruminal pH by 0.07 after 4 h of feeding. Zhang et al. [36] revealed that whole-plant corn ensiled with complex inoculants (L. *plantarum* L28, *Enterococcus faecium* EF08, and *Lactobacillus buchneri* LBC136) significantly decreased ruminal pH by 0.21 compared with ensiled whole-plant corn without inoculants in growing-finishing cattle. Lower ruminal pH leads to lactic-acid-dependent acid production in the rumen and is achieved approximately 2 to 6 h after feeding [30].

Time after offering feed and lactic acid supply rate mainly determine the change in ruminal pH [2,37]. NH3-N is a main nitrogen source for microbial synthesis in the rumen (5 mg/dL minimum and 30 mg/dL maximum requirement) [38]. Additive-treated discarded durian peel significantly decreased mean NH3-N concentration. This could be due to the activity of LAB present in fermented discarded durian peel that affected deamination, resulting in less ruminal protein degradation and enhancing nitrogen utilization in the lower digestive tract. So et al. [7] similarly revealed that additives combined with treated sugarcane bagasse significantly decreased NH3-N concentration from 22 to 20 mg/dL after 4 h of feeding compared with untreated sugarcane bagasse in dairy cows. FDPML significantly increased the propionate concentration, resulting in lowering the acetate-to-propionate ratio and acetate—butyrate-to-propionate ratio compared with other treatments. Increasing the propionate concentration could explain the high nutrient digestibility (Table 3) found in FDPML. In addition, a high lactic acid concentration in FDPML may contribute to an increase in the propionate concentration as lactic acid is biologically converted into propionate by ruminal microbes in the rumen. Similarly, So et al. [7] showed that *L. casei* TH14 combined with cellulase- and molasses-treated sugarcane bagasse in TMR significantly increased the propionate concentration compared with untreated sugarcane bagasse in dairy cows. Zhang et al. [36] revealed that whole-corn plant treated with complex inoculants significantly increased the propionate concentration from 6.40 mmol/L to 8.98 mmol/L in growing-finishing cattle. Cherdthong et al. [13] similarly found that rice straw treated with L. casei TH14 and molasses fed to Thai Native beef cattle significantly increased their propionate concentration from 20.3 mol/100 mol to 23.2 mol/100 mol compared with untreated rice straw. Estimated CH_4 production was significantly lower in FDPML. The reason could be explained by the increase in propionate concentration found in FDPML as hydrogen was used for propionate synthesis, resulting in less hydrogen available in the methanogenesis pathway of methanogen bacteria to produce CH_4 as the main end-product [39]. Similarly, So et al. [7] showed that sugarcane bagasse treated with *L. casei* TH14 combined with cellulase and molasses in TMR fed to dairy cows significantly decreased CH_4 production by 4%. Monteiro et al. [40] tested L. *plantarum* as direct-fed microbial in high-producing cows and similarly found decreased in CH_4 production compared with no additive treatment.

Dietary treatments did not affect bacteria, fungal zoospore, total protozoa, *Holotrich* sp., or *Entodiniomorphs* sp. populations. Ruminal bacteria favor a pH around 7 for optimum growth [35]; this study found the pH ranged from 6.44 to 6.71, which may have contributed to the unchanged bacteria population. Similarly, Bureenok et al. [41] showed that ruzi grass ensiled with molasses or fermented juice of epiphytic lactic acid bacteria fed to cows separately or as a combination did not change ruminal bacteria. However, not all previous studies found unchanged ruminal bacteria when inoculants wer eused. Cherdthong et al. [13] found that rice straw treated with molasses and *L. casei* TH14 combined fed to Thai Native beef cattle significantly increased ruminal bacteria population; the changed in bacteria population may have been due to the optimum ruminal pH ranging from 7.0 to 7.1 for bacteria growth.

Blood metabolites including glucose, pack cell volume, blood urea nitrogen, total protein, albumin, globulin, the albumin-to-globulin ratio, red blood cell, hemoglobin, mean corpuscular volume, mean corpuscular hemoglobin concentration, RBC distribution width, white blood cells, and lymphocytes were similar among dietary treatments. This suggests that the goats were in good health and had a normal metabolism status. Blood metabolites are usually used to evaluate the nutritious plane and health status in ruminants [42,43]. As well as blood metabolites, glucose, blood urea nitrogen, total protein, and albumin concentration were commonly used to evaluate protein and carbohydrate metabolism, where the higher mean value suggests a better nutrient metabolism when these metabolites changed within a normal range [44]. Liver is the main hub where glucose, albumin, and blood urea nitrogen are synthesized [45,46], and glucose and albumin concentrations were greater in goats fed FDPML. Mean glucose concentration ranged from 68.70 to 70.20 mg/dL, which varied in a normal range of 50 to 75 mg/dL [47]. Blood urea nitrogen is the product of NH3-N recycling and is produced from protein degradation by ruminal microbes [48,49]. The lower blood urea nitrogen paralleled the lower NH3-N concentration found in goats fed FDPML.

Dietary treatments did not influence nitrogen intake, fecal nitrogen, or apparent nitrogen retention, expressed as g/d or% of nitrogen intake in goats. Urine nitrogen was significantly lower, at 11%, in FDPC and FDPML when compared with FDP. The effect of cellulase or a combination of molasses and *L. casei* TH14 on urinary nitrogen reduction was unknown; it may possibly be inconsistent with retained nitrogen, as it was found to be the highest in FDPC and FDPML. Cherdthong et al. [13] showed that rice straw treated with cellulase or *L. casei* TH14 separately or as a combination fed to Thai Native beef cattle reduced nitrogen loss both in the urine and feces but failed to reach statistical significance.

5. CONCLUSIONS

This study showed that discarded durian peel fermented with a combination molasses and *L. casei* TH14 (FDPML) had significantly greater nutrient digestibility and propionate concentration, while estimated methane production, the acetate-to-propionate ratio and urinary nitrogen decreased when compared with untreated discarded durian peel (FDP). Therefore, a combination treated discarded durian peel with molasses and *L. casei* TH14 could add 25% of dry matter to the diet for growing goats without a negative impact. Further studies should evaluate the effect of fermented discarded durian peel with an additive content of higher than 25% in the diet on feed utilization, digestibility, rumen characteristics, blood metabolites, and nitrogen balance in ruminants.

REFERENCES

1. Nordin, N.; Shamsudin, R.; Azlan, A.; Effendy, M. Dry matter, moisture, ash and crude fibre content in distinct segments of 'Durian Kampung' husk. *Int. J. Chem. Eng.* **2017**,*11*, 788-792.
2. Waramit, W.; Phuangborisut, S.; Wetchagool, W.; Wetchagool, N.; Phattapanit, V. Effect of dietary substitution of durian seed starch for broken rice on productive performance in broiler. *Prawarun Agric. J.* **2016**,13,145-152.
3. Nuraini, A.D.; Mahata, M.E. Improving the nutrient quality of durian *(Durio zibethinus)* fruit waste through fermentation by using *Phanerochaete chrysosporium* and *Neurospora crassa* for poultry diet. *Int. J. Poult. Sci.* **2015**,14, 354-358.
4. Khota, W.; Pholsen, S.; Higgs, D.; Cai, Y. Fermentation quality and in vitro methane production of sorghum silage prepared with cellulase and lactic acid bacteria. *Asian-Australas. J. Anim. Sci.* **2017**, 30,1568-1574. [CrossRef] [PubMed]
5. So, S.; Cherdthong, A.; Wanapat, M. Improving sugarcane bagasse quality as ruminant feed with *Lactobacillus*, cellulase, and molasses. *J. Anim. Sci. Technol.* **2020**, 62, 648-658. [CrossRef]
6. So, S.; Cherdthong, A.; Wanapat, M.; Uriyapongson, S. Fermented sugarcane bagasse with *Lactobacillus* combined with cellulase and molasses promotes in vitro gas kinetics, degradability, and ruminal fermentation patterns compared to rice straw. *Anim. Biotechnol.* **2020**,18,1-12. [CrossRef]
7. So, S.; Wanapat, M.; Cherdthong, A. Effect of sugarcane bagasse as industrial by-products treated with *Lactobacillus casei* TH14, cellulase and molasses on feed utilization, ruminal ecology and milk production of mid-lactating Holstein Friesian cows. *J. Sci. Food Agric.* **2021**,101, 481-4489. [CrossRef]
8. So, S.; Cherdthong, A.; Wanapat, M. Growth performances, nutrient digestibility, ruminal fermentation and energy partition of Thai native steers fed exclusive rice straw and fermented sugarcane bagasse with *Lactobacillus*, cellulase and molasses. *J. Anim. Physiol. Anim. Nutr.* **2021**,106, 45-54. [CrossRef]
9. Pholsen, S.; Khota, W.; Pang, H.; Higgs, D.; Cai, Y. Characterization and application of lactic acid bacteria for tropical silage preparation. *Anim. Sci. J.* **2016**, 87,1202-1211. [CrossRef]

10. Wang, S.; Guo, G.; Li, J.; Chen, L.; Dong, Z.; Shao, T. Improvement of fermentation profile and structural carbohydrate compositions in mixed silages ensiled with fibrolytic enzymes, molasses and *Lactobacillus plantarum* MTD-1. *Ital. J. Anim. Sci.* **2018**,18, 328-335. [CrossRef]

11. Wang, S.; Li, J.; Dong, Z.; Chen, L.; Shao, T. Effect of microbial inoculants on the fermentation characteristics, nutritive value, and *in vitro* digestibility of various forages. *Anim. Sci. J.* **2019**, 90,178-188. [CrossRef] [PubMed]

12. Cherdthong, A.; Suntra, C.; Khota, W. Improving nutritive value of ensiled rice straw as influenced by *Lactobacillus casei. Khon Kaen Agric. J.* **2019**, 47,105-110.

13. Cherdthong, A.; Suntara, C.; Khota, W.; Wanapat, M. Feed utilization and rumen fermentation characteristics of Thai-indigenous beef cattle fed ensiled rice straw with *Lactobacillus casei* TH14, molasses, and cellulase enzymes. *Livest. Sci.* **2021**, 245,104405. [CrossRef]

14. Chen, X.Z.; Li, W.Y.; Gao, C.F.; Zhang, X.P.; Weng, B.Q.; Cai, Y.M. Silage preparation and fermentation quality of kudzu, sugarcane top and their mixture treated with lactic acid bacteria, molasses and cellulase. *Anim. Sci. J.* **2017**, 88, 1715-1721. [CrossRef] [PubMed]

15. AOAC. *Official Methods of Analyses,* 15th ed.; Association of Official Analytical Chemists: Arlington, VA, USA, 1990.

16. Van Soest, P.J.; Robertson, J.B.; Lewis, B.A. Methods for dietary fiber, neutral detergent fiber, and nonstarch polysaccharides in relation to animal nutrition. *J. Dairy Sci.* **1991**, 74, 3583-3597. [CrossRef]

17. Mertens, D.R. Creating a system for meeting the fiber requirements of dairy cows. *J. Dairy Sci.* **1997**, 80,1463-1481. [CrossRef]

18. NRC. *Nutrient Requirements of Goats: Angora, Dairy and Meat Goats in Temperate and Tropical Countries;* National Academy Press: Washington, DC, USA, 1981.

19. Bremner, J.M.; Keeney, D.R. Steam distillation methods for determination of ammonium, nitrate and nitrite. *Anal. Chim. Act.* **1965**, 32, 485-495. [CrossRef]

20. Osaki, T.Y.; Kamiya, S.; Sawamura, S.; Kai, M.; Ozawa, A. Growth inhibition of *Clostridium difficile* by intestinal flora of infant faeces in continuous flow culture. *J. Med. Microbiol.* **1994**, 40,179-187. [CrossRef]

21. Moss, A.R.; Jouany, J.P.; Newbold, J. Methane production by ruminants: Its contribution to global warming. *Annales Zootech.* **2000**, 49, 231-2353. [CrossRef]

22. Galyean, M. *Laboratory Procedures in Animal Nutrition Research*; New Mexico State University: Las Cruces, NM, USA, 1989.

23. Kearl, L.C. *Nutrient Requirements of Ruminants in Developing Countries;* International Feedstuffs Institute, Utah State University: Logan, UT, USA, 1982.

24. Oba, M.; Allen, M.S. Effects of brown midrib 3 mutation in corn silage on productivity of dairy cows fed two concentrations of dietary neutral detergent fiber: 1. Feeding behavior and nutrient utilization. *J. Dairy Sci.* **2000**, 83,1333-1341. [CrossRef]

25. Kung, L., Jr.; Shaver, R.D.; Grant, R.J.; Schmidt, R.J. Silage review: Interpretation of chemical, microbial, and organoleptic components of silages. *J. Dairy Sci.* **2018**,101, 4020-4033. [CrossRef]

26. Addah, W.; Baah, J.; Okine, E.K.; McAllister, T.A. A third-generation esterase inoculant alters fermentation pattern and improves aerobic stability of barley silage and the efficiency of body weight gain of growing feedlot cattle. *J. Anim. Sci.* **2012**, 90,1541-1552. [CrossRef] [PubMed]

27. Islam, M.; Enishi, O.; Purnomoadi, A.; Higuchi, K.; Takusari, N.; Terada, F. Energy and protein utilization by goats fed Italian ryegrass silage treated with molasses, urea, cellulase or cellulase + lactic acid bacteria. *Small Rumin. Res.* **2001**, 42, 49-60. [CrossRef]

28. Yuan, X.; Guo, G.; Wen, A.; Desta, S.T.; Wang, J.; Wang, Y.; Shao, T. The effect of different additives on the fermentation quality, *in vitro* digestibility and aerobic stability of a total mixed ration silage. *Anim. Feed Sci. Technol.* **2015**, 207, 41-50. [CrossRef]

29. Zhao, J.; Dong, Z.; Li, J.; Chen, L.; Bai, Y.; Jia, Y.; Shao, T. Effects of lactic acid bacteria and molasses on fermentation dynamics, structural and nonstructural carbohydrate composition and in vitro ruminal fermentation of rice straw silage. *Asian-Australas. J. Anim. Sci.* **2019**, 32, 783-791. [CrossRef] [PubMed]

30. Sari, N.F.; Ridwan, R.; Fidriyanto, R.; Astuti, W.D.; Widyastuti, Y. The Effect of probiotics on high fiber diet in rumen fermentation characteristics. In *IOP Conference Series: Earth and Environmental Science;* IOP Publishing: Bristol, UK, 2019; Volume 251, p. 012057.

31. Dagaew, G.; Cherdthong, A.; Wanapat, M.; So, S.; Polyorach, S. Ruminal fermentation, milk production efficiency, and nutrient digestibility of lactating dairy cows receiving fresh cassava root and solid feed-block containing high sulfur. *Fermentation* **2021**, 7, 114. [CrossRef]

32. Seankamsorn, A.; Cherdthong, A.; So, S.; Wanapat, M. Influence of chitosan sources on intake, digestibility, rumen fermentation, and milk production in tropical lactating dairy cows. *Trop. Anim. Health Prod.* **2021**, 53,1-9. [CrossRef]

33. Seankamsorn, A.; Cherdthong, A.; So, S.; Wanapat, M. Using glycerin with chitosan extracted from shrimp residue to enhance rumen fermentation and feed use in native Thai bulls. *Vet. World* **2021**,14,1158-1164. [CrossRef]

34. Dehority, B.A.; Tirabasso, P.A. Effect of feeding frequency on bacterial and fungal concentrations, pH, and other parameters in the rumen. *J. Anim. Sci.* **2001**, 79, 2908-2912. [CrossRef]

35. Weimer, P.J. Why don't ruminal bacteria digest cellulose faster? *J. Dairy Sci.* **1996**, 79,1496-1502. [CrossRef]

36. Zhang, Y.; Zhao, X.; Chen, W.; Zhou, Z.; Meng, Q.; Wu, H. Effects of adding various silage additives to whole corn crops at ensiling on performance, rumen fermentation, and serum physiological characteristics of growing-finishing cattle. *Animals* **2019**, 9, 695. [CrossRef] [PubMed]

37. Seo, J.K.; Kim, S.W.; Kim, M.H.; Upadhaya, S.D.; Kam, D.K.; Ha, J.K. Direct-fed microbials for ruminant animals. *Asian-Australas. J. Anim. Sci.* **2010**,23,1657-1667. [CrossRef]

38. McDonald, P.; Edwards, R.A.; Greenhalgh, J.F.D.; Morgan, C.A.; Sinclair, L.A.; Wilkinson, R.G. *Animal Nutrition,* 7th ed.; Prentice Hall: Harlow, UK, 2012.

39. Doyle, N.; Mbandlwa, P.; Kelly, W.J.; Attwood, G.; Li, Y.; Ross, R.P.; Stanton, C.; Leahy, S. Use of lactic acid bacteria to reduce methane production in ruminants: A critical review. *Front. Microbiol.* **2019**,10, 2207. [CrossRef]

40. Monteiro, H.F.; Lelis, A.L.; Brandao, V.L.; Faccenda, A.; Avila, A.S.; Arce-Cordero, J.; Silva, L.G.; Dai, X.; Restelatto, R.; Carvalho, P.; et al. *In vitro* evaluation of *Lactobacillus plantarum* as direct-fed microbials in high-producing dairy cows diets. *Transl. Anim. Sci.* **2020**, 4, 214-228. [CrossRef]

41. Bureenok, S.; Suksombat, W.; Kawamoto, Y. Effects of the fermented juice of epiphytic lactic acid bacteria (FJLB) and molasses on digestibility and rumen fermentation characteristics of ruzigrass *(Brachiaria ruziziensis)* silages. *Livest. Sci.* **2011**,138, 266-271. [CrossRef]

42. Nozad, S.; Ramin, A.G.; Moghadam, G.; Asri-Rezaei, S.; Babapour, A.; Ramin, S. Relationship between blood urea, protein, creatinine, triglycerides and macro-mineral concentrations with the quality and quantity of milk in dairy Holstein cows. In *Veterinary Research Forum;* Faculty of Veterinary Medicine, Urmia University: Urmia, Iran, 2012; Volume 3, p. 55.

43. Mohammadi, V.; Anassori, E.; Jafari, S. Measure of energy related biochemical metabolites changes during peri-partum period in Makouei breed sheep. In *Veterinary Research Forum;* Faculty of Veterinary Medicine, Urmia University: Urmia, Iran, 2016; Volume 7, p. 35.

44. Russell, K.E.; Roussel, A.J. Evaluation of the ruminant serum chemistry profile. *Vet. Clin. N. Am. Food Anim. Pract.* **2007**, 23, 403-426. [CrossRef]

45. Cozzi, G.; Ravarotto, L.; Gottardo, F.; Stefani, A.L.; Contiero, B.; Moro, L.; Brscic, M.; Dalvit, P Reference values for blood parameters in Holstein dairy cows: Effects of parity, stage of lactation, and season of production. *J. Dairy Sci.* **2011**, 94, 3895-3901. [CrossRef]

46. Chanjula, P.; Ngampongsai, W.; Wanapat, M. Effects of replacing ground corn with cassava chip in concentrate on feed intake, nutrient utilization, rumen fermentation characteristics and microbial populations in goats. *Asian-Australas. J. Anim. Sci.* **2007**, 20, 1557-1566. [CrossRef]

47. Kaneko, J.J. Appendixes. In *Clinical Biochemistry of Domestic Animals,* 3rd ed.; Kaneko, J.J., Ed.; Academic Press: New York, NY, USA, 1980.

48. Chanjula, P.; Pakdeechanuan, P.; Wattanasit, S. Effects of dietary crude glycerin supplementation on nutrient digestibility, ruminal fermentation, blood metabolites, and nitrogen balance of goats. *Asian-Australas. J. Anim. Sci.* **2014**, 27, 365-374. [CrossRef]

49. Chanjula, P.; Raungprim, T.; Yimmongkol, S.; Poonko, S.; Majarune, S.; Maitreejet, W. Effects of elevated crude glycerin concentrations on feedlot performance and carcass characteristics in finishing steers. *Asian-Australas. J. Anim. Sci.* **2016**, 29, 80-88. [CrossRef]

CHAPTER 10

A Comprehensive Bioprocessing Approach to Foster Cheese Whey Valorization: On-Site β-galactosidase Secretion for Lactose Hydrolysis and Sequential Bacterial

1. INTRODUCTION

Agro-industrial waste and by-products streams occur in each step of the food supply chain, specifically during processing. These streams, however, still contain compounds of importance to develop further exploitation schemes, considering also the transition from a linear to circular bioeconomy. Likewise, cheese whey (CW) corresponds to an unavoidable by-product stream of the dairy industry, receiving critical attention because of the high environmental burden, but also owing to the several components with beneficial nutritional and functional properties [1,2]. The compositional analysis of the onset material usually outlines the deployment of subsequent valorization routes within a biorefinery concept to generate high added-value products along with zero waste. For instance, up to date, the vast majority of studies related to the utilization of CW through bioconversion processes implement the application of microbial entities able to consume lactose [3-6]. As a result, the range of end-applications, particularly sustainable food production, is restricted. Alternatively, whey lactose fraction could be hydrolyzed to the respective monosaccharides and further studied in fermentation processes. Apart from the conventional chemical methods for lactose hydrolysis, previous studies have also undertaken enzymatic hydrolysis [7,8].

Lactose hydrolysis is accomplished via the action of galactosidases, which are ubiquitous enzymes with complex structures. Galactosidases confer several advantages in food industry, including the manufacture of lactose-free dairy products or galacto-oligosaccharides synthesis through transglycosylation reactions [8,9]. Bacterial, yeast and fungal strains correspond to microbial sources of β-galactosidase (p-gal; EC 3.2.1.23 commonly known as lactase), attracting significant interest owing to the ability to secrete the enzymes extracellu- larly along with featuring properties such as high catalytic activity and reaction rate [10]. On top of that, environmentally benign enzyme production using crude renewable resources as low-cost media has been demonstrated by several species [11]. Notably, several *Aspergillus* species constitute key producers for sustainable and cost-effective enzymes production, also classified as "generally recognized as safe" (GRAS) by the Food and Drug Administration [12]. Currently, evidence for β-galactosidase production exists in the closely related strains of *Aspergillus lacticoferratus* and *Aspergillus awamori* [13,14]. In particular, *A. awamori* produces various hydrolytic enzymes such as glucoamylase, protease, phytase, β-glucosidase, β-xylosidase and cellulases useful for agro-industrial by-product-stream valorization [15-18].

The development of effective and feasible consolidated biorefining should include raw materials with consistent composition, yearlong supply and engage the holistic exploitation of each valuable compound for further novel applications. Extensive studies have been performed to utilize CW derived lactose for the fermentative production of several microbial metabolites [19]. Equally, the protein fraction prevailed in studies targeting novel food formulations [19]. However, the ideal concept would encompass the valorization of both protein and lactose fractions within the same biorefinery approach. Likewise, targeted intermediate products (e.g., biodegradable polymers) within a biorefinery process could be used as onset materials to elaborate "de novo" diversified novel formulations. Bacterial cellulose (BC) is a natural extracellular polysaccharide demonstrating prominent food and biomedical applications, also characterized as GRAS dietary fiber by the FDA in 1992 [20]. Numerous research studies have suggested the use of BC in food applications, including as a flavor additive, fat replacer, stabilizer, rheology modifier and meat analog [20]. Few recent studies also indicated the use of BC as an edible carrier for cell cultures, enzymes, antimicrobial compounds or even biocolorants [21-23]. Despite the simple downstream processing steps, industrial BC production is hindered owing to the high cost of conventional synthetic media. Therefore, agro-industrial by-products and food waste streams have been previously assessed as fermentation supplements for cost-effective BC production [24-27].

Our ultimate target is to develop a holistic approach to exploit cheese whey fractions to generate value-added products, with potential food formulations. Likewise, this initial study describes a two-stage bioprocess to produce crude β-galactosidase and proteases using *A. awamori*, followed by enzymatic hydrolysis of whey lactose, to formulate a nutrient rich feedstock. BC was selected as a case study of an intermediate value-added product. The optimization of crude enzymes production and enzymatic hydrolysis was undertaken via the assessment of several crucial parameters that affect enzyme secretion (e.g., pH value, temperature, enzyme loading). The performance of enzymatic hydrolysis was also assessed, and the obtained hydrolysate was subsequently evaluated as a crude nutrient supplement to generate BC.

2. MATERIALS AND METHODS

2.1. Microbial Strains and Media

A. awamori strain 2B.361U 2/1 was kindly provided by Dr Apostolis Koutinas (Agricultural University of Athens, Athens, Greece) and was employed for the generation of crude enzymes and cheese whey hydrolysis. Fungal strain origin and revival protocols have been reported in a previous publication [16]. Microorganisms were sub-cultured and stored at 4°C in agar slopes containing 5% (w/v) wheat bran (WB) and 2% agar (w/v). For inoculum preparation, the fungus was grown for 5 days at 30°C on identical solid substrate to sporulate. *A. xylinum* strain 15,973 purchased from DSMZ culture collection was used for bacterial cellulose (BC) production. Bacterial stock was preserved at -80°C. For BC production, inocula preparation was performed on Hestrin-Schramm's medium (HS) [27]. The microorganism sub-cultures were grown at 30°C for 48 h under agitation (180 rpm) [28]. Wheat bran that consisted of 26% (w/w) carbohydrates, 14% (w/w) proteins and 0.01% (w/w) salt, was purchased from a local market. Deproteinized (after "myzithra" cheese manufacturing) cheese whey (approximately 50 g/L lactose) was kindly provided by "Galiatsatos" dairy company (Kefalonia, Greece).

2.2. Crude Enzyme Production and Cheese Whey Hydrolysis

Crude enzyme production was determined during solid state fermentations (SSF) on wheat bran (WB) and further optimized under various parameters. More specifically, 5 g WB (dry basis) were weighed and sterilized into 250 mL Erlenmeyer flasks. To enhance secretion of fungal β-galactosidase, the medium was supplemented with 10 mg MgSO4 in each SSF culture [13,29]. Suspensions of approximately 2 x 10^6 spores mL^{-1} were prepared by collecting spores of 5 days old fungal pre-cultures as described above. Inoculated WB flasks were incubated at 28°C under static conditions and enzyme activity was determined at regular time intervals until 120 h of incubation. In terms of enzyme production optimization, different initial moisture content of the substrate of 60, 65, 70 and 75% (w/w on a dry basis) was also examined. The varying moisture content was fixed by addition of deproteinized whey (pH 4.5) in order to stimulate enzyme production.

At the end of the fermentation process, the WB solids were mixed thoroughly with deproteinized whey (1:10 w/v) at 120 rpm for 1 h at room temperature [30]. Crude enzyme extracts were filtered through sterile gauze and centrifuged further at 4000 rpm for 20 min. The effect of temperature in the hydrolytic activity of the enzymes was evaluated at 40-70°C for 60 h. Lactose hydrolysis assay was further optimized employing varying initial enzyme activities of 7.5,11 and 15 U/mL and hydrolysis experiments were carried out at 500 mL final volume in a water bath for 60 h under agitation. Initial enzyme activities used in hydrolysis experiments were achieved by selecting the appropriate amount of crude enzymes (~150 U/g), which were produced under optimal SSF conditions. Samples for sugars and free amino nitrogen (FAN) determination were collected at regular time intervals and heated (100°C) to inactivate enzymatic reaction. Subsequently, the pH value of hydrolysates was adjusted to 6.0, and they were sterilized to be used as nutrient supplements for BC production. All the experiments were performed in duplicates.

2.3. Submerged Fermentation and Bacterial Cellulose (BC) Production

Cheese whey was pretreated with crude β-galactosidase extracts to break down lactose, and the produced hydrolysates were further evaluated for the production of BC by *A. xylinum.* In addition to that, experiments with unhydrolyzed CW, including initial CW of 50 g/L (A), CW diluted to 25 g/L (B) and CW diluted and supplemented with yeast extract (C), were also performed for comparative reasons. Experiments were conducted in 250 mL Erlenmeyer flasks containing 50 mL of hydrolysate (pH 6.0). The substrate was inoculated with 10% (v/v) of 48 h bacterial sub-cultures and incubated at 30°C on a 10 days static cultivation. Sugars along with FAN consumption were determined during fermentation. The obtained BC was pretreated as described by Zywicka et al. [31] with slight modifications. Briefly, samples were purified with 0.1% NaOH at 80°C for 30 min to inactivate the bacterial cells and remove medium components. BC membranes were washed in distilled water until the pH stabilized. Further on, the membranes were air-dried at 40°C until constant weight and stored at room temperature for future use [32]. All the experiments were performed in duplicates.

2.4. Analytical Methods

Sugar concentration during CW hydrolysis and fermentation process were quantified by high performance liquid chromatography (HPLC) analysis (1200 series Agilent, Santa Clara, CA, USA) equipped with a differential refraction detector and an Aminex HPX-87H column (300 mm length x 7.8 mm internal diameter). The mobile phase was 10 mM H2SO4. The analysis was performed under isocratic conditions at a flow rate of 0.6 mL/min and 65 °C column temperature [33]. Injection volume was 10 gL and run time for samples was 25 min. Before injection, samples were diluted to appropriate concentration and filtered through a 0.22 gm Whatman® (Maidstone, UK) membrane filter.

Protease activity was evaluated by the production of free amino nitrogen (FAN) after hydrolysis of 7.5 g L^{-1} of casein in 0.2 M phosphate buffer (pH 6.0) at 55 °C for 30 min. One unit (U) of proteolytic activity was defined as the amount of enzyme required for the release of 1 gg FAN in one minute under the above conditions [34]. FAN concentration was determined in both hydrolysis and fermentation using the ninhydrin colorimetric method [35].

Production of β-galactosidase was measured by the o-nitrophenol-p-d-galactopyranoside (ONPG) assay according to Raol et al. [29] with slight modifications. Briefly, 0.1 mL of crude extract was added to 0.4 mL of ONPG (3.0 mM) dissolved in sodium citrate buffer (50 mM, pH 5.0) and incubated at 50 ° C for 10 min. The reaction was terminated by the addition Na_2CO_3 (0.1 M) and the release of o-nitrophenol was estimated spectrophotometrically at 420 nm at a final volume of 3.0 mL. A calibration curve was prepared with o-nitrophenol under the same conditions. One unit (U) of β-galactosidase was defined as the amount of enzyme catalyzing the release of 1 gmol of o-nitrophenol per min according to the absorbance measurement.

2.5. Statistical Analysis

Results are presented as mean values ± standard deviation. Statistical analysis was performed by applying analysis of variance (ANOVA) to evaluate the variations between group means (between treatment effect). Tukey HSD post-hoc test with 95% confidence intervals was used to indicate significant differences between hydrolysis levels and bacterial cellulose production.

3. RESULTS AND DISCUSSION

3.1. Solid State Fermentation (SSF) and Crude β-Galactosidase Production

The leading target of this study was to evaluate the hydrolytic activity of *A. awamori* on CW to obtain a nutrient-rich supplement deriving from lactose hydrolysis, that will substitute synthetic media in a following bioconversion process. Therefore, SSF optimization to enhance β-galactosidase production using WB as a single substrate was initially undertaken, based also on previous studies that have outlined that WB reinforced β-galactosidase production [36]. This has been attributed to the appropriate ratio of hemicellulose to sugars, that is defined as a stimulus factor for galactosidase production [37]. Figures 1 and 2 demonstrate the effect of initial moisture content, ranging from 60 to 75%, along with incubation time (1-5 days). Maximum production of β-galactosidase reached 148 U/g (db) at 70% of initial moisture after 70 h of fermentation. Earlier reports highlighted that increased moisture levels enhanced β-galactosidase yield in *A. tubingensis* [29]. The latter usually associates with the fact that moisture crucially affects nutrient solubility within the substrate [38]. As it can be easily observed, the production rate exhibits an increasing trend (Figure 1), during the first three days of fermentation followed by a decrease after approximately 70 h (three days) of incubation. Similar results were also obtained in studies using *A. tubigensis* and *A. awamori*, respectively [14,29], whereby prolonged fermentation times entailed higher β-galactosidase activities. For instance, Nizamuddin et al. [30] demonstrated optimum β-galactosidase production by *A. oryzae* after seven days of incubation, Raol et al. [29] found maximum enzyme activity by *A. tubingensis* at seven days, whereas Cardoso et al. [13] performed SSF for six days to produce β-galactosidase production by *A. lacticoffeatus*.

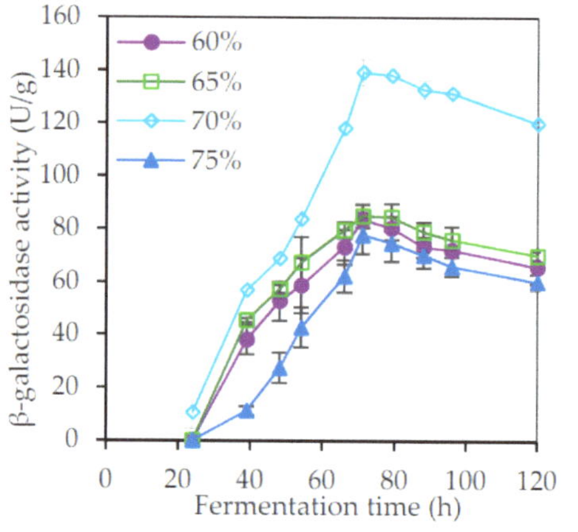

Figure 1. Effect of solid state fermentation (SSF) time in crude β-galactosidase production by A. awamori, at different initial moisture contents.

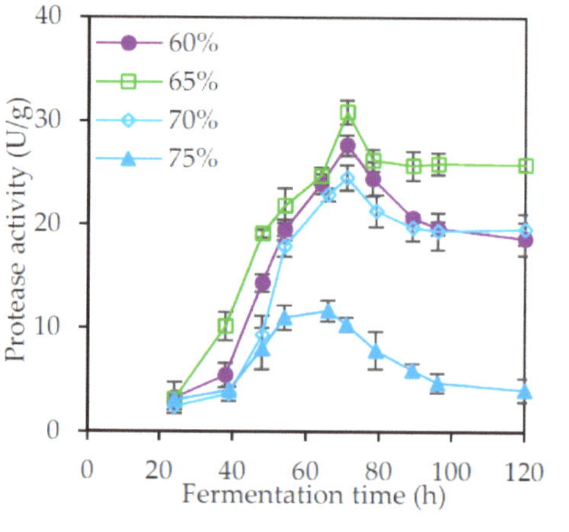

Figure 2. Effect of solid state fermentation (SSF) time in crude protease production by A. awamori, at different initial moisture contents.

Recently, Vidya et al. [14] studied α- and β-galactosidase production from *A. awamori* (MTCC 548), whereby the purified enzyme exhibited 25.5-176.5 U/mg of activity, respectively. On top of that, the authors reported β-xylosidase and β-glucosidase activities, tuggesting the ample substrate specificity. Several preceding studies had also suggested multi-enzyme produdtion by *A. awamori* including glucoamylase and protease [16,39]. Therefore, proteolytic activity was also undertaken (Figure 2), reaching the highest value after 70 h of fermentation (30.9 U/g). Similarly, Wang et al. [40] reported profease activities up to 40 U/g, (db) after 120 h employing similar SSF conditions. Evidently, ic could be speculated that the addition of CW in SSF cultures, induced the secretion of β-galactosidases considering that fungal strains tend to adapt in the environmental niches and develop mechanisms for the production of specific enzymes. Moreover, this could be attributed to the low pH during fermentation, that could potentially enhance *Aspergillus* β-galactosidase production [11,13,41] Ultimately, SSF time for crude [5-galactosidase and pro tease was standardized at 70 h to obtain maximal activities, that would be implemented in subs equent hydro lytic react ions of OW.

3.2. Cheese Whey Hydrolysis Study

CW hydrolysis was performed using the crude enzymatic extracts obtained from SSF cultures. Figure 3a illustrates the results obtained from different hydrolysis temperatures, whereby it can be observed that 60-65°C was the optimum hydrolysis temperature of *A. awamori.* Ultimately', at the end of -the bioprocess, crude enzymes hydrolyzed >90% of the initial whey' lactose (Figure 3a). On the other hand, the optimum proteolyt'c activity was observed at 55°C, as it h as been earlier indicated by Tsakona et al. [1 6]. More particular, as displayed in Figure 3b, FAN production increased along with ths increase in temperature up to 55°C, followed by a grad ua l reduction with further temperature increments (Figure 3b). Based on our results, significant differences ($p < 0.05$) were observed on the performed hydrolyses, at almost all evaluated temperatures. Likewise, no significant differences ($p > 0.05$) were observed on hydrolysis experiments carried out at 60 and 65°C. Previous studies have also demonstrated processing of cheese whey via the implementation of microbial β-galactosidase to generate value-added products [4:2,43]. Generally, temperatures; ranging between 50 and 60°C and acidic pH values (3.5-41.5) have been reported a s the optimal condi tions for fungal β-ga lactosidase activity [13]. Additionally, Silverio et at. [44] recenlly studied β-galactosidase production in several Aspergillus species, aiming to synthesize potential prebiotics, whereby tan increased enzyme activity in the range of 50-60°C was noted. The current observation highlights the significant potential of the enzymes, since thermal stability is of imperative practical use for diverce bioprocesses, preventing various contaminations [45,46]. Furthermore, the results obtained postulate that the enzyme is more accessible during the first hours of hydrolysis. More specifically, a higher hydrolysis rate during the first 1d h entailed 30-50°% of lactose hydrolysis, fol- Cowed by a decreased rate at prolonged incubation time. Several studies also coincide with such fundings where hydrolysis products decreased or even restricted lactose hydrolysis reaction [41,47,48]. Indeed, it has been previously established that at high galactose concentrations, β -galactosidase activity is imp aired since the conformational mo dification of the enzyme's active site reduces the affinity for its substrate [49,50]. Moreover, galactose could also act as a competitive inhibitor of β-galactosidase via the formation of galactosyl-enzyme intermediate products [51].

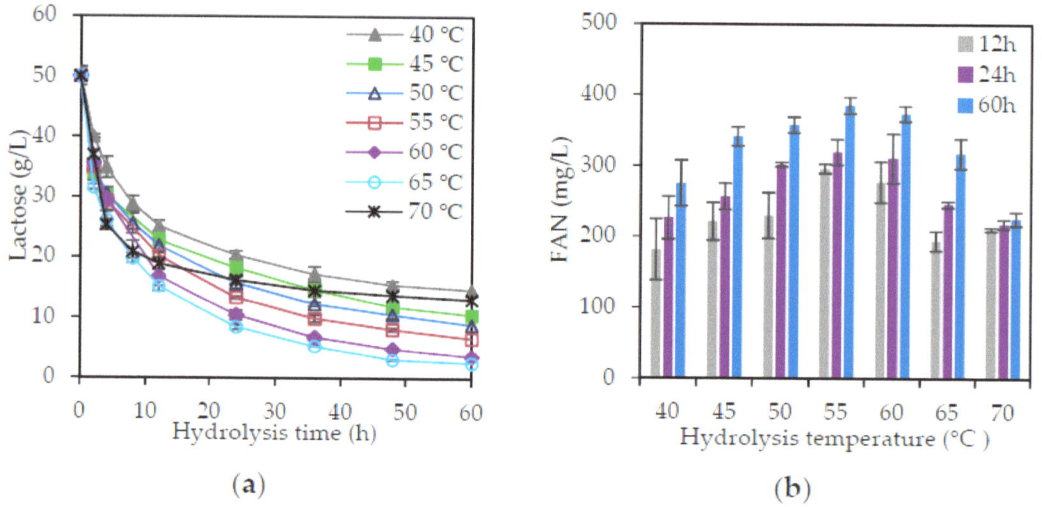

Figure 3. Effect of temperature on cheese whey (CW) hydrolysis using crude β-galactosidase and proteases. Kinetics of (a) lactose hydrolysis and (b) free amino nitrogen (FAN) production.

In an effort to further optimize whey hydrolysis, trials were also performed to evaluate the effect of different initial enzymatic activities on lactose breakdown and FAN production. Initial enzymatic activities of 7.5,11,15 U/mL were employed, and the results are illustrated in Figure 4. Figure 4a presents the kinetic profile of lactose hydrolysis, whereas Figure 4b presents FAN production in specific timepoints. Evidently, the use of 15 U/mL resulted in accelerated rates and complete lactose hydrolysis at 36 h and the production of 583.13 mg/L FAN. On the other hand, initial enzymatic activities of 7.5 and 11 U/mL yielded 87 and 93% of hydrolysis, respectively, at the same time point, providing lower productivities. Even though the degree of hydrolysis seems to follow a dose-dependent trend, apparently much higher concentrations do not significantly alter the hydrolysis efficiency, although complete hydrolysis is performed significantly earlier at higher initial enzymatic activities. Worth noting, FAN production increased almost *two-fold* at higher inifial enzymatic activities. Rosolen et al. [52] also presented similar efficiency levels on whey lactose hydrolysis by *A. oryzae,* regardless ot the enzyme concentrations used (3, 6 and 9 U/mL). Thi s observation probably also indicates the satu ration of lac tose at high β-galactosidase concentrations [53]. Thus, as in previous studies, our results imply that CW lactose hydrolysis is not strictly proportional with enzyme concentration [7,54]. However, complete hydrolysis was performed in almost half the time, using 15 U/mL, compared with the case of 7.5 U/mL. Nonetheless, in the event that scale up should be considered, lactose hydrolysis efficiency and FAN production should coincide with the feasibility of the process to highlight the most favorable operating condifions, which will be designated by the end target products.

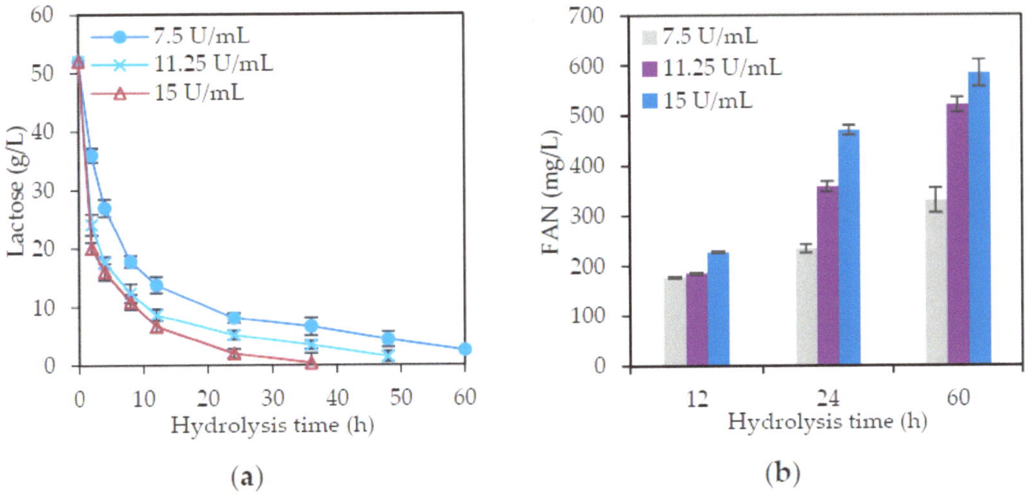

Figure 4. Effect of different initial enzymatic activity of crude β-galactosidase on cheese whey (CW) hydrolysis. Kinetics of (a) lactose hydrolysis and (b) free amino nitrogen (FAN) production.

3.3. Bacterial Cellulose Production

CW constitutes a renewable, zero-cost substrate suitable for microbial fermentation, mostly requiring minimal pretreatment. However, often lactose does not undergo fermentation by several microorganisms including acetic acid bacteria. Previous reports demonstrated low BC-production from unhydrolyzed CW, thus hindering further implementation. Thus, pretreatment is often essential to overcome such limitations. Besides this, only limited studies have evaluated CW for BC production [55,56]. Based on similar literature reports, BC production is species and strain dependent. Evidently, the results of the current work confirmed the ability of *A. xylinum* to use CW hydrolysate under three different fermentation schemes. Hydrolysates derived from 11.25 (Hydrolysate A) and 7.5 U/mL (Hydrolysate B) crude β-galactosidase, respectively, were used to evaluate BC production, and the results are prefsen ted in Table 1. Different nitrogen concentrations were used , based on previous observations where elevated levels of nitrogen content induced cell proliferation at the expense of BC production [32]. As it can be seen in Figure 5a, the consumption of 13.91 g/L of glucose and 224.79 mg/L of FAN resulted in the production of 7.05 g/L BC (Hydrolysate A). Hydrolysate B followed a similar trend with respect to glucose consumption rate. The consumption of 13.41 g/L of glucose and 132.64 mg/L FAN, resulted in 5.78 g/L of BC production (Figure 5b). However, it is worth noting that in both experiments a considerable amount of sugars remained unfermented by *A. xylinum*. Therefore, a third treatment was deployed using diluted hydrolysate (hydrolysate C) (1:1 CW:H_2O) in order to evaluate the BC production yield on approximately 25 g/L total sugar content and 260 mg/L FAN concentration (Figure 5c). In fact, in these experimental conditions, almost complete glucose consumption was attained along with the consumption of 114.94 mg/L FAN, achieving a final BC concentration of 5.59 g/L.

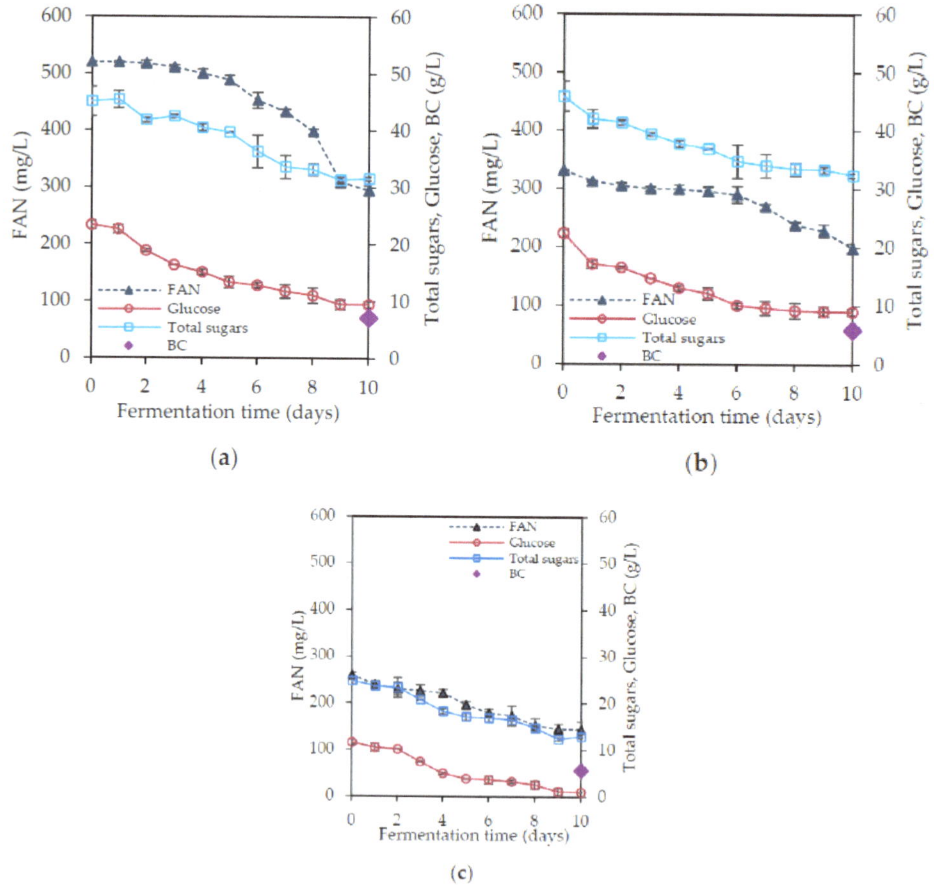

Figure 5. Bacterial cellulose (BC) production and kinetics of sugars and free amino nitrogen (FAN) consumption, using different cheese whey (CW) hydrolysates. (a) Hydrolysate A; (b) Hydrolysate B; (c) Hydrolysate C.

Table 1. Experimental schemes of cheese whey and cheese whey hydrolysates fermentation by A. xylinum.

Fermentation Media	Initial Total Sugars (g/L)	Initial Glucose (g/L)	Residual Glucose (g/L)	Initial FAN (mg/L)	FAN Consumption (mg/L)	BC Production * (g/L)	BC Productivity (g/L/d)
Hydrolysate A	45.04 ± 2.60	23.31 ± 0.77	9.40 ± 0.24	520.05 ± 0.34	224.79 ± 8.39	7.05 ± 0.14 [A]	0.71
Hydrolysate B	45.80 ± 0.77	22.35 ± 0.46	8.93 ± 0.19	331.36 ± 12.96	132.64 ± 5.21	5.78 ± 0.35 [A_B]	0.58
Hydrolysate C	24.68 ± 0.69	11.48 ± 0.48	1.00 + 0.08	259.25 ± 5.70	114.94 ± 4.73	5.59 ± 0.22 [B]	0.56
Cheese whey A	50.00 ± 1.22	2.51 ± 0.23	-	56.00 ± 3.35	19.49 ± 2.78	0.58 ± 0.01 [a]	0.06
Cheese whey B	24.45 ± 1.03	1.28 ± 0.10	-	22.98 ± 5.70	22.98 ± 0.00	0.71 ± 0.05 [a_b]	0.07
Cheese whey C	24.29 ± 1.18	1.39 ± 0.14	-	250.00 ± 10.06	81.33 ± 4.65	1.07 ± 0.09 [b]	0.11

** Different letters (A, B, a, b) within each group (hydrolysates and cheese whey) indicate significant differences ($p < 0.05$). FAN: free amino nitrogen; BC: bacterial cellulose.*

The above results are in accordance with other studies describing the utilization of several monosaccharides and disaccharides as carbon sources to generate BC by various *Acetobacter* spp. strains. Semjonovs et al. [55] reported a high BC yield with CW hydrolysate (20 g/L reducing sugars) as the sole carbon source using the strain *Komagataeibacter rhaeticus* P 1463. Additionally, Salari et al. [57] recently referred to a BC production of 3.5 g/L within 14 days of fermentation by *Gluconacetobacter xylinum* PTCC 1734 in static cultures, using an equimolar glucose/galactose mixture from hydrolyzed CW. In all the conducted experiments, a considerable increase in BC production was observed when compared with the results obtained by media with lower amount of FAN concentration. On the other hand, BC production levels by unhydrolyzed whey were quite close to those previously reported [58]. More specifically, as it is presented in Table 1, *A. xylinum* consumed 19.49 mg/L of FAN, producing 0.58 g/L of BC in unhydrolyzed CW (cheese whey A) (Table 1). Likewise, significant differences ($p < 0.05$) on BC production were observed, when different fermentation media were applied, whereas significantly higher concentrations were produced using all types of CW hydrolysates, compared to sole CW (Table 1).

Recently, Kumar et al. [59] demonstrated the production of 1.4 g/L of BC under static culture conditions in whey medium by *Acetobacter pasteurianus*. The formation of BC in these cases is mostly due to the presence of several other compounds such as the residual carbon present in the initial inocula. In addition to this, higher

BC production was observed using diluted CW, which is consistent which similar studies [60]. In specific, *A. xylinum* produced 0.71 g/L and 1.07 g/L BC, when diluted CW (cheese whey B) and diluted CW supplemented with yeast medium (cheese whey C) were, respectively, applied (Table 1). In general, lactose as a sole carbon source is reported as a weak substrate for BC production leading to 0.04-0.07 g/L [39,61], while BC production by unhydrolyzed CW is recorded slightly higher ranging from 0.15 to 0.78 g/L [58,60]. Our results (using unhydrolyzed CW) are in agreement with those previously reported, whereas BC production was significant higher using CW hydrolysates. Overall, in this study, high production of BC was achieved using CW hydrolysates compared even to BC production using conventional synthetic HS medium. These findings are exceptionally promising pointing out potential for a cost-effective bioprocess.

3.4. Technological Consideration of the Study

The principal target of this study was the development of a holistic exploitation approach for cheese whey, that will engage with sustainability and generate value-added products via the aligned food waste reduction and by-product streams treatment, as cornerstones of the circular economy concept. Likewise, an efficient fungal-based, two-stage bioprocess was employed to produce a nutrient rich feedstock for subsequent upstream bacterial bioconversions. Cost effective production of crude enzymes, without further purification steps was undertaken using food industry by-products, specifically cheese whey. The significant hydrolytic activity of this novel biocatalyst was demonstrated, leading to the formulation of a suitable feedstock for bacterial cellulose (BC) production. The results of our study confer an insight for the fermentative production of BC using whey lactose hydrolysates, which effectively sustained the nutrient requirements of *A. xylinum,* displaying high production yields. Evidently, enhanced feasibility could be established through the development of suitable bioprocesses to mediate BC production costs via the replacement of conventional fermentation media. The consolidated bioprocess presented hereof is currently further extended within the concept of holistic refining of cheese whey streams (lactose and protein). In particular, in our forthcoming research, novel probiotic starter cultures will be developed, and BC generated in this study will be implemented as a carrier for lactic acid bacteria starter to be reintroduced into dairy products, thereby closing the loop. Ultimately, the combined proposed approach conforms to the pillars of circular bioeconomy, encompassing environmentally benign processes, zero waste generation in parallel with novel food product development and potential health benefits.

4. CONCLUSIONS

The results of the present study indicate the successful development of a novel cheese whey valorization approach within the concept of circular bioeconomy. More specifically, a two-stage operation was established to generate crude enzymatic consortia via fungal solid state fermentations with A. *awamori.* Fermentation conditions were optimized and a novel biocatalyst was effectively secreted, and subsequently implemented to hydrolyze whey lactose formulating a nutrient substrate for fermentative bioconversions. BC production was conceptualized as a transitional compound for subsequent functional food formulations, along with the protein fraction to complement sustainability and circularity of the process.

REFERENCES

1. Prazeres, A.R.; Carvalho, F.; Rivas, J. Cheese whey management: A review. *J. Environ. Manag.* **2012**,110, 48-68. [CrossRef]
2. Yadav, J.S.S.; Yan, S.; Pilli, S.; Kumar, L.; Tyagi, R.D.; Surampalli, R.Y. Cheese whey: A potential resource to transform into bioprotein, functional/nutritional proteins and bioactive peptides. *Biotechnol. Adv.* **2015**, 33, 756-774. [CrossRef] [PubMed]
3. Okamoto, K.; Nakagawa, S.; Kanawaku, R.; Kawamura, S. Ethanol production from cheese whey and expired milk by the brown rot fungus *Neolentinus lepideus. Fermentation* **2019**, 5, 49. [CrossRef]
4. Mehri, D.; Perendeci, N.A.; Goksungur, Y. Utilization of whey for red pigment production by *Monascus purpureus* in submerged fermentation. *Fermentation* **2021**, 7, 75. [CrossRef]
5. Marcus, J.F.; DeMarsh, T.A.; Alcaine, S.D. Upcycling of whey permeate through yeast- and mold-driven fermentations under anoxic and oxic conditions. *Fermentation* **2021**, 7,16. [CrossRef]
6. Costa, S.; Summa, D.; Semeraro, B.; Zappaterra, F.; Rugiero, I.; Tamburini, E. Fermentation as a strategy for bio-transforming waste into resources: Lactic acid production from agri-food residues. *Fermentation* **2021**, 7, 3. [CrossRef]
7. Ghosh, B.C.; Prasad, L.N.; Saha, N.P. Enzymatic hydrolysis of whey and its analysis. *J. Food Sci. Technol.* **2017**, 54,1476-1483. [CrossRef]
8. Liu, P.; Xie, J.; Liu, J.; Ouyang, J. A novel thermostable |3 -galactosidase from *Bacillus coagulans* with excellent hydrolysis ability for lactose in whey. *Int. J. Dairy Sci.* **2019**,102, 9740-9748. [CrossRef] [PubMed]
9. Saqib, S.; Akram, A.; Halim, S.A.; Tassaduq, R. Sources of |3-galactosidase and its applications in food industry. *3 Biotech* **2017**, 7, 79. [CrossRef] [PubMed]

10. Deng, Y.; Xu, M.; Ji, D.; Agyei, D. Optimization of ^-galactosidase production by batch cultures of *Lactobacillus leichmannii* 313 (ATCC 7830™). *Fermentation* **2020**, 6, 27. [CrossRef]

11. Martarello, R.D.; Cunha, L.; Cardoso, S.L.; de Freitas, M.M.; Silveira, D.; Fonseca-Bazzo, Y.M.; Homem-de-Mello, M.; Filho, E.X.F.; Magalhaes, P.O. Optimization and partial purification of beta-galactosidase production by *Aspergillus niger* isolated from Brazilian soils using soybean residue. *AMB Express* **2019**, 9, 81. [CrossRef]

12. FDA. Partial List of Microorganisms and Microbial-Derived Ingredients Used in Foods. 2015. Available online: http://www.fda.gov/Food/IngredientsPackagingLabeling/GRAS/MicroorganismsMicrobialDerivedIngredients/default.htm (accessed on 20 June 2021).

13. Cardoso, B.B.; Silverio, S.C.C.; Abrunhosa, L.; Teixeira, J.A.; Rodrigues, L.R. |3-galactosidase from *Aspergillus lacticoffeatus:* A promising biocatalyst for the synthesis of novel prebiotics. *Int. J. Food Microbiol.* **2017**,257, 67-74. [CrossRef] [PubMed]

14. Vidya, C.H.; Kumar, B.S.; Chinmayee, C.V.; Singh, S.A. Purification, characterization and specificity of a new GH family 35 galactosidase from *Aspergillus awamori. Int. J. Biol. Macromol.* **2020**,156, 885-895. [CrossRef]

15. Koutinas, A.A.; Arifeen, N.; Wang, R.; Webb, C. Cereal-based biorefinery development: Integrated enzyme production for cereal flour hydrolysis. *Biotechnol. Bioeng.* **2007**, 97, 61-72. [CrossRef]

16. Tsakona, S.; Kopsahelis, N.; Chatzifragkou, A.; Papanikolaou, S.; Kookos, I.K.; Koutinas, A.A. Formulation of fermentation media from flour-rich waste streams for microbial lipid production by *Lipomyces starkeyi. J. Biotechnol.* **2014**,189, 36-45. [CrossRef]

17. de Sousa Paredes, S.R.; de Barros, R.R.; Inoue, H.; Yano, S.; Bon, E.P. Production of xylanase, a-l-arabinofuranosidase, [3-xylosidase, and [3-glucosidase by *Aspergillus awamori* using the liquid stream from hot-compressed water treatment of sugarcane bagasse. *Biomass Conv. Bioref.* **2015**, 5, 299-307. [CrossRef]

18. Nishida, V.S.; de Oliveira, R.F.; Brugnari, T.; Correa, R.C.G.; Rosely, A.; Peralta, R.A.; Castoldi, R.; de Souza, C.G.; Bracht, A.; Peralta, R.M. Immobilization of *Aspergillus awamori* |3-glucosidase on commercial gelatin: An inexpensive and efficient process. *Int. J. Biol. Macromol.* **2018**,111,1206-1213. [CrossRef] [PubMed]

19. Lappa, I.K.; Papadaki, A.; Kachrimanidou, V.; Terpou, A.; Koulougliotis, D.; Eriotou, E.; Kopsahelis, N. Cheese whey processing: Integrated biorefinery concepts and emerging food applications. *Foods* **2019**, 8, 347. [CrossRef]

20. Azeredo, H.M.C.; Barud, H.; Farinas, C.S.; Vasconcellos, V.M.; Claro, A.M. Bacterial cellulose as a raw material for food and food packaging applications. *Front. Sustain. Food Syst.* **2019**, 3. [CrossRef]

21. Fijalkowski, K.; Peitler, D.; Rakoczy, R.; Zywicka, A. Survival of probiotic lactic acid bacteria immobilized in different forms of bacterial cellulose in simulated gastric juices and bile salt solution. *LWT Food Sci. Technol.* **2016**, 68, 322-328. [CrossRef]

22. Bayazidi, P.; Almasi, H.; Aslm, A.K. Immobilization of lysozyme on bacterial cellulose nanofibers: Characteristics, antimicrobial activity and morphological properties. *Int. J. Biol. Macromol.* **2018**,107, 2544-2551. [CrossRef] [PubMed]

23. Mohammadalinejhad, S.; Almasi, H.; Moradi, M. Immobilization of *Echium amoenum* anthocyanins into bacterial cellulose film: A novel colorimetric pH indicator for freshness/spoilage monitoring of shrimp. *Food Control* **2020**,113,107169. [CrossRef]

24. Cacicedo, M.L.; Castro, M.C.; Servetas, I.; Bosnea, L.; Boura, K.; Tsafrakidou, P.; Dima, A.; Terpou, A.; Koutinas, A.; Castro, G.R. Progress in bacterial cellulose matrices for biotechnological applications. *Bioresour. Technol.* **2016**, 213,172-180. [CrossRef]

25. Revin, V.; Liyaskina, E.; Nazarkina, M.; Bogatyreva, A.; Shchankin, M. Cost-effective production of bacterial cellulose using acidic food industry by-products. *Braz. J. Microbiol.* **2018**, 49,151-159. [CrossRef] [PubMed]

26. Abol-Fotouh, D.; Hassan, M.A.; Shokry, H.; Roig, A.; Azab, M.S.; Kashyout, A.B. Bacterial nanocellulose from agro-industrial wastes: Low-cost and enhanced production by *Komagataeibacter saccharivorans* MD1. *Sci. Rep.* **2020**,10, 3491. [CrossRef]

27. Son, H.J.; Kim, H.G.; Kim, K.K.; Kim, H.S.; Kim, Y.G.; Lee, S.J. Increased production of bacterial cellulose by *Acetobacter* sp. V6 in synthetic media under shaking culture conditions. *Bioresour. Technol.* **2003**, 86, 215-219. [CrossRef]

28. Paximada, P.; Tsouko, E.; Kopsahelis, N.; Koutinas, A.A.; Mandala, I. Bacterial cellulose as stabilizer of o/w emulsions. *Food Hydrocoll.* **2016**, 53, 225-232. [CrossRef]

29. Raol, G.G.; Raol, B.V.; Prajapati, V.S.; Bhavsar, N.H. Utilization of agro-industrial waste for ^-galactosidase production under solid state fermentation using halotolerant *Aspergillus tubingensis* GR1 isolate. *3 Biotech* **2015**, 5, 411-421. [CrossRef]

30. Nizamuddin, S.; Sridevi, A.; Narasimha, G. Production of |3-galactosidase by *Aspergillus oryzae* in solid-state fermentation. *Afr. J. Biotech.* **2008**, 7,1096-1100.

31. Zywicka, A.; Wenelska, K.; Junka, A.; Chodaczek, G.; Szymczyk, P.; Fijalkowski, K. Immobilization pattern of morphologically different microorganisms on bacterial cellulose membranes. *World J. Microbiol. Biotechnol.* **2019**, 35,11. [CrossRef]

32. Tsouko, E.; Kourmentza, C.; Ladakis, D.; Kopsahelis, N.; Mandala, I.; Papanikolaou, S.; Paloukis, F.; Alves, V.; Koutinas, A. Bacterial cellulose production from industrial waste and by-product streams. *Int. J. Mol. Sci.* **2015**,16,14832-14849. [CrossRef] [PubMed]

33. Papadaki, A.; Papapostolou, H.; Alexandri, M.; Kopsahelis, N.; Papanikolaou, S.; de Castro, A.M.; Freire, D.M.G.; Koutinas, A.A. Fumaric acid production using renewable resources from biodiesel and cane sugar production processes. *Environ. Sci. Pollut. Res.* **2018**, *25*, 35960-35970. [CrossRef] [PubMed]

34. Kachrimanidou, V.; Kopsahelis, N.; Chatzifragkou, A.; Papanikolaou, S.; Yanniotis, S.; Kookos, I.; Koutinas, A.A. Utilisation of by-products from sunflower-based biodiesel production processes for the production of fermentation feedstock. Waste *Biomass Valorization* **2013**, 4, 529-537. [CrossRef]

35. Lie, S. The EBC-ninhydrin method for determination of free alpha amino nitrogen. *J. Inst. Brew.* **1973**, 79, 37-41. [CrossRef]

36. Hatzinikolaou, D.G.; Katsifas, E.; Mamma, D.; Karagouni, A.D.; Christakopoulos, P.; Kekos, D. Modeling of the simultaneous hydrolysis-ultrafiltration of whey permeate by a thermostable |3-galactosidase from *Aspergillus niger. Biochem. Eng. J.* **2005**, *24*, 161-172. [CrossRef]

37. Merali, Z.; Collins, S.R.A.; Elliston, A.; Wilson, D.R.; Kasper, A.; Waldron, K.W. Characterization of cell wall components of wheat bran following hydrothermal pretreatment and fractionation. *Biotechnol. Biofuels* **2015**, 8, 23. [CrossRef]

38. Shah, A.R.; Madamwar, D. Xylanase production under solid-state fermentation and its characterization by an isolated strain of *Aspergillus foetidus* in India. *World J. Microbiol. Biotechnol.* **2005**,21, 233-243. [CrossRef]

39. Wang, J.; Tavakoli, J.; Tang, Y. Bacterial cellulose production, properties and applications with different culture methods—A review. *Carbohydr. Polym.* **2019**, 219,63-76. [CrossRef]

40. Wang, R.; Godoy, L.C.; Shaarani, S.M.; Melikoglu, M.; Koutinas, A.; Webb, C. Improving wheat flour hydrolysis by an enzyme mixture from solid state fungal fermentation. *Enzyme Microb. Technol.* **2009**, 44, 223-228. [CrossRef]

41. Ansari, S.A.; Husain, Q. Lactose hydrolysis from milk/whey in batch and continuous processes by concanavalin A-Celite 545 immobilized *Aspergillus oryzae* P galactosidase. *Food Bioprod. Process.* **2012**, 90, 351-359. [CrossRef]

42. Geiger, B.; Nguyen, H.M.; Wenig, S.; Nguyen, H.A.; Lorenz, C.; Kittl, R.; Mathiesen, G.; Eijsink, V.G.H.; Haltrich, D.; Nguyen, T.H. From by-product to valuable components: Efficient enzymatic conversion of lactose in whey using |3-galactosidase from *Streptococcus thermophilus. Biochem. Eng. J.* **2016**,116, 45-53. [CrossRef] [PubMed]

43. Sampaio, F.C.; de Faria, J.T.; da Silva, M.F.; de Souza Oliveira, R.P.; Converti, A. Cheese whey permeate fermentation by *Kluyveromyces lactis:* A combined approach to wastewater treatment and bioethanol production. *Environmen. Technol.* **2019**,1-9. [CrossRef] [PubMed]

44. Silverio, S.C.; Macedo, E.A.; Teixeira, J.A.; Rodrigues, L.R. New ^-galactosidase producers with potential for prebiotic synthesis. *Bioresour. Technol.* **2018**,250,131-139. [CrossRef] [PubMed]

45. Turner, P.; Mamo, G.; Karlsson, E.N. Potential and utilization of thermophiles and thermostable enzymes in biorefining. *Microb. Cell Factories* **2007**, 6, 9. [CrossRef] [PubMed]

46. Rigoldi, F.; Donini, S.; Redaelli, A.; Parisini, E.; Gautieri, A. Review: Engineering of thermostable enzymes for industrial applications. *APLBioeng.* **2018**,2, 011501. [CrossRef]

47. Park, A.R.; Oh, D.K. Effects of galactose and glucose on the hydrolysis reaction of a thermostable ^-galactosidase from *Caldicellu- losiruptor saccharolyticus. Appl. Microbiol. Biotechnol.* **2010**, 85,1427-1435. [CrossRef]

48. Gosling, A.; Stevens, G.W.; Barber, A.R.; Kentish, S.E.; Gras, S.L. Recent advances refining galactooligosaccharide production from lactose. *Food Chem.* **2010**,121, 307-318. [CrossRef]

49. Kim, C.S.; Ji, E.S.; Oh, D.K. A new kinetic model of recombinant |3-galactosidase from *Kluyveromyces lactis* for both hydrolysis and transgalactosylation reactions. *Biochem. Biophys. Res. Commun.* **2004**, 316, 738-743. [CrossRef]

50. Neri, D.F.; Balcao, V.M.; Carneiro-da-Cunha, M.G.; Carvalho, L.B.; Teixeira, J.A. Immobilization of |3-galactosidase from *Kluyveromyces lactis* onto a polysiloxane-polyvinyl alcohol magnetic (mPOS-PVA) composite for lactose hydrolysis. *Catal. Commun.* **2008**, 9, 2334-2339. [CrossRef]

51. Ansari, S.A.; Husain, Q. Lactose hydrolysis by P galactosidase immobilized on concanavalin A-cellulose in batch and continuous mode. *J. Mol. Catal. B Enzym Enzymatic* **2010**, 63, 68-74. [CrossRef]

52. Dutra Rosolen, M.; Gennari, A.; Volpato, G.; Volken de Souza, C.F. Lactose hydrolysis in milk and dairy whey using microbial B-galactosidases. *Enzyme Res.* **2015**, 806240. [CrossRef]

53. Akgul, F.B.; Demirhan, E.; Ozbek, B. A Modelling study on skimmed milk lactose hydrolysis and B-galactosidase stability using three reactor types. *Int. J. Dairy Technol.* **2012**, 65, 217-231. [CrossRef]

54. Horner, T.W.; Dunn, M.L.; Eggett, D.L.; Ogden, L.V. B-galactosidase activity of commercial lactase samples in raw and pasteurized milk at refrigerated temperatures. *J. Dairy Sci.* **2011**, 94, 3242-3249. [CrossRef]

55. Semjonovs, P.; Ruklisha, M.; Paegle, L.; Saka, M.; Treimane, R.; Skute, M.; Rozenberga, L.; Vikele, L.; Sabovics, M.; Cleenwerck, I. Cellulose synthesis by *Komagataeibacter rhaeticus* strain P 1463 isolated from Kombucha. *Appl. Microbiol. Biotechnol.* **2017**, *101*, 1003-1012. [CrossRef] [PubMed]

56. Bekatorou, A.; Plioni, I.; Sparou, K.; Maroutsiou, R.; Tsafrakidou, P.; Petsi, T.; Kordouli, E. Bacterial cellulose production using the Corinthian currant finishing Side-Stream and Cheese Whey: Process Optimization and Textural Characterization. *Foods* **2019**, *8*, 193. [CrossRef]

57. Salari, M.; Khiabani, M.S.; Mokarram, R.R.; Ghanbarzadeh, B.; Kafil, H.S. Preparation and characterization of cellulose nanocrystals from bacterial cellulose produced in sugar beet molasses and cheese whey media. *Int. J. Biol. Macromol.* **2019**,122, 280-288. [CrossRef] [PubMed]

58. Battad-Bernardo, E.; McCrindle, S.L.; Couperwhite, I.; Neilan, B.A. Insertion of an *E. coli lacZ* gene in *Acetobacter xylinus* for the production of cellulose in whey. *FEMS Microbiol. Lett.* **2004**, 231, 253-260. [CrossRef]

59. Kumar, V.; Sharma, D.K.; Sandhu, P.P.; Jadaun, J.; Sangwan, R.S.; Yadav, S.K. Sustainable process for the production of cellulose by an *Acetobacter pasteurianus* RSV-4 (MTCC 25117) on whey medium. *Cellulose* **2021**, 28,103-116. [CrossRef]

60. Carreira, P.; Mendes, J.A.; Trovatti, E.; Serafim, L.S.; Freire, C.S.; Silvestre, A.J.; Neto, C.P. Utilization of residues from agro-forest industries in the production of high value bacterial cellulose. *Bioresour. Technol.* **2011**,102, 7354-7360. [CrossRef]

61. Nguyen, V.T.; Flanagan, B.; Gidley, M.J. Characterization of cellulose production by a *Gluconacetobacter xylinus* strain from Kombucha. *Curr. Microbiol.* **2008**, 57, 449. [CrossRef]

CHAPTER

11

Concentrated Buffalo Whey as Substrate for Probiotic Cultures and as Source of Bioactive Ingredients: A Local Circular Economy Approach towards Reuse of Wastewaters

1. INTRODUCTION

Whey is the main and most polluting by-product obtained from cheese manufacturing processes due to its organic load consisting of lactose, lactic acid, proteins and salts. However, the substantial production of whey worldwide, estimated to be around 180-190 x 106 ton/year [1], and the consideration that 1 or 2 kg of cheese yields 8 to 9 kg of whey [2] is fostering its valorization. In fact, the discovery of its potential use as a functional food with nutritional applications is transforming it from a waste [3] into an added value product. Numerous studies have attributed several biological actions to these by-products which are important in the medical, pharmaceutical and food industries for their properties with potential benefits to human health [4]. The biological components of whey, including lactoferrin, beta-lactoglobulin, alpha-lactalbumin, glycomacropeptide, and immunoglobulins, demonstrate a range of immune-enhancing properties [5]. In addition, whey has the ability to act as an antioxidant, antihypertensive, antitumor, hypolipidemic and chelating agent. A number of clinical trials have successfully been performed using whey in the treatment of cancer, HIV and hepatitis B [6-8]. Moreover, today, whey is a popular dietary protein supplement that may provide antimicrobial activity, immune modulation, improved muscle strength and body composition, and prevent cardiovascular disease and osteoporosis [8,9]. The commercial success of whey proteins has led to the development of high-quality protein-based supplements manufactured as primary products, and not as a by-product, of cheese manufacturing. In fact, currently, proteins are processed with less aggressive treatments, for example under low temperatures and controlled pH, to avoid denaturing their native structures [5].

Due to the high lactose content in addition to proteins, whey, very often in combination with other medium components, was also used as a substrate for the cultivation of diverse microorganisms [10]. Whey-based media were in fact investigated for the production of value-added chemicals (e.g., succinic acid, lactic acid), antimicrobial peptides and probiotic biomasses [11-16].

The use of membrane technologies in order to obtain bioactive molecules from whey is a topic of growing interest. The main advantages are low energy requirement, no need for additives, separation efficiency and easy scale-up, and temperature control that prevents denaturation of recoverable added value products [5]. Several treatment technologies based on membranes were proposed. In particular, four basic types of membrane filtrations present potential applications for the dairy industry, i.e., microfiltration (MF), ultrafiltration (UF), nanofiltration (NF) and reverse osmosis (RO) [17,18]. The application of filtration processes to produce clean effluents thereby reducing wastewater and generating a purified stream (e.g., machinery washing, irrigation) transforms a difficult to manage and polluting effluent into a resource. Moreover, the use of membrane devices with different cut-offs allows the separation of compounds (proteins, peptides and lactose) present in whey into differentially enriched fractions that are, therefore, suitable for different applications.

Cheese manufacturing is one of the main industrial activities in the food sector present in the Campania region. The scope of this study was to promote a locally integrated bio refinery approach fully exploiting discarded whey. Therefore, the permeate and retentate of ultrafiltered whey, both provided by a local dairy factory, were evaluated in this work. The permeate was further processed to investigate a potential downstream approach to obtain reusable water with a low organic load. The retentate was evaluated to identify other potential biotechnological applications of whey from buffalo milk. In particular, it was investigated as the main substrate for the growth of *Lactobacillus fermentum,* a probiotic with several potential biomedical usages [19]. Moreover, it was also assessed for the presence of molecules active on tissue repair induction by using wound healing assays on mammalian cells.

2. MATERIALS AND METHODS

2.1. Materials

The buffalo whey used in this work was provided by the dairy factory "La Perla del Mediterraneo" (Battipaglia, Italy) in the framework of research and development processes in collaboration with the Dept. of Experimental Medicine, Bioteknet and the manufacturing facility. The company performed ultrafiltration on polyethersulphone 20 kDa spiral membranes (filtering surface 40 m^2) and provided (i) a sample (15–20 L) of the retentate concentrated fraction that was spray-dried in the framework of this research and used as substrate for the growth of L. fermentum and for wound healing assays (ii) a permeate that was nanofiltered and characterized.

L. fermentum was isolated from buffalo milk (data not shown). All salts and medium components for bottle and fermentation experiments were supplied by Sigma-Aldrich (St. Louis, MO, USA). Yeast extract was furnished by Organotechnie (La Corneuve, France), while sulphuric acid was purchased by Biochem s.r.l. (Turin, Italy). Dulbecco's Modified Eagle Medium (DMEM), Fetal bovine serum(FBS), penicillin–streptomycin, Phosphate Buffer Solution, (PBS) and Trypsin are provided by Gibco Invitrogen (Milan, Italy). HaCaT cells, a spontaneously transformed non-tumorigenic human keratinocytes cell line, were provided by the Zooprophylactic Institute (Brescia, Italy).

2.2. Downstream Process

NF processes were performed using a polyethersulfone spiral membrane with a nominal cut-off of 150-200 Da, respectively, with a total filtering area of 0.3 m^2 (Fluxa Filtri, Milano, Italy). The system used for the membrane process was a UF-NF system equipped with a 10 L volume steel tank, pressure gauges on the inlet and retentate lines, and a thermostatic bath to keep the temperature constant (Idea 3 Engineering, Lessona, Italy). The whey fraction provided by the dairy factory had been ultrafiltered in site on 20 kDa polyethersulfone membranes with a filtering surface of about 40-50 m^2 and flux of 2500-3000 L/h. The concentration factor reported was 13-14 fold, on the basis of 10,000 L used at the inlet for each treatment. In this study, 10 L of the permeate of the above mentioned UF on 20 KDa were nanofiltered and concentrated. All recovered fractions were characterized by analyzing lactate, lactic acid and total proteins as described in the following paragraph.

2.3. Spray Drying

Spraying was carried out on the UF retentate sample (UF_Ret20) provided by dairy company. The spray drier used was a Mobile MinorTM (GEA Process Engineering, Dani- marca). Samples were loaded into the specific chamber by means of a peristaltic pump with a flow rate of 2-2.5 Kg/h, the internal temperature of the chamber (in which the sample remained in contact with the gas for drying for about 10 s) was of about 160-170°C, the outlet temperature 80-85°C; compressed air was used as the drying gas. One g of spray- dried sample indicated as UF_Ret20Pow was dissolved in 100 mL of sterile bi-distilled water and characterized to determine the protein, sugars, lactic acid and insoluble solid content. Water content was determined by drying 100 mg of powder on a thermobalance (Mettler Toledo HR 83 Halogen). The sample was heated at 105°C and temperature was maintained for 6 h until a constant weight was recorded.

For ashes determination, 10 g of sample were placed in a tared crucible. The crucible was placed in a muffle furnace for 18 h at about 550°C. When the temperature dropped below 250°C the sample was quickly transferred to a desiccator until cooled before weighing. Calculation for dry ash content [20]:

% ash (dry basis) = (wt after ashing — tare wt of crucible)/(original sample wt x dry matter coefficient) x 100.

2.4. Analytical Methods

The various fractions and samples were analyzed for lactose, galactose, glucose, lactic acid, acetic acid and ethanol content using a UHPLC Dionex Ultimate 3000+ chromatograph (Thermofisher, Italy) equipped with a UV/Vis and RI detector. The standards and the samples, previously ultrafiltered on Centricon systems with 3 kDa cut-off, were injected in a Shodex sugar SH1011 300 x 8 mm, 6 g column according to the following operating conditions: isocratic elution with 0.1% sulfuric acid in water at a flow rate of 0.8 mL/min; temperature: 40°C; concentration range: 20-0.01 mg/mL; acquisition time 25 min [21].

2.5. Total Protein Quantification

The total protein content was obtained by analyzing the samples by UV/Vis spectrophotometry at 595 nm with a colorimetric method by using the Kit Protein assay Biorad (Bio-Rad Laboratories Inc., Hercules, CA, USA) [22] and the bovine serum albumin (BSA) as standard (Bio-Rad Laboratories Inc., Hercules, CA, USA). A Beckmann DU-800 spectrophotometer was used.

2.6. Bottle and Bioreactor Experiments

Frozen stocks of *L. fermentum* were prepared from cells growing exponentially on Mann, Rogosa and Sharpe (MRS) broth and stored at -80°C after the addition of a 20% *v*/*v* glycerol solution. Bottle experiments were performed in 100 mL screw-cap bottles with a working volume of 90 mL and incubated at 37°C and 150 rpm in a rotary shaker incubator (model Minitron, Infors, Bottmingen, Switzerland) for 24 h. Modified semi-define medium SGSL [23] contained the following per liter: 10 g yeast extract; 10 g soy peptone; 2 g Na3C6H5O7; 0.5 g L-ascorbic acid; 0.5 mL Tween80; 0.25 g MgSO4 • 7 H2O; 0.2 g NaCl; 0.05 g $MnSO_4 \cdot H_2O$. The concentrated and spray-dried UF retentate (UF_Ret20Pow) was used for growth experiments. In particular, the medium consisted of 1 ± 0.1, 2 ± 0.2 and 4 ± 0.2% UF_Ret20Pow reconstituted in SGSL medium. UF_Ret20Pow 4 ± 0.2% reconstituted in distilled water was also tested. Control experiments on SGSL supplemented with 30 g/L glucose or lactose were also performed. Samples were withdrawn at time 0, and after 8 and 24 h of growth to analyze optical density (600 nm) carbon sources consumption, and acid and ethanol production. All bottle experiments were performed in triplicate. Viability was evaluated by serially diluting the samples and plating on MRS-agar medium. Plates were incubated at 37°C for 36 h before counting viable cells. Each sample was analyzed in triplicate.

Bioreactor experiments were performed in a Biostat CT plus (Sartorius Stedim, Gottingen, Germany) bioreactor with a working volume of about 2.2 L. Temperature was controlled at 37°C, pH at 6.1 and agitation was fixed at 150 rpm. Fermentation medium containing UF_Ret20Pow reconstituted in water or in water supplemented with salts present in the SGSL medium and 2 g/L of yeast extract and soy peptone (1/5th of that present in SGLS) were used for bioreactor growth. Before each experiment, a concentrated stock solution of *L. fermentum* was inoculated in 0.25 L of SGSL medium at 37°C and 150 rpm and grown for 8 h. The pre-culture was then transferred to the bioreactor with a peristaltic pump (model 313 U, Watson-Marlow, England) to reach up to 10% (v/v) of the working volume inside the fermenter. Stirring was set to 150 rpm and air was sparged at a constant flow of 0.44 vvm. A constant pH of 6.1 was maintained by addition of NaOH 10 M and 30% *v/v* H_2SO_4 solutions. Experiments lasted up to 24 h. Samples were withdrawn during the experiments to analyze cell density (600 nm), cell viability, carbon source consumption, and acid and ethanol production. Viability was evaluated as previously described. Batch experiments on UF_Ret20Pow only were repeated 4 times whereas those on UF_Ret20Pow supplemented with salts and complex nitrogen sources were performed in duplicate.

2.7. Cell Cultures and Treatments

Human Keratynocytes cell lines (HaCat) were grown in Dulbecco's Modified Eagle Medium DMEM, supplemented with 10% (v/v) heat-inactivated FBS, penicillin 100 U/mL and streptomycin 100 gg/mL (Sigma Aldrich, MI, USA). The cells were grown on tissue culture plates (BD Bioscience-Falcon, San Jose, CA, USA), in a humidified atmosphere (95% air and 5% CO_2, v/v) at 37°C.

UF_Ret20Pow whey powder was dissolved in ultrapure (bi-distilled) water at a concentration of about 20 g/L (w/v) and sterilized by microfiltration on 0.22 gm filter devices. The sample was diluted with PBS to a final titer of 2 and 4 g/L before being added to the medium. Opportunely diluted PBS was used as control in the medium. During the wound healing assay, the concentration of FBS in the medium was reduced to 1% v/v to slow down the migration phenomena that could then be better evaluated.

2.8. In Vitro Scratch Test and Time-Lapse Video Microscopy (TLVM)

Briefly, HaCaT cells were seeded in 12-wells until complete cellular confluence was reached. Successively the confluent monolayer was scratched with a sterile tip (0 = 0.1 mm) and the diluted UF_Ret20 samples were added to the medium.

The in vitro cell migration was analyzed by a video microscopy time-lapse station (TLVM) (OKOLAB, Pozzuoli, Italy), assembled with an inverted microscope (AxioVision200, Zeiss, Germany), a CCD-gray-camera (ORCA ER, Hamamatsu Photonics, Hamamatsu Ci-tyo Japan) that records the images, a motorized stage? incubator that logs the position and maintains the in vitro condition of the cell culture (37 ° C, 5% CO2 in humidified air) and the custom-tailored software OKO-Vision 4.3 software that follows the overall process and allows image analysis. Thee TVLM tracks the wound repair process in real-time for 48-72 h, due to the selection and recording; of representative images of the experiments. TRe quantitative analycis of wound closure rates is calculated as [(Area t0 — Area t)/Area t0] x 100 directly by the software or alternatively by manual mode tracking for each image of the wound area over time. For each well a minimum of 5 fields of view was used for deriving the overall averaged curves of wound closures (%) as a function of the time, thus ensuring the statistical significance of the experiment.

2.9. Statistical Analysis

All data were analyzed by means of two-tailed non homoscedastic Student's t-test, and $p < 0.05$ was considered as statistically significant.

3. RESULTS AND DISCUSSION

In line with current circular economy approaches, a waste material from one of the production sectors of excellence of the Campania region, namely the dairy industry, was processed and recovered to obtain on the one side purified water for cleaning[1] use within the manufacturing facility, and on the other a spray-dried concentrate that could be tested as substrale for the growth of *L. fermentum* and as bioactive compound usoful in p romotfng tissue regeneration. An overview of the process approach is presented in Figure 1.

3.1. Clean Water Generation, Spray Drying of Concentrated Ultrafiltered Whey and Characterization of the Obtained Fractions

The first part of the project regarded the evaluation of a simple membrane-based process based on sequential UF and NF steps for the reduction of the organic load (BOD and COD) of whey that represents a substantial pollution source. The use of filtration is easily applicable in small/medium-sized companies that can not only reduce costs due to the disposal of numerous tons of discarded whey produced daily, but also reuse it for the separation of fractions enriched in lactose, proteins and peptides, and for the recovery of clean water. Results of the NF process are reported in Table 1a,b.

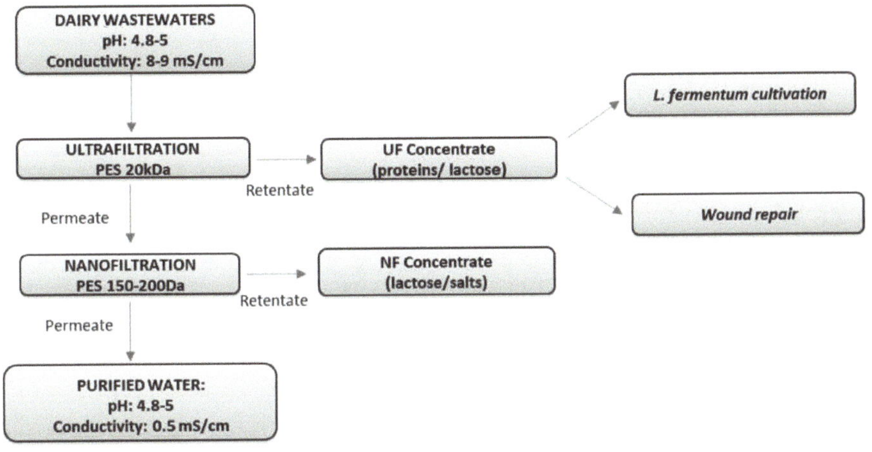

Figure 1. Overall downstaeam process of discarded whey and obtainment of added value fractions.

Table 1. Downstream processing of whey (a) Nanofiltration on 150-200 Da membranes of buffalo whey previously ultrafiltered on 20 kDa membranes. TMP, transmembrane pressure; LMH, L/m2-h; (b) Composition change of buffalo whey during ultrafiltration and nanofiltration. * Indicates the volume of permeate used for the nanofiltration experiments in the present study.

(a) Nanofiltration Parameters	Initial Vol. (L)	Final Vol. (L)	Concentration Factor	Initial Flux (LMH)	Final Flux (LMH)	Initial TMP (bar)	Final TMP (bar)
150-200 Da cut-off	10	1.25	8	33	28	10	12

(b) Sample	Protein (g/L)	Lactose (g/L)	Lactic Acid (g/L)	Vol (L)
Buffalo whey	1.10	29.8	5.0	10,000
UF ret	5.19	42.9	6.7	800
UF per	0.05	31.7	5.5	9200 (10 *)
NF ret	0.09	120.3	10.9	1.25
NF per	n.d.	0.03	2.3	8

Thirteen liters of UF_Ret20 were spray-dried in 6 h and resulted in the recovery of about 525 g of powder (Table 2). The residual water present in the sample resulted equal to 3.50 ± 0.50%. The powder was of thin and palpable grain size and contained prevalently lactose (Table 2). When suspended at 20 g/L, pH in bidistilled water was equal to 5.15 ± 0.05 and a conductivity of 3.18 ± 0.10 mS/cm was measured. Spray drying of the volume used in this work was affected by the void volume within the equipment, thus the yield was lower than 70%. However, the treatment of greater volumes on an industrial scale typically improves process yields, as the amount of solids lost in the spray dryer remains constant once a steady-state is achieved, and only the very fine powder that cannot be separated in the cyclone defines the actual process yield on solids.

Table 2. Characterization of spray-dried powder UF_Ret20Pow. Lac, lactose; Gal, galactose; Glu, glucose; LA, lactic acid. Ins. Solids, insoluble solids.

Sample	Protein w/v (%)	Lac w/v (%)	Gal w/v (%)	Glu w/v (%)	LA w/v (%)	Ins. Solids w/v (%)	Water w/w (%)	Ash w/w (%)
Dried powder	6	44	8	6	5	7	4	11

3.2. Evaluation of Ultrafiltered Spray-Dried Whey as Substrate for the Growth of L. fermentum

The use of whey proteins for the growth of biotechnologically interesting microorganisms is well established [16]. For example, the probiotic strain *L. casei* is a well-known case study for the production of biomass [11] and other antimicrobial products such as nisin and bacteriocins [12,13]. Since lactic acid bacteria present specific and critical nutritional requirements, often supplementation with growth factors, vitamins and amino acids is necessary. *L. fermentum* DSM 20,049 was previously grown on whey with the addition of hydrolyzed lupin flour as an auxiliary nitrogen source in flask experiments, showing a shorter lag phase and a 70% higher biomass yield as compared to growth on MRS in the same conditions [24].

In the present study, buffalo milk-derived whey was evaluated as a substrate for the growth of an *L. fermentum* strain isolated from buffalo milk. In particular, the powder obtained from the spray-dried ultrafiltered retentate (UF_Ret20Pow) was used in fermentation experiments to evaluate its potential as a one-pot medium; this would in fact strongly simplify cultivation medium preparation and overall upstream processes. Since this fraction was not diafiltered it contained a large amount of sugars, in particular 44%, 8% and 6% of lactose, glucose and galactose, respectively, and about 6% of protein, a necessary nitrogen source for bacterial growth. Strain viability, sugars consumed and metabolic products produced (lactic acid, acetic acid and erhanol) were initially evaluated in bottle experiments. *L. fermentum* was cultivated on semi defined SGSL medium supplemented with different amounts of UF_Ret20Pow. SGLS supplemented with glucose, the carbon source most efficiently used by *L. fermentum* [25], or lactose, the main sugar present in whey powder, were used as controls. As shown in Figure 2, a higher concentration of viable cells and related metabolic products (e.g., lactic acid, ethanol) were observed in relation to higher initial concentrations of ultrafiltered whey; in particular, in the presence of 4% UF_Ret20Pow, the final average concentration of viable cells (8.85 ± 0.17 Log10 CFU/mL) was comparable to that obtained in control experiments on glucose (8.97 ± 0.14 Log10 CFU /mL), whereas it was significantly higher compared to results obtained in control experiments with lactose (8.43 ± 0.17 Log10 CFU/mL). Additional expeolments un UF_Ret20Pow dissolved in water (in the aUsence of SGLS medium components) showed that growth was still supported by the organic compounds present in the ultrafiltered whey fraction, although a significantly lower sugar consumption and biomass concentration were achieved (Figure 2), probably due to the lower amount of nitrogen source compared to thar present in SGSL medium [25].

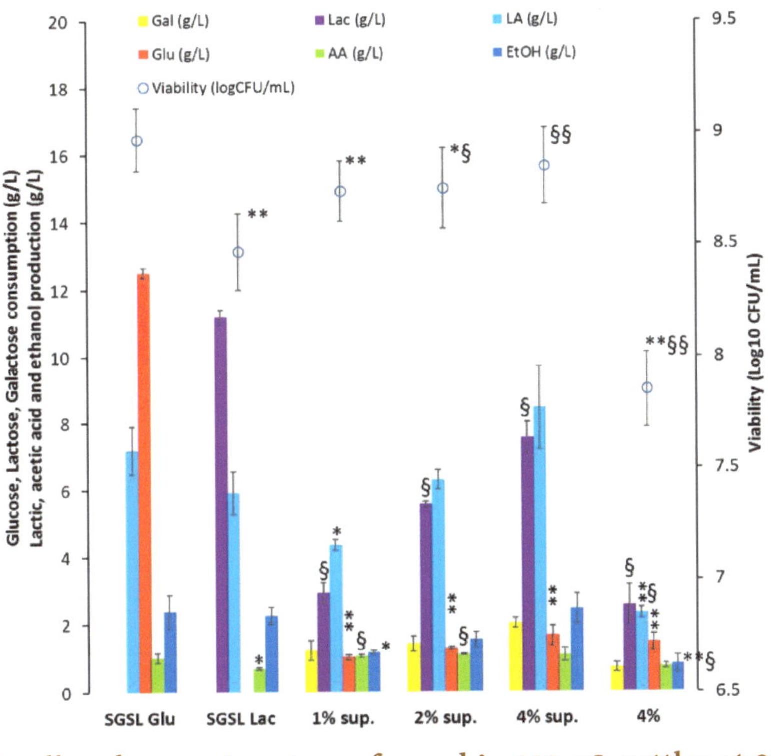

Figure 2. Small-scale experiments performed in 100 mL pottles at 37°C and 150 rpm. Gal, galactose; Lac, lactose; Glc, glucose; LA, lactic acid; AA, acetic acid; EtOH, ethanol. Sup, indicates media in which UF_Ret20Pow wns reconstituted in SGLS medium. Data were analyzed by two-tailed non homoscedastic Student's t-test. * indicates $p < 0.05$ compared to results obtained on SGLS Glu; ** indicates $p < 0.01$ compared to results obtained on SGLS Glu; § indicates $p < 0.05$ compared to results obtained on SGLS Lac; §§ indicates $p < 0.01$ compared to results obtained on SGLS Lac.

With the aim of simplifying upstream and downstream procedures on an industrial scale by using buffalo milk waste as the only component of fermentation media, bioreactor experiments in controlled conditions were performed. *L. fermentum* probiotic biomass production was therefore investigated on UF_Ret20 as a one-pot medium (powder reconstituted in water) and on UF_Ret20 supplemented with SGSL salts, yeast extract and soy peptone (1/5th of that present in SGLS). Each experiment was performed at least in duplicate. Table 3 shows the results obtained. Controlled pH and constant air sparging improved viability that reached 8.1 ± 0.2 Log10 CFU/mL on the medium containing concentrated whey only. The addition of salts and of low amounts of complex N sources (soy peptone and yeast extract), yielded similar results, indicating an impact only on sugar consumption and LA production which increased (Table 3); LA, in particular, showed a 3.3 fold titer increase with a final concentration of about 10.0 ± 0.3 g/L and a final yield of 0.61 ± 0.03 g/g. Apparently, due to the lower amount of nitrogen source, the sugars were addressed to acid instead of biomass production [26].

Table 3. Data obtained by growing L. fermentum in batch in a Biostat CT plus (3 L) bioreactor reported as mean ± s.d. UF_Ret20Pow indicates that the fermentation medium was obtained by reconstituting the UF_Ret20Pow spray-dried powder in water; UF_Ret20Pow sup indicates the additional presence of SGLS salts, yeast extract and soy peptone (2 g/L) in the medium. Lac, initial lactose; Gal, initial galactose; Glu, initial glucose; LA, lactic acid; AA, acetic acid; EtOH, ethanol. Data were analyzed by two-tailed non homoscedastic Student's t-test: * $p < 0.05$; ** $p < 0.01$.

	Viability (Log10 CFU/mL)	Lac (g/L)	Glu (g/L)	Gal (g/L)	Sugars Cons. (g/L)	LA (g/L)	AA (g/L)	EtOH (g/L)	^{Y}LA/s (g/g)
UF_ Ret20Pow	8.1 ± 0.2	3.3 ± 0.8**	2.5 ± 0.8*	1.5 ± 0.2	7.2 ± 1.4 **	3.0 ± 0.6**	1.3 ± 0.4	1.4 ± 0.5	0.42 ± 0.01
UF_ Ret20Pow sup	8.0 ± 0.2	9.4 ± 0.3	5.1 ± 0.3	2.0 ± 0.2	16.4 ± 0.3	10.0 ± 0.3	1.5 ± 0.4	1.9 ± 0.2	0.61 ± 0.03

3.3. Whey Valorization as Wound Repair Agent

Buffalo milk whey has been shown to contain various proteins with immunomodulatory and antitumor activity, and a series of bioactive peptides with antimicrobial, antioxidant, antihypertensive and remineralizing properties [5]. For example, the effect of whey proteins in improving the inflammatory status during wound healing in diabetic rats, in particular by reducing the expression of specific cytokines involved in the reparation, was previously reported [27]. Kalinina and collaborators [28] identified whey proteins, and in particular WFDC12, as a specific marker for the last stage of keratinocytes differentiation, probably enhancing the occurrence of epidermal homeostasis.

Our research aimed to evaluate the ability of ultrafiltered and concentrated whey from buffalo milk, to prompt cell migration and regeneration of the human keratinocyte monolayers by using a scratch assay [29]. This assay is generally run to obtain a preliminary but robust evaluation of the regenerative potential of biomolecules.

The sprayed powder was diluted in the cell growth medium and tested by an in vitro wound healing assay, as reported in the material and methods section. Representative fields of view of wound closure reported in Figure 3a are clearly showing that the addition of UF_Ret20Pow prompts keratinocytes migration, confirming the role of whey proteins in dermal repair. Quantitative analyses (Figure 3b) indicate that in the presence of the sprayed powder at a concentration of 4 g/L, 40% wound closure occurred within 15 h, as compared to the control that, at the same time, reached about 20% of repair. Moreover, the sample induced complete healing within 30 h, whereas the closure area of the control was lower than 40%. Data analysis by means of a two-tailed non homoscedastic Student t-test indicates that the wound closure rate significantly improves ($p < 0.05$) in the supplemented sample from 9 h onwards, as compared to the control (Figure 3b). The treatment with a less concentrated sample (2 g/L) did not improve significantly the natural occurrence of wound healing. Overall these data clearly establish a beneficial effect of diluted fractionated whey powder, suggesting a potentially promising effect in topical products for skin treatments.

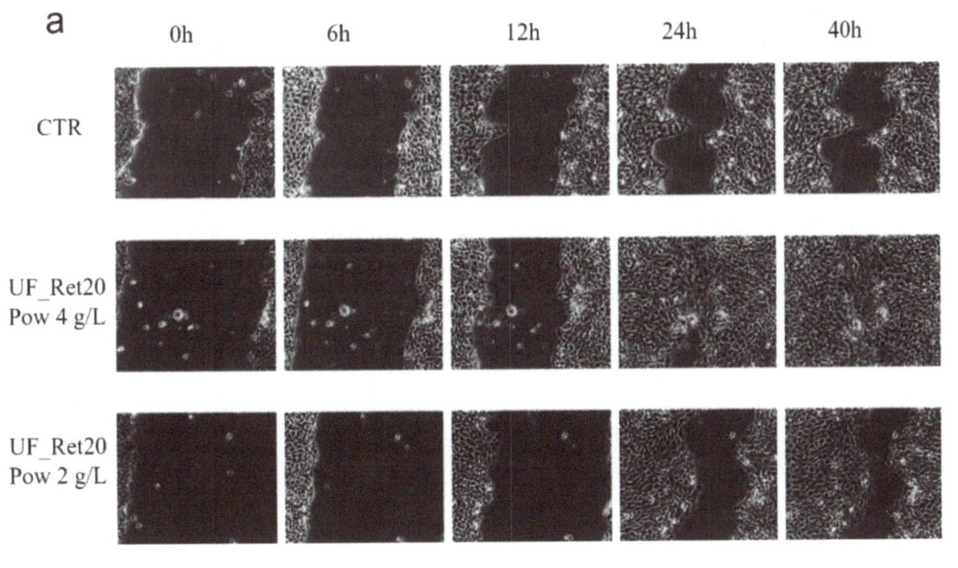

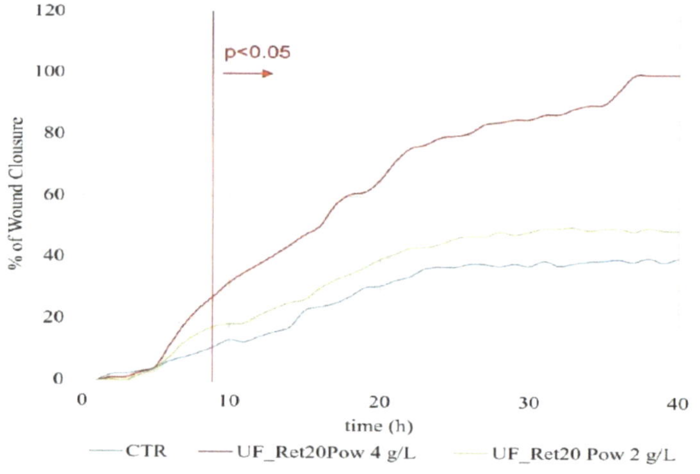

Figure 3. Wound healing assay (a) Representative field of view of HaCat scratch assays during time course of the experiments. At least five fields of view for each sample were analyzed; (b) Quantitative analyses of wound closure as percentage over time. Data significance was analyzed by two-tailed non homoscedastic Student t-tests. The vertical line in the graph indicates the time point from which a significant difference of wound closure % was observed between the CTR and the UF_Ret20Pow 4 g/L treated sample. CTR, control

4. CONCLUSIONS

Overall, the present study provides a small local circular economy example in which one of the most abundant industrial wastes productd by small and medium regional companies could be easily valorized by configuring processes that generate value-added producfs using a bio-refining concept.

In particular, the implementation of a membrane-based process involving a ultrafiltration and a nanofiltration step was demonstrated to be sufficient to reduce the organic load of whey resulting in a clean and reusable effluent.

Maybe even more importantly, this downstream process generates fractions differentially enriched with potentially interesting compounds. The concentrated UF retentate was investigated here and demonstrated (i) to be by itself a sufficient source of sugars and proteins to support the growth of a probiotic strain with known biomedical applications, and (ii) to stimulate epidermis (keratinocyte) regeneration and therefore meaning potential applicability as an ingredient in skincare products.

REFERENCES

1. Baldasso, C.; Barros, T.C.; Tessaro, I.C. Concentration and purification of whey proteins by ultrafiltration. *Desalination* **2011**, *278*, 381-386. [CrossRef]

2. Cuartas-Uribe, B.; Alcaina-Miranda, M.I.; Soriano-Costa, E.; Bes-Pia, A. Comparison of the behavior of two nanofiltration membranes for sweet whey demineralization. *J. Dairy Sci.* **2007**, *90*,1094-1101. [CrossRef]

3. Baisali, S.; Chakrabarti, P.P.; Vijaykumar, A.; Kale, V. Wastewater treatment in dairy industries—Possibility of reuse. *Desalination* **2006**,195,141-152.

4. Dullius, A.; Goettert, M.I.; Volken de Souza, C.F. Whey protein hydrolysates as a source of bioactive peptides for functional foods—Biotechnological facilitation of industrial scale-up. *J. Func. Foods* **2018**, 42, 58-74. [CrossRef]

5. Pires, A.F.; Marnotes, N.G.; Rubio, O.D.; Garcia, A.C.; Pereira, C.D. Dairy By-Products: A Review on the Valorization of Whey and Second Cheese Whey. *Foods* **2021**,*10*,1067. [CrossRef] [PubMed]

6. Bounous, G. Whey protein concentrate (WPC) and glutathione modulation in cancer treatment. *Anticancer Res.* **2000**, *20*, 4785-4792.

7. Teixeira, F.J.; Santos, H.O.; Howell, S.L.; Pimentel, G.D. Whey protein in cancer therapy: A narrative review. *Pharmacol. Res.* **2019**, 144, 245-256. [CrossRef]

8. Agin, D.; Gallagher, D.; Wang, J.; Heymsfield, S.B.; Pierson, R.N.; Jr Kotler, D.P. Effects of whey protein and resistance exercise on body cell mass, muscle strength, and quality of life in women with HIV. *AIDS* **2001**,15, 2431-2440. [CrossRef]

9. Marshall, K. Therapeutic applications of whey protein. *Altern. Med. Rev.* **2004**, 9,136-156.

10. Zotta, T.; Solieri, L.; Iacumin, L.; Picozzi, C.; Gullo, M. Valorization of cheese whey using microbial fermentations. *Appl. Microbiol. Biotechnol.* **2020**,104, 2749-2764. [CrossRef]

11. Aguirre-Ezkauriatza, E.J.; Aguilar-Yanez, J.M.; Ramirez-Medrano, A.; Alvarez, M.M. Production of probiotic biomass *(Lactobacillus casei)* in goat milk whey: Comparison of batch, continuous and fed-batch cultures. *Bioresour. Technol.* **2010**,101, 2837-2844. [CrossRef]

12. Kumar, M.; Jain, A.K.; Ghosh, M.; Ganguli, A. Industrial whey utilization as a medium supplement for biphasic growth and bacteriocin production by probiotic *Lactobacillus casei* LA-1. *Probiotics Antimicrob. Proteins* **2012**, 4,198-207. [CrossRef]

13. Malvido, M.C.; Gonzalez, E.A.; Bazan Tantalean, D.L.; Bendana Jacome, R.J.; Guerra, N.P. Batch and fed-batch production of probiotic biomass and nisin in nutrient-supplemented whey media. *Braz. J. Microbiol.* **2019**, *50*, 915-925. [CrossRef]

14. Meng, W.; Zhang, Y.; Cao, M.; Zhang, W.; Lu, C.; Yang, C.; Gao, C.; Xu, P.; Ma, C. Efficient 2,3-butanediol production from whey powder using metabolically engineered *Klebsiella oxytoca. Microb. Cell Factories* **2020**,19,162-171. [CrossRef]

15. Pais, J.; Serafim, L.S.; Freitas, F.; Reis, M.A. Conversion of cheese whey into poly(3-hydroxybutyrate-co-3-hydroxyvalerate) by *Haloferax mediterranei. New Biotechnol.* **2016**, 33, 224-230. [CrossRef]

16. Prasad, S.; Srikanth, K.; Limaye, A.M.; Sivaprakasam, S. Homofermentative production of D-lactic acid by *Lactobacillus* sp. employing casein whey permeate as a raw feed-stock. *Biotechnol. Lett.* **2014**, 36,1303-1307. [CrossRef]

17. Atra, R.; Vatai, G.; Bekassy-Molnar, E.; Balint, A. Investigation of ultrα- and nanofiltration for utilization of whey protein and lactose. *J. Food Eng.* **2005**, 67, 325-332. [CrossRef]

18. Yorgun, M.S.; Balcioglu, I.A.; Saygin, O. Performance comparison of ultrafiltration, nanofiltration and reverse osmosis on whey treatment. *Desalination* **2008**, 229, 204-216. [CrossRef]

19. Naghmouchi, K.; Belguesmia, Y.; Bendali, F.; Spano, G.; Seal, B.S.; Drider, D. *Lactobacillus fermentum:* A bacterial species with potential for food preservation and biomedical applications. *Crit. Rev. Food Sci. Nutr.* **2020**, 60, 3387-3399. [CrossRef] [PubMed]

20. Baldini, M.; Fabietti, F.; Giammarioli, S.; Onori, R.; Orefice, L.; Stacchini, A. *Determinazione delle Ceneri nel Formaggio, nel Formaggio Fuso e Nella Ricotta. G. U. Supplemento al n. 229 del 02/10/1986. Istituto Superiore di Sanitd.* Available online: https://www.iss.it/ (accessed on 10 October 2021).

21. Jin, J.; Lao, J.; Zhou, R.; He, W.; Qin, Y.; Zhong, C.; Xie, J.; Liu, H.; Wan, D.; Zhang, S.; et al. Simultaneous identification and dynamic analysis of saccharides during steam processing of rhizomes of *Polygonatum cyrtonema* by HPLC-QTOF-MS/MS. *Molecules* **2018**,*11*, 2855. [CrossRef] [PubMed]

22. Bradford, M.M. A rapid and sensitive method for the quantitation of microgram quantities of protein utilizing the principle of protein-dye binding. *Anal. Biochem.* **1976**, 72, 248-254. [CrossRef]

23. Wayah, S.B.; Philip, K. Characterization, yield optimization, scale up and biopreservative potential of fermencin SA715, a novel bacteriocin from *Lactobacillus fermentum* GA715 of goat milk origin. *Microb. Cell Factories* **2018**,17,125-142. [CrossRef]

24. Hanoune, S.; Djeghri-Hocine, S.; Kassas, S.; Derradji, S.; Boudour, S.; Boukhemis, S. Optimization of *Lactobacillus fermentum* DSM 20049 growth on whey and lupin based medium using response surface methodology. *J. Adv. J. Food Sci. Technol.* **2015**, 9, 679-685. [CrossRef]

25. Gao, X.; Qiao, S.Y.; Lu, W.Q. Determination of an economical medium for growth of *Lactobacillus fermentum* using response surface methodology. *Lett. Appl. Microbiol.* **2009**, 49, 556-561. [CrossRef]

26. Kariluoto, S.; Aittamaa, M.; Korhola, M.; Salovaara, H.; Vahteristo, L.; Piironen, V. Effects of yeasts and bacteria on the levels of folates in rye sourdoughs. *Int. J. Food Microbiol.* **2006**,106,137-143. [CrossRef] [PubMed]

27. Ebaid, H.; Salem, A.; Sayed, A.; Metwalli, A. Whey protein enhances normal inflammatory responses during cutaneous wound healing in diabetic rats. *Lipids Health Dis.* **2011**,10, 235. [CrossRef] [PubMed]

28. Kalinina, P.; Vorstandlechner, V.; Buchberger, M.; Eckhart, L.; Lengauer, B.; Golabi, B.; Mildner, M. The whey acidic protein WFDC12 is specifically expressed in terminally differentiated keratinocytes and regulates epidermal serine protease activity. *J. Investig. Dermatol.* **2021**,141,1198-1206. [CrossRef]

29. D'Agostino, A.; Pirozzi, A.V.A.; Finamore, R.; Grieco, F.; Minale, M.; Schiraldi, C. Molecular mechanisms at the basis of pharmaceutical grade *Triticum vulgare* extract efficacy in prompting keratinocytes healing. *Molecules* **2020**, 25, 431. [CrossRef] [PubMed]

CHAPTER 12

Utilization of Whey for Red Pigment Production by Monascus purpureus in Submerged Fermentation

1. INTRODUCTION

Color introduces data to the class of products and particular brands, creating effective visual recommendations to show attractiveness, tastes, distinction, and novelty in packages. Food producers have preferred to use natural pigments instead of artificial pigments in response to the developing consumer perception that natural pigments are safer. Plant-originated pigments have some disadvantages, such as low solubility in water, being unstable against heat and light [1,2], and high prices along with the required large agricultural areas. However, microbial pigments have many advantages such as being independent of seasonal changes, use of low-cost raw materials, being biodegradable, having different color tones depending on the conditions of culture and higher efficiency compared to chemical synthesis [1,3]. Thus, microbial pigments identified as bio-pigments are the safer preference for the food industry. The production of microbial pigments has been gradually increasing since health concerns such as carcinogenicity and teratogenicity are related more to the synthetic pigments [4]. Furthermore, environmental policies are other main drivers for the diffusion of biopigments in the market.

Among the various pigment-producing organisms, Monascus purpureus, is a fungus that is isolated from red-fermented rice in Indonesia [5]. This filamentous fungus has been extensively preferred in the production of fermented foods in Asian countries, southern China, Japan, and Southeastern Asia [6]. Monascus pigments (MPs) as secondary metabolites have many therapeutic effects such as antioxidant, anti-inflammatory, antimicrobial, anticarcinogenic and antimutagenic [7]. The main color pigments produced by *Monascus* spp. in polyketone structure are red, orange and yellow. The red ones of those six major pigments are called rubropunctamine ($C_{21}H_{26}NO_4$) and monascorubramine ($C_{23}H_{27}NO_4$); orange ones are called rubropunctatin ($C_{21}H_{22}O_5$) and monascorubrin ($C_{23}H_{26}O_5$) and yellow pigments are called as monascin ($C_{21}H_{26}O_5$) and ankaflavin ($C_{23}H_{30}O_5$) [8,9].

Monascus pigments are used as food additives, color thickener, or nitrite substitute in different types of foods (red wine, tofu, surimi, sausage, ham, different sauces, noodle products, ready meals, meat products, etc.) especially in East Asian countries. There are also application areas in the dairy, textile and cosmetics industries [1,4,10-12].

The wastes generated by the food industry cause serious environmental pollution and global warming. The major criteria of sustainable industrial production is the recovery and reuse of these wastes as a resource within the cycle of circular economy [13]. Hence, researchers in recent years have concentrated on studies about the use of food industry wastes in the production of high value biotechnological products. The utilization of wastes containing carbon and nitrogen in the bioprocesses is important in terms of reducing environmental pollution and also building low cost, robust and sustainable production schemes.

Various food industry wastes were used to produce *Monascus* pigments in the literature. These are; hydrolyzed rice straw [14], waste beer [15], brewer's spent grain [9], orange peels [16], chicken feather [17], sugarcane bagasse [18,19] bakery wastes [20], rice water based medium [21], sweet potato [22], corn cob [23,24], potato powder [1], the grape pulp [10], corn step liquor [25], Jack fruit wastes [26], wheat [5] and prickly pear juice [27].

Whey, which is a by product of cheese, casein and yogurt manufacturing, is considered as waste of the dairy industry [28]. When processed further, whey becomes a high value by product that is used as substrate in microbial fermentations utilizing lactose. However, it is defined as an ecologically harmful and most polluting waste when released directly into water receiving bodies [29]. With the rapid industrialization observed in the last century [30] and the growth rate of milk production (around 2.8% per annum), dairy processing is usually considered as the largest industrial food wastewater source, especially in Europe [31,32]. Depending on cheese type and production method, 150-200 kg of cheese is produced from a ton of milk, while 800-850 kg of whey is generated. Approximately 180-190 million tons of whey are produced annually in the world. It is a great threat to the environment [33] due to its very high biological oxygen demands (BOD > 35,000 ppm) and chemical oxygen demands (COD > 60,000 ppm) along with its low pH [34]. For a long time in the 20th century, the industry worked on an inexpensive removal strategy for whey, which included release into waterways, the ocean, municipal sewage treatment works, and/or onto fields. Today, these disposal methods are prevented by strict environmental regulations. Treatment and/or recovery of whey for its use in the production of value- added products have become a major concern. Since whey contains 4.5-5.0% lactose, 0.6-0.8% protein, 0.4-0.5% lipid, vitamins and minerals, there are many studies on its use as substrate in bioprocesses [34-36]. Since some microorganisms can not utilize lactose in whey as the carbon source, some bioprocesses require the hydrolysis of lactose into its monomers, glucose and galactose by the enzyme β-galactosidase.

The aim of this study is to investigate whey as an alternative low-cost sustainable substrate in the fermentation of *Monascus purpureus* CMU 001 strain to produce red color pigment for the food industry. Raw, demineralized, deproteinized whey as a substrate, fermentation pH, initial lactose concentration, monosodium glutamate (MSG) concentration as the nitrogen source, inoculation ratio, mycelial development and pigment synthesis kinetics of *Monascus purpureus* were studied. In contrast to the production of red pigments from different waste resources as a fermentation substrate, to the best of our knowledge no previous study in the literature has systematically investigated the simultaneous hydrolysis and fermentation of whey for the production of red pigment by *Monascus purpureus* from the point of sustainable resource recovery.

2. MATERIALS AND METHODS

2.1. Microorganisms and Culture Media

The microorganism used in this study was the *Monascus purpureus* CMU 001 kindly provided by the Department of Biology, Chiang Mai University of Thailand. Our previous work [37] has showed that this strain did not produce the mycotoxin citrinin, which is a secondary metabolite of *Monascus* species. The strain was kept on potato dextrose agar (PDA; pH:5.6 ± 0.2, Merck, Darmstadt, Germany) at 4°C and sub-cultured every 4 weeks to fresh PDA slants incubated at 30°C for 7 days.

Raw whey powder, demineralized whey powder and deproteinized whey powder were supplied by Enka Sut ve Gida Mamulleri Sanayi ve Ticaret A.S. located in Konya city, Turkey. All the whey samples were sweet whey-based samples as declared by the supplier. β-galactosidase (Saphera 2600L) enzyme was kindly donated by Novozymes A/S (Bagsvaerd, Denmark) for the hydrolysis of lactose in whey. Saphera 2600L is a bacterial β-galactosidase enzyme from *Bacillus licheniformis* with a specific activity of 2600 LAU-B/g (enzyme activity was stated as β-galactosidase that hydrolyzes terminal non-reducing β-D-galactosides releasing beta-D-galactose residues). Whey powder was diluted with distilled water to contain 50 g/L lactose concentration (unless otherwise stated) for fermentation experiments.

2.2. Preparation of Inoculum and Fermentation Medium

The inoculum was prepared by collecting spores from the surface of the PDA dishes under aseptic conditions with 10 mL of sterile distilled water. The spore suspension was used to prepare the inoculum culture containing 1.25 x 10^6 spores/mL. The spore concentration was estimated using a Neubauer chamber (Marienfeld-Superior, Laudα- Konigshofen, Germany). The inoculation medium (50 mL in 250 mL Erlenmeyer flasks) was prepared by reconstituting whey powder in water to contain lactose at 50 g/L concentration and pH 6.0. Since *Monascus purpureus* can not utilize lactose as the carbon source, β-galactosidase enzyme (Saphera 2600L) was added to the fermentation medium at a rate of 0.1% (v/v) (simultaneous hydrolysis and fermentation, SHF). The fermentation for inoculum preparation was performed in a shaking incubator operated at 200 rpm, 30°C for 4 days.

Fifty milliliters of demineralized whey medium (pH 6.0, unless otherwise stated) containing 50 g/L lactose was the substrate in SHF. The pH of the whey medium was adjusted to pH 6.0 by using 1 N potassium hydroxide and sterilized at 121°C for 15 min. The inoculation ratio was 2% (v/v) and the SHF's were carried out in a shaking incubator at 200 rpm, 30°C for 8 days. Inoculation and fermentation media were supplemented with 5 g/L of MSG (MSG; Sigma Aldrich, St Louis, MO, USA) as the nitrogen source. Samples were collected at equal time intervals for the determination of biomass growth rate and red pigment synthesis.

Prehydrolysis and a separate fermentation (PHSF) method was also used for the production of the red pigment. In this method, the prehydrolysis of lactose in demineralised whey medium was done at 50°C (pH 6.0) for 8 h enzymatically before fermentation with *M. purpureus*.

2.3. The Effect of Whey Type on Pigment Production

Whey powder (W), demineralized whey powder (DM), deproteinized whey powder (DP), treated deproteinized raw whey powder (DPW), and treated deproteinized demineralized whey powder (DPDM) were used as fermentation substrate. Initial concentration of lactose was 50 g/L for each fermentation. DM and W samples were subjected to deproteinization to obtain DPW and DPDM. The process of deproteinization was adapted from Roukas et al., [38]; the pH of W and DM were adjusted to 5.0 and protein precipitation was induced by heating W and DM solutions at 90°C for 20 min. The solutions were kept at room temperature for 1 h and precipitated proteins were removed by centrifugation (Hettich Universal 320 R, Tuttlingen, Germany) at 6000 x *g* for 15 min.

2.4. Optimization of Fermentation Parameters

The experiments were conducted to find optimum fermentation conditions and to determine the effect of fermentation parameters on red pigment production. Different initial pH values (5.0, 6.0, 7.0, 8.0, 9.0), initial lactose concentrations (25, 50, 75,100 g/L), MSG concentrations (2.5, 5.0, 7.5,10.0,12.5,15.0,20.0, 25.0, 30.0 g/L) as nitrogen source, and inoculation ratios of *Monascus* (1,2,3,4,5, 6%; (v/v)) were investigated for the pigment production. The pigment production kinetics and mycelial development were examined at the optimal conditions of fermentation determined from the experiments.

2.5. Analytical Methods

The biomass was measured gravimetrically at the end of fermentation by filtering the mycelia through Whatman No: 1 filter paper, washing three times with distilled water and drying at 65°C to constant weight in an oven (Memmert, UN 55, Schwabach, Germany). The biomass concentration was expressed as mycelial dry weight per unit volume of culture medium [39].

The supernatant obtained by centrifugation of the fermentation medium in the centrifuge (Hettich Universal 320 R, Tuttlingen, Germany) at 5400 rpm/20 min was filtered with Whatman No. 1 filter paper. The filtrate obtained was used for the quantification of the synthesized red pigments using a spectrophotometer (Thermo Scientific, Genesys 10S UV-VIS, Paisley, UK) at 510 nm and the readings were given in absorbance units (UA_{510}) [40]. Sterile fermentation medium was used as the blank sample since the medium itself may have some absorbance at 510 nm wavelength. The pH values were determined with the pH-meter (WTW Inolab 7110, Weilheim, Germany). Lactose concentration in the medium was determined spectrophotometrically at 540 nm by the phenol-sulfuric acid method [41]. Total nitrogen content was determined by the Kjeldahl method [42] and the protein content of demineralized whey was calculated by multiplying the nitrogen content by the factor 6.25 in order to convert the nitrogen content into protein content. Moisture content of demineralized whey was determined by drying the sample to constant weight at 80°C. The ash content of demineralized whey was determined by burning the sample for 12-18 h at 550°C in a muffle furnace followed by cooling to room temperature and weighing.

Each experiment was performed in triplicate and the results were expressed as the mean ± standard deviation. The experiments were carried out in two repetitions and the analyses were carried out in three parallel samples. The results were expressed as the mean ± standard deviation. All statistical analyses were performed using the software IBM SPSS (v. 22). All data obtained were analyzed by one way analysis of variance, and tests of significant differences were determined by using Duncan's test at $p < 0.05$. In all the figures, mean values for all the factors given in x-axis that are not followed by the same letter (a-g) are significantly different ($p < 0.05$).

2.6. Cost Analysis

In this study, red pigment production cost was determined. The production cost includes the handling of raw materials (whey), chemicals (MSG and enzyme), electricity for fermentation, downstream processes (autoclaving, centrifuging, and drying) and the cost of water. Energy consumption values for fermentation and downstream processes were measured by a TT-Technic PM 001 plug power meter device. Prices of whey, chemicals, electricity, and water were obtained from the suppliers for cost analysis.

3. RESULTS AND DISCUSSION

3.1. The Effect of Whey Type as Fermentation Substrate on Pigment Production

Raw whey powder (W), demineralized whey powder (DM), deproteinized whey powder (DP), treated deproteinized raw whey powder (DPW) and treated deproteinized demineralized whey powder (DPDM) were prepared to investigate the effect of whey type on pigment production by *Monascus purpureus.* The simultaneous hydrolysis and fermentation (SHF) method was used in the fermentation experiments. The chemical composition of whey type has an important role in red pigment production since different types of whey have different protein andmineral contents [34].

As seen from Figure 1, the highest pigment concentration (20.8 UA$_5$1$_0$ nm) was produced from demineralized whey (DM) medium. Similar pigment yield value of 20.2 UA$_5$10 nm was obtained with deproteinized and demineralized whey powder medium (DPDM). No significant difference ($p > 0.05$) in pigment ytelds were? observed with DM and DPDM medie. The lowest pigment yield was obtained from the DP medium (5.3 UA$_5$1$_0$ m)- While DP and DPW have yielded very/^{1} low red pigment production values, DM and DPDM produced higher values. Lower pigment synthesis values observed in DP and PDW samples rmght have originated from the inhibitory effect of high concentration of cations (Na+, NH4+, K+, Mg^{+2}, and Ca^{+2}) and anions (Cl$^-$, SO.|^2, PO4^{-3}, and citrate) present in whey medium [43,44]. Since the demineralization process decreased monovaleent, divalent cation and anion levels in whey, pigment production values obtained from DIff and DPDM are satisfactory compared to the literature [45]. Raw whey powder (W) yielded 12.1 UA$_5$1$_{0run}$ pigment production value which was slightly higher than the values of DP and DPW as a substrate of pigment synthesis. The biomass concentration valurs observed in all types of whey were very close to each other and no significant correlation was observed between pigment and biomass production vaues (Figure 1). This indicates that biomass can reproduce in all topes of whey, but the conditions for pigment production canbe achieved at low anion and cation concentfations.

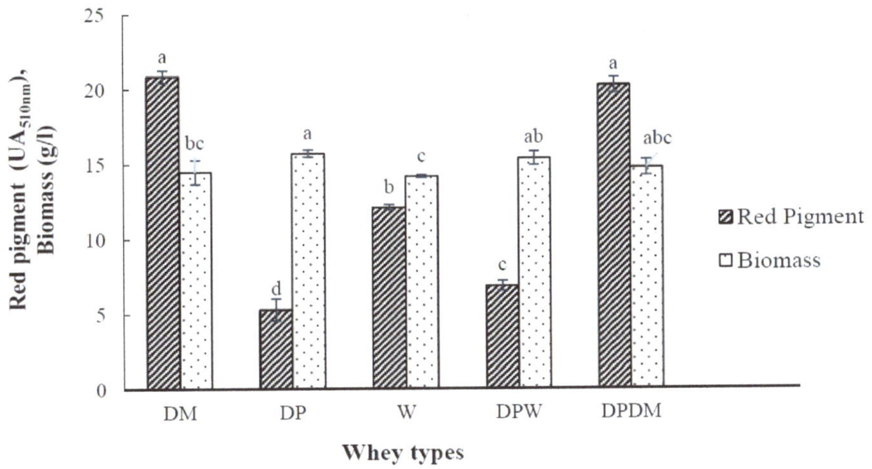

Figure 1. The effect ot whey types on the: red pigment and biomass production by M. purpureas CMU 001 (Fermentation conditions: 30°C, 8 days, 200 rpm, pH 6.0, 5 g/L monosodium glutamate (MSG), 50 g/L lactose).

Demineralised whey powder with the highest pigment yield had the following composition (%): total sugar 70.46 ± 0.24, protein 5.32 ± 0.05, ash 3.32 ± 0.36, total nitrogen 0.86 ± 0.01, moisture 2.72 ± 0.3. As long as the lactose present in whey is enzymatically hydrolyzed, DM whey with its high lactose sugar content (70.5%) is a suitable substrate for *M. purpureas* in the synthesis of the red pigment.

3.2. Effects of Different Fermentation Methods on Pigment Production

Raw demineralised whey powder (DM) containing lactose as the carbon source in unhydrolyzed form was used for pigment production to determine the effect of lactose on growth and pigment production by *M. purpureus*. As seen in Figure 2, lactose found in DM was not utilized efficiently by *M. purpureus* for growth and pigment production. Other researcher also pointed out the difficulties of using lactose as the carbon source for the production of biopigments [45-48]. Therefore, it is necessary to hydrolyze lactose with [3-galactosidase enzyme to its monomers of glucose and galactose for its proper utilization by *M. purpureus* as an energy source. In this study, two fermentation methods were compared for lactose hydrolysis; simultaneous lactose hydrolysis and fermentation (SHF) and enzymatic pre-hydrolysis and separate fermentation (PHSF) for pigment production by *M. purpureut.* Demineralised whey powder (DM) containing 50 g/L lactose was used as the substrate: in tire fermentation eoperiments. As shown in Figure 2, the highest pigment production valua (21.3 $UA_{5}i_{0\ nm}$) was obtained wit[a] the SHI[7] application. When lactose was hydrolyzed by the enzyme prior to fermentation (PHSF), lower red pigment synthesis (16.8 $UA_{510\ nm}$) was obtai ned compared to S HF. Similar b iomass yiel ds of 1[e].3 and 14.9 g/L were obtained for PHSF and SHF, respectively'.

Hence, SHF yielded higher pigment production values. Da Costa and Vendruscolo [45], determined pigment production by *Monatcut rubor* CCT 3802 in the presence of several carbon sources such as gluaose (20 g/L), lactose (20 g/L), and hydrolyzed lactose (20 g/L). Parallel to the results of our study, they found that the average production of yellova and orange pigmenfs were higher *in* the hydrolyzed lactose medium by *M. ruber* compared to lactose.

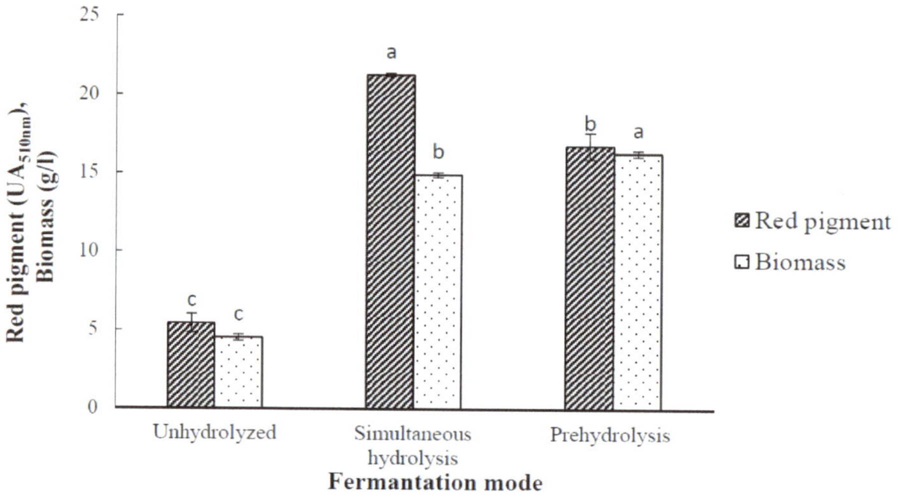

Figure 1. The effect ot whey types on the: red pigment and biomass production by M. purpureas CMU 001 (Fermentation conditions: 30 °C, 8 days, 200 rpm, pH 6.0, 5 g/L monosodium glutamate (MSG), 50 g/L lactose).

3.3. Effect of Initial pH on Pigment Production

The pH of fermentation medium is an important factor in red pigment synthesis by *Monascus* species sincehigh phi values and the existence of suitable nitrogen source leads to the chemical modification of orange pigments changing into axtracellular and water-soluble red pigments [49]. The effect of initial pH of fermentation medium was studied for pH range 5.0-9.0 using demineralised whey powder containing 50 g/L of lactose as the substrate. SHF (30°C/8 days) was performed in 250 mL Erlenmayer flasks containing 50 mL of fermentation medium. As seen in Figure 3, the highest pigment synthesis of 25.3 $UA510_{nm}$ was observed at the initial pH value of 7.0. A low amount of pigment formation was observed at high and low pH values. In the fermentation medium with an initial pH of 9.0, growth and pigment synthesis were found to be the lowest (4.4 $UA510_{nm}$). Other researchers have also reported that pH of fermentation medium was highly important for red pigment synthesis by *M. purpureus* [48,50]. Orozco and Kilikian [50] investigated the synthesis of pigments by *Monascus purpureus* and they obtained 11.3 U of red pigments at pH 8.5. They stated that high pH medium facilitates the transfer of intracellular pigments to the fermentation medium. Parallel to the findings of our study, Mukherjee and Singh [39] stated that *M. purpureus* produced more red pigments at pH values of 6.0-8.0. Lee et al. [51] also indicated that the pH range of 5.5-8.5 was suitable for red pigment production, whereas pH values higher than 8.5 and lower than 5.5 led to a decrease in red pigment production. Prapajati et al. [48] achieved maximum *Monascus* pigment synthesis at pH 8.5. It has been reported that water-soluble extracellular red pigment production increased at high pH values and at high MSG concentrations and the transition of pigments from the cell to fermentation medium was reatricteal at low pH values [49].

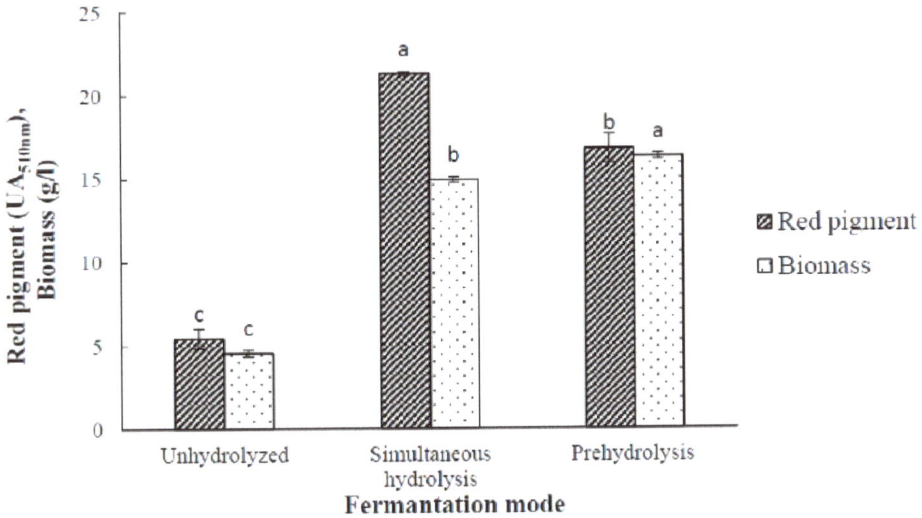

Figure 2. The effect of fermentation mode on the red pigment and laiomass production by M. purpureut CMU 001 (Fermentation conditions: 30 °C, 8 days, 200 rpm, pH 6.0, 5 g/L MSG, 50 g/L lactose).

3.4. Effect of Initial Lactose Concentration on Pigment Production

The effect of lactose concentration on pigment production by *M. purpureus* CMU 001 was investigated with DM whey containi ng different initial lactose c one entrations as 25, 50, 75, arid 100 g/L by SHF at 30°C fot 8 days. The initial pH of the fermentation medium was 7.0. Tlte highest pigment densitywas determined as 27.7 $UA_{5}i_{0\ nm}$ in the fermentation medium containing 75 g/L of lactose (Figure 4). Red pigment concenitrations of 10.2, 25.[a], and 25.3 UA510 $_{nm}$ were found for the fermentation media containing initial lactose eaneentrationa of 25, 50, and "100 g/L, respectively. Chen and Johns [52] reparttd that; high glucose concentration of 50 g/L led to low biomass, pigment synthesis and ethanol production, while high maltose concentration of 50 g /L increased red pigment production by three fold compared to tlie same concentration of glucoae. Liu et al., [53] studied pigment production by Ad. *purpureus* M183 and found the optimum glucose csncentration as 80 g/L in terms of maximum efficiency, pigment yield and cost efficiency for industrial applications. Parallel to our findings, they stated that the initial substrate concentration has a negligible effect on biomass (dry cell weight). The high glucose concentration in the fermentation medium can be an advantage for mycelium growth, but as fermentation progresses, the fermentation medium becomes more acidic and this can lead to low pigment yields [54].

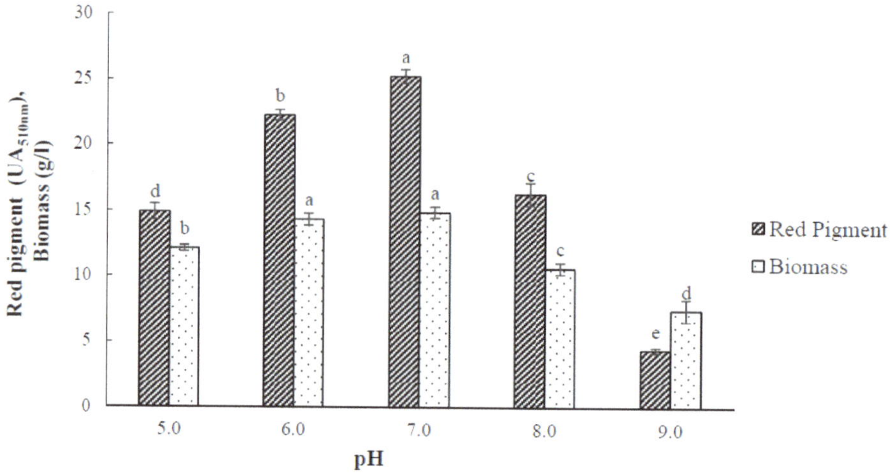

Figure 3. The effect of initial pH on the red pigment and biomass production by M. purpureus CMU 001 (Fermentation conditions: 30 °C, 8 days, 200 rpm, 5 g/L MSG, 50 g/L lactose).

3.5. Effect of Monosodium Glutamate (MSG) Concentration on Pigment Production

The previous studies showed that *M. purpureus* had higher red pigment production efficiency in the presence of MSG than other nitrogen sources [9,15]. Pigments produced by *Monascus* species are usually intracellular and insoluble in water. These pigments turn into extracellular and water-soluble red pigments as a result of a non-enzymatic reactions in the presence of MSG at neutral pH values. MSG replaces the ammonia in the orange pigment to produce the red pigment derivatives [51,55-57]. To investigate the effect of MSG concentration on red pigment production, SHF was performed with DM whey containing 75 g/L of lactose at pH 7.0 and 30°C temperature during 8 days. As seen in Figure 5, MSG concentration had a significant effect on red pigment synthesis by *M. purpureus.* The pigment production increased steadily and linearly up to 25 g/L of MSG concentration resulting in a maximum pigment synthesis of 45.7 $UA510_{nm}$ and then declined at higher MSG concentrations. Atalay et al. [15] tested different nitrogen sources (monosodium glutamate, malt sprouts, corn steep liquor, peptone, urea, ammonium sulfate and yeast extract) for *Monascus* pigment production using residual beer as the fermentation medium and found that fermentation medium containing 7.5 g/L of MSG yielded the highest red pigment production of 18.5 UA510 $_{nm}$. Sharmila et al. [1] obtained the maximum pigment synthesis by using 6 g/L of MSG as the nitrogen source. Babitha et al. [26] stated that Jackfruit seeds could not produce water-soluble pigments without using additional nitrogen sources. Zhang et al. [57] determined that glutamate and glycine was the most suitable source for growth of *Monascus.* Silbir and Goksungur [9] and Lee et al. [51] also used a submerged culture technique to determine the effect of various nitrogen sources on red pigment synthesis by *M. purpureus* and obtained maximum pigment production when MSG was used. bee et al. [51] stated that increased MSG concentrations decreased pigment production while increasing biomass concentrations. Hamano and Kilikian [25] stated that the highest pigment production (20.7 U) was obtained in a fermentation medium containing 7.6 g/L of MSG concentration. Our results showed that the optimal MSG concentration found for DM is different and higher than the concentration values found for other substrates in the literature for the production of red pigment by *M. purpureus.* Demineralized whey medium might contain low concentration of nitrogen as stated before. When higher MSG concentrations are used to overcome this problem, *Monascus purpureus* gave higher pigment production values.

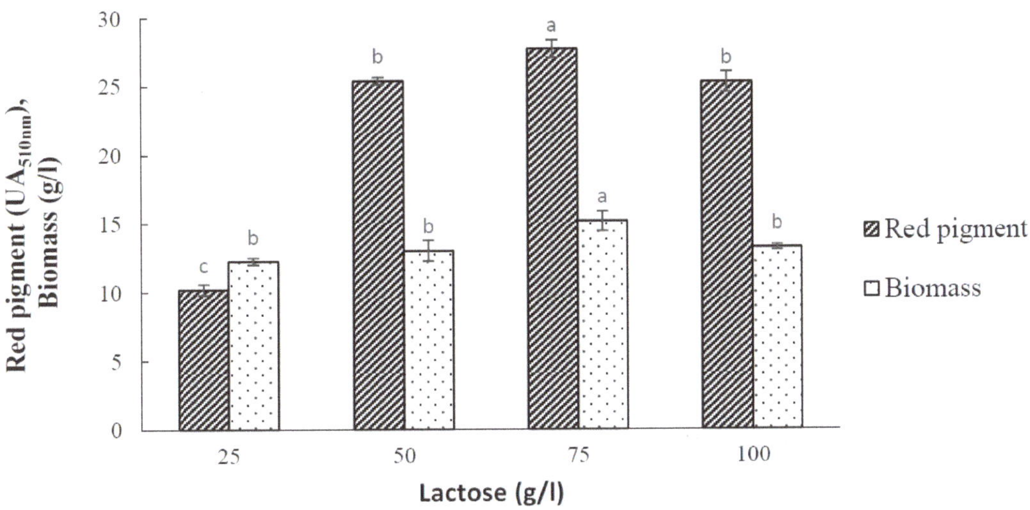

Figure 4. The effect of initial lactose concentration on red pigment and biomass production by M. purpureus CMU 001 (Fermentation conditions: 30°C, 8 days, 200 rpm, pH 7.0, 5 g/L MSG).

3.6. Effect of Inoculation Ratio on Pigment Production

To investigate the effect of inoculation ratio on red pigment production, DM medium was inoculated with 1 , 2, 3, 4, 5, and 6% (v/v) spore suspension solutions. SHF was performed at 30°C *for* 8 days (pH 7.0) with initial lacaoseand MSG concentrations oa 75 g/L and 5 g/L, respectively (Figure 6). Although maximum pigmena synthesis was obtained at an MSG concentration of 25 g/L, an MSG concentration of 5 g/L was used in the fermentation experiments cine to economical considerations and feasibility of tire process. It was observed that the highest pigment production of 38.4 UA510$_{nm}$ was obtained when DM fermentation medium was inoculated with 5^{0}% of spore suspension. The results of this study showed that different inoculum ratios influenced the synthesis of the red pigment bye *M. purpureus,* while- 2 to 5%t inoculum ratios had very little effect on the- biomass concentration. Babitha et al. [260 studied pigment production by *M. purpureus* from jackfruit seed using solid-state fermentation and reported th at low inoculum concentration produced insufficient biomass whale high inoculum conrentrations ted to excessive biomass formation depleting the nutrients in the fermentation medium that are necessary for the product formation. Atalay et al. [15] studied pigment production from residual beer using *M. purpureus* and obtained the highest pigment production of 18.5 UA$_{510\ nm}$ with the fermentation medium inoculated with 2% *(v/v)* of culture medium. Silbir and Goksungur [9] also studied pigment production from brewer's spent grain in submerged fermentation system and found that 2% (v/v) spore suspension yielcit'd the highest p^{i}gment production of 22.3 UA$_{500nm}$.

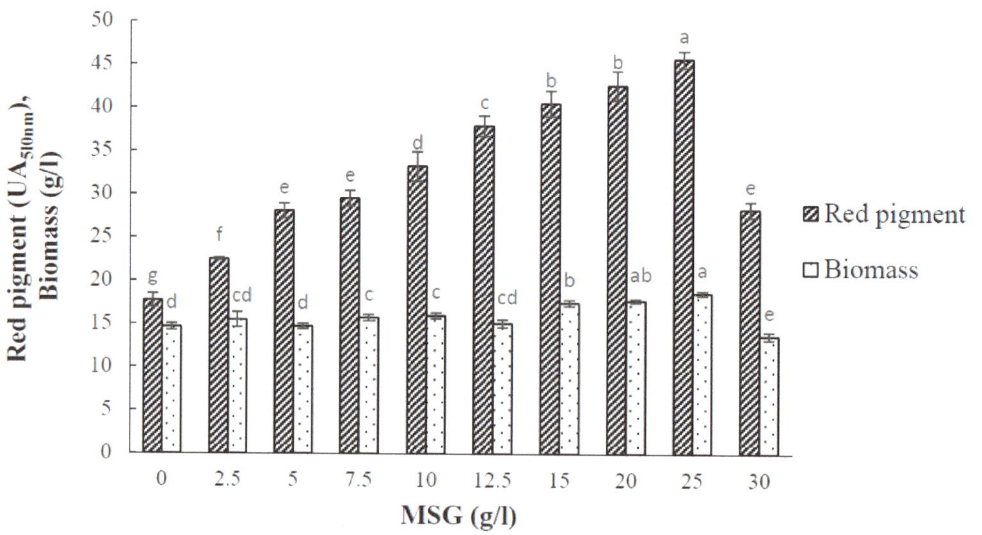

Figure 5. The effect of monosodium glutamate (MSG) concentration on the red pigment and biomass production by M. purpureus CMU 001 (Fermentation conditions: 30°C, 8 days, 200 rpm, pH 7.0, 75 g/L lactose).

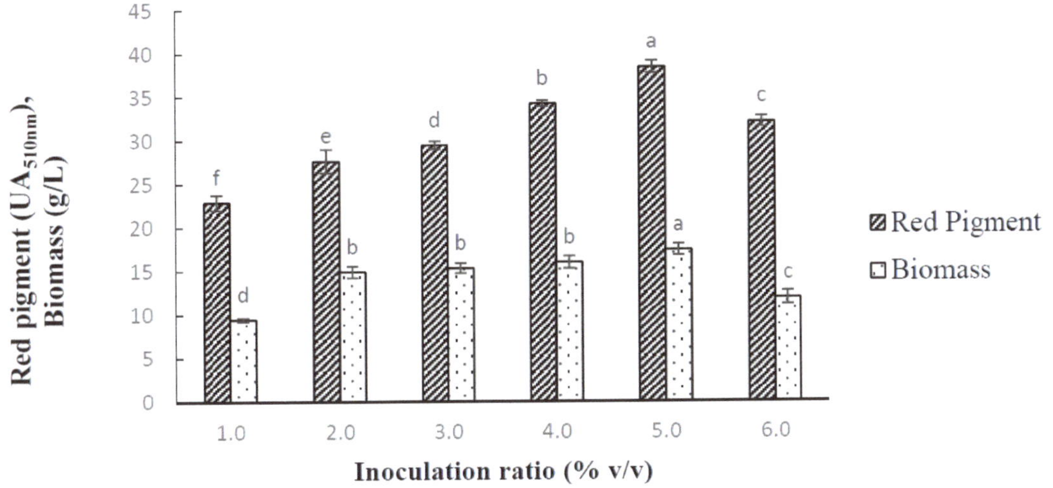

Figure 6. The effect of inoculation ration on the red pigment and biomass production by M. purpureus CMU 001 (Fermentation conditions: 30°C, 8 days, 200 rpm, pH 7.0, 5 g/L MSG, 75 g/L lactose.

3.7. Kinetics of Red Pigment Production by M. purpureus

The mycelial deve lopment and pigment synohesis kinetics ot *M. purpureus* were investigated by SHF under the optimized conditions (demineralized whey diluted to contain 75 g/L lactose, 5 g/L MSG concentrations, pH 7.0,30°C, 200 rpm, and 8 days). As shown in Figure 7, red pigment production started at the beginning of the exponential growth phase and reached a maximum value of 37 $UA510_{nm}$ on the 8th days of fermentation. The red pigment concentration decreased after the 8th day of fermentation most probably due to the substrate limitation, possible chemical decomposition of the pigment, conversion to other products or oxidation by the micboorganisms. The decrease in to tail sugar concentration during fermentation proved that hydrolyznd lactose was used by *M. purpureus* for growth and pigment synthesis. The pH value dropped during the first 3 days of fermentation and increased afterwards to 8.0 at tire end of the fermentation. The pH increase in the last: stage oh fermentation was probably due to *M. purpureus* producing ammonia as a result of the deamination of amino acids. Dry biomass weight steadily increased during the 8 days of fermentation period and then decreased.

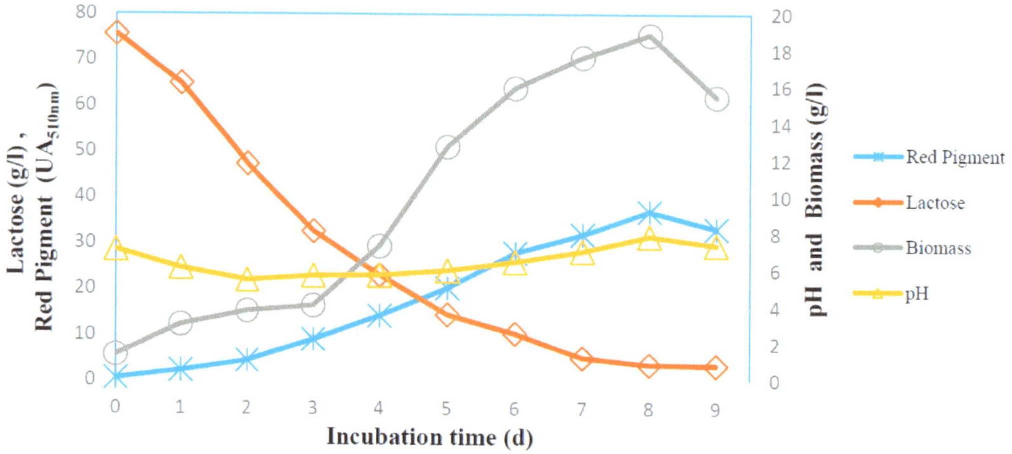

Figure 7. Simultaneous hydrolysis and fermentation (SHF) profiles of mycelial growth of M. purpureus CMU 001, red pigment synthesis, lactose consumption, and pH under the optimum conditions (Fermentation conditions: 30 o C, 8 days, 200 rpm, pH 7.0, 5 g/L MSG, 75 g/L lactose). The standard deviation of each experimental point ranged from 1.5 to 4.8.

Maximum pigment productivity and specific growth rate of *M. purpureus* were determined as 4.55 UAh^{-1} and 0.023 h^{-1}, respectively for SHF of DM whey at the optimized conditions in this study. Da Costa and Vendrusculo [45] investigated the use of hydrolyzed lactose, glucose, and lactose as a substrate for pigment production. While productivity values were calculated as 0.059 AU510 h^{-1},0.072 AU510 h^{-1}, and 0.032 AU510 h^{-1}, specific growth rate values were determined as 0.031 h^{-1},0.042 h^{-1}, and 0.017 h^{-1} for hydrolyzed lactose, glucose, and lactose, respectively. Atalay et al. [15] produced 18.5 UA510 red pigment from residual beer by *M. purpureus* after 192 h of fermentation. Productivity and specific growth rate values were found as 2.3 UAh^{-1} and 0.03 h^{-1} respectively. Even though the calculated specific growth rate value of *M. purpureus* grown on DM whey was in the range reported by the other authors, pigment productivity was found to be higher than the other substrates. These findings make DM whey a productive and sustainable substrate for the production of red pigment by SHF of *M. purpureus.*

3.8. Assesment of Production Cost of Pigment Production

The utilization of agricultural residues and organic waste from agro-industries as carbon sources for the production of value-added products reduce the production cost and provide environmental sustainability. On the other hand, product yield, transporting, handling and pretreatment costs should be considered when wastes are used as inexpensive organic substrates for fermentation processes. The operational cost includes the handling and pretreatment processes of raw materials, fermentation, downstream processes, labor, and maintenance along with the plant and administrative costs. Furthermore, the pretreatment process of the waste substrate should be appropriate in terms of process and environmental sustainability. Physical, chemical and hydrothermal processes are among the several pretreatments applied to agricultural waste materials [58].

As can be seen from Table 1, the amount of pigment produced from soybean, coconut and bagasse is low, therefore transportation and handling costs for utilizing large volumes of waste materials for pigment production will be very high. The concentration of pigment produced from corn meal supplemented with glucose (6%) and whey (1%) is higher than pigment concentrations when other waste materials are used. The addition of whey to the fermentation medium may have provided the necessary nitrogen and mineral sources for cell proliferation. Since coconut and bagasse contain glucose sources in the form of cellulose, the pigment yield will also be low unless the substrate is pretreated. Hence, the use of cellulosic agricultural waste in fermentation requires a costly pretreatment process and more waste will be generated after the fermentation limiting the environmental sustainability of the process. However, instead of lignocellulosic waste, the use of agro industrial wastes such as whey that does not require many pre-treatment unit operations prior to the fermentation seems to be a more economical and sustainable alternative for the red pigment production process.

Table 1. Pigment yields and necessary waste material for red pigment production.

Agricultural Waste Material	Pigment Yield (mg/g Dry Substrate)	Waste Material (g Dry Substrate/g Pigment)	Cost ($/kg)	References
Soy bean residues (Supp. with 6% glucose)	1.65	606.06		[2]
Coconut residues (Supp. with 6% glucose)	5.65	176.99		[2]
Bagasse (Supp. with 6% glucose)	7.5	133.33		[2]
Corn meal (Supp. with 6% glucose)	20.86	47.94		[2]
Corn meal (Supp. with 6% glucose + 1% whey)	47.42	21.09		[2]
Demineralized whey (6% glucose equivalent)	133.3	7.50	14.92	This Study
Glucose (6%)	166.67	6.00	14.84	Theoretical

In this study, the operational cost of red pigment production by *M. purpureus* is calculated as 14.92 dollars/kg when commercially processed, demineralized whey was used for pigment production. This calculated cost is close to the pigment production process cost when glucose is used as the carbon source, since both are processed products. Pigment production cost can be reduced by the use of raw liquid whey coming directly from cheese production, but the membrane filtration process needed to concentrate the sugar will possibly increase the production cost of the process. Therefore, it is important to concentrate more on the economic feasibility of the process for future work.

4. CONCLUSIONS

Suitable substrates rich in nutrients are necessary for microbial growth in the fermentation processes to produce economical, sustainable high value-added bioproducts. In this study, whey from the dairy industry was investigated as a fermentation medium for red pigment production by *M. purpureus.* To optimize fermentation conditions and maximize the yield, the effects of whey type, fermentation methods, initial lactose concentration, MSG concentration, initial pH, and inoculation ratio on red pigment synthesis by *M. pur- pureus* were systematically investigated. The results showed that DM whey is a suitable substrate for the red pigment synthesis by *M. purpureus* and SHF was highly affected by the tested process parameters. This research provides new insights into the utilization of whey produced in massive amounts and presents a possible solution for serious environmental pollution problems. The highest pigment synthesis was obtained with treated samples of whey (DM and DPDM) in this study. Since the nutrients or inhibitory compounds are different in each form of whey, they might have different effects on biomass formation and product synthesis. Hence, future studies will focus on the full characterization of all whey types and the positive and negative effects of the major and minor constituents on pigment synthesis by *Monascus purpureus.*

REFERENCES

1. Sharmila, G.; Nidhi, B.; Muthukumaran, C. Sequential statistical optimization of red pigment production by Monascus pur-pureus (MTCC 369) using potato powder. *Ind. Crops Prod.* **2013**, *44*,158-164. [CrossRef]

2. Nimnoi, P.; Pongsilp, N.; Lumyong, S. Utilization of agro-industrial products for increasing red pigment production of *Monascus purpureus* (AHK12). *Chiang Mai J. Sci.* **2015**, 42, 331-338.

3. Aberoumand, A. A review article on edible pigments properties and sources as natural biocolorants in foodstuff and food industry. *World J. Dairy Food Sci.* **2013**, 6, 71-78.

4. Panesar, R.; Kaur, S.; Panesar, P.S. Production of microbial pigments utilizing agro-industrial waste: A review. *Curr. Opin. Food Sci.* **2015**,1, 70-76. [CrossRef]
5. Dominguez-Espinosa, R.M.; Webb, C. Submerged fermentation in wheat substrates for production of Monascus pigments. *World J. Microbiol. Biotechnol.* **2003**,19,329-336. [CrossRef]
6. Dufosse, L. Microbial production of food grade pigments. *Food Technol. Biotechnol.* **2006**, 44, 313-321.
7. Kumura, H.; Ohtsuyama, T.; Matsusaki, Y.; Taitoh, M.; Koyanagi, H.; Kobayashi, K.; Hayakawa, T.; Wakamatsu, J.; Ishizuka, S. Application of red pigment producing edible fungi for development of a novel type of functional cheese. *J. Food Process. Preserv.* **2018**, 42, e13707. [CrossRef]
8. Agboyibor, C.; Kong, W.-B.; Chen, D.; Zhang, A.-M.; Niu, S.-Q. Monascus pigments production, composition, bioactivity and its application: A review. *Biocatal. Agric. Biotechnol.* **2018**,16, 433-447. [CrossRef]
9. Silbir, S.; Goksungur, Y. Natural red pigment production by *Monascus purpureus* in submerged fermentation systems using a food industry waste: Brewer's spent grain. *Foods* **2019**, 8,161. [CrossRef]
10. Silveira, S.T.; Daroit, D.J.; Brandelli, A. Pigment production by *Monascus purpureus* in grape waste using factorial design. *LWT Food Sci. Technol.* **2008**, 41,170-174. [CrossRef]
11. Feng, Y.; Shao, Y.; Chen, F. Monascus pigments. *Appl. Microbiol. Biotechnol.* **2012**, 96,1421-1440. [CrossRef] [PubMed]
12. Vendruscolo, F.; Muller, B.L.; Moritz, D.E.; Oliveira, D.; Schmidell, W.; Ninow, J.L. Thermal stability of natural pigments pro-duced by Monascus ruber in submerged fermentation. *Biocatal. Agric. Biotechnol.* **2013**,2, 278-284. [CrossRef]
13. Gunerhan, U.; Us, E.; Dumlu, L.; Yilmaz, V.; Carrere, H.; Perendeci, A. Impacts of chemical-assisted thermal pretreatments on methane production from fruit and vegetable harvesting wastes: Process optimization. *Molecules* **2020**, 25, 500. [CrossRef]
14. Liu, J.; Luo, Y.; Guo, T.; Tang, C.; Chai, X.; Zhao, W.; Bai, J.; Lin, Q. Cost-effective pigment production by *Monascus purpureus* using rice straw hydrolysate as substrate in submerged fermentation. *J. Biosci. Bioeng.* **2020**,129, 229-236. [CrossRef] [PubMed]
15. Atalay, P.; Sargin, S.; Goksungur, Y. Utilization of residual beer for red pigment production by *Monascus purpureus* in submerged fermantation. *Fresenius Environ. Bull.* **2020**, 29,1025-1034.

16. Kantifedaki, A.; Kachrimanidou, V.; Mallouchos, A.; Papanikolaou, S.; Koutinas, A. Orange processing waste valorisation for the production of bio-based pigments using the fungal strains *Monascus purpureus* and Penicillium purpurogenum. *J. Clean. Prod.*

17. Orak, T.; Caglar, O.; Ortucu, S.; Ozkan, H.; Taskin, M. Chicken feather peptone: A new alternative nitrogen source for pigment production by *Monascus purpureus. J. Biotechnol.* **2018**,271, 56-62. [CrossRef]

18. Silveira, S.T.; Daroit, D.J.; Sant'Anna, V.; Brandelli, A. Stability Modeling of red pigments produced by *Monascus purpureus* in submerged cultivations with Sugarcane Bagasse. *Food Bioprocess Technol.* **2011**, 6,1007-1014. [CrossRef]

19. Hilares, R.T.; de Souza, R.A.; Marcelino, P.F.; da Silva, S.S.; Dragone, G.; Mussatto, S.I.; Santos, J.C. Sugarcane bagasse hy-drolysate as a potential feedstock for red pigment production by Monascus ruber. *Food Chem.* **2018**, 245, 786-791. [CrossRef] [PubMed]

20. Haque, A.; Kachrimanidou, V.; Koutinas, A.; Lin, C.S.K. Valorization of bakery waste for biocolorant and enzyme production by *Monascus purpureus. J. Biotechnol.* **2016**, 231, 55-64. [CrossRef]

21. Singh, N.; Goel, G.; Singh, N.; Pathak, B.K.; Kaushik, D. Modeling the red pigment production by *Monascus purpureus* MTCC 369 by Artificial Neural Network using rice water based medium. *Food Biosci.* **2015**,11,17-22. [CrossRef]

22. Srivastav, P.; Yadav, V.K.; Govindasamy, S.; Chandrasekaran, M. Red pigment production by *Monascus purpureus* using sweet potato-based medium in submerged fermentation. *Nutrafoods* **2015**,14,159-167. [CrossRef]

23. Velmurugan, P.; Hur, H.; Balachandar, V.; Kamala-Kannan, S.; Lee, K.-J.; Lee, S.-M.; Chae, J.-C.; Shea, P.J.; Oh, B.-T. Monascus pigment production by solid-state fermentation with corn cob substrate. *J. Biosci. Bioeng.* **2011**,112, 590-594. [CrossRef] [PubMed]

24. Zhou, Z.; Yin, Z.; Hu, X. Corncob hydrolysate, an efficient substrate for Monascus pigment production through submerged fermentation. *Biotechnol. Appl. Biochem.* **2014**, 61, 716-723. [CrossRef] [PubMed]

25. Hamano, P.S.; Kilikian, B.V. Production of red pigments by Monascus ruber in culture media containing corn steep liquor. *Braz. J. Chem. Eng.* **2006**, 23, 443-449. [CrossRef]

26. Babitha, S.; Soccol, C.R.; Pandey, A. Solid-state fermentation for the production of Monascus pigments from jackfruit seed. *Bioresour. Technol.* **2007**, *98*,1554-1560. [CrossRef]

27. Hamdi, M.; Blanc, P.J.; Goma, G. Effect of aeration conditions on the production of red pigments by *Monascus purpureus* growth on prickly pear juice. *Process. Biochem.* **1996**, 31, 543-547. [CrossRef]

28. Guven, G.; Perendeci, A.; Ozdemir, K.; Tanyolac, A. Specific energy consumption in electrochemical treatment of food industry wastewaters. *J. Chem. Technol. Biotechnol.* **2012**, 87, 513-522. [CrossRef]

29. Hausjell, J.; Miltner, M.; Herzig, C.; Limbeck, A.; Saracevic, Z.; Saracevic, E.; Weissensteiner, J.; Molitor, C.; Halbwirth, H.; Spadiut, O.; et al. Valorisation of cheese whey as substrate and inducer for recombinant protein production in E. coli HMS174(DE3). *Bioresour. Technol. Rep.* **2019**, 8,100340. [CrossRef]

30. Nadais, M.H.; Capela, M.I.; Arroja, L.M.; Hung, Y.T. Anaerobic treatment of milk processing wastewater. In *Environmental Bioengineering;* Humana Press: Totowa, NJ, USA, 2010; Volume 11, pp. 555-618.

31. Britz, J.T.; van Schalwyk, C.; Hung, Y.T. Treatment of dairy processing wastewaters. In *Waste Treatment in the Food Processing Industry;* Wang, L.K., Hung, Y.T., Lo, H.H., Yapijakis, C., Eds.; CRC Press: Boca Raton, FL, USA, 2006; pp. 1-25.

32. Karadag, D.; Koroglu, O.E.; Ozkaya, B.; Cakmakci, M. A review on anaerobic biofilm reactors for the treatment of dairy industry wastewater. *Process. Biochem.* **2015**, 50,262-271. [CrossRef]

33. Yadav, J.S.; Yan, S.; Pilli, S.; Kumar, L.; Tyagi, R.D.; Surampalli, R.Y. Cheese whey: A potential resource to transform into bio-protein, functional/nutritional proteins and bioactive peptides. *Biotechnol. Adv.* **2015**, 33, 756-774. [CrossRef]

34. Smithers, G.W. Whey-ing up the options—Yesterday, today and tomorrow. *Int. Dairy J.* **2015**, 48, 2-14.

35. Prazeres, A.R.; Carvalho, F.; Rivas, J. Cheese whey management: A review. *J. Environ. Manag.* **2012**,110, 48-68. [CrossRef]

36. Das, M.; Raychaudhuri, A.; Ghosh, S.K. Supply chain of bioethanol production from whey: A review. *Procedia Environ. Sci.* **2016**, 35, 833-846. [CrossRef]

37. Silbir, S. Bira Atigindan Monascus Renk Pigmentleri Uretimi ve Stabilitesinin Belirlenmesi. Ph.D. Thesis, Ege University, Bornova/Izmir, Turkey, 2019; p. 590713. (In Turkish)

38. Roukas, T.; Mantzouridou, F.; Boumpa, T.; Vafiadou, A.; Goksungur, Y. Production of |3-carotene from beet molasses and deproteinized whey by Blakeslea trispora. *Food Biotechnol.* **2003**,17, 203-215. [CrossRef]

39. Mukherjee, G.; Singh, S.K. Purification and characterization of a new red pigment from *Monascus purpureus* in submerged fermentation. *Process. Biochem.* **2011**, 46,188-192. [CrossRef]

40. Shi, K.; Song, D.; Chen, G.; Pistolozzi, M.; Wu, Z.; Quan, L. Controlling composition and color characteristics of Monascus pigments by pH and nitrogen sources in submerged fermentation. *J. Biosci. Bioeng.* **2015**,120,145-154. [CrossRef]

41. Dubois, M.; Gilles, K.A.; Hamilton, J.K.; Rebers, P.A.; Smith, F. Colorimetric method for determination of sugars and related substances. *Anal. Chem.* **1956**, 28, 350-356. [CrossRef]

42. Kjeldahl, J.G.C. En ny Methode til Kvaelstofvestemmelse i organiske Stoffer. *Z. Anal. Chem.* **1883**,22, 366-382. (In German) [CrossRef]

43. Cataldi, T.R.I.; Angelotti, M.; D'Erchia, L.; Altieri, G.; Di Renzo, G.C. Ion-exchange chromatographic analysis of soluble cations, anions and sugars in milk whey. *Eur. Food Res. Technol.* **2003**, 216, 75-82. [CrossRef]

44. Duke, M.; Vasiljevic, T. Whey demineralization with membrane operations. In *Encyclopedia of Membranes;* Droli, E., Giorno, L., Eds.; Springer: Berlin/Heidelberg, Germany, 2015.

45. Da Costa, J.P.V.; Vendruscolo, F. Production of red pigments by Monascus ruber CCT 3802 using lactose as a substrate. *Biocatal. Agric. Biotechnol.* **2017**,11, 50-55. [CrossRef]

46. Lopes, F.C.; Tichota, D.M.; Pereira, J.Q.; Segalin, J.; Rios, A.D.O.; Brandelli, A. Pigment production by filamentous fungi on agro-industrial byproducts: An eco-friendly alternative. *Appl. Biochem. Biotechnol.* **2013**,171, 616-625. [CrossRef] [PubMed]

47. Pisareva, E.; Kujumdzieva, A. Taxonomic investigation and growth characteristics of citrinin free Monascus pilosus C1 strain. *Biotechnol. Biotechnol. Equip.* **2006**,20, 88-96. [CrossRef]

48. Prajapati, V.S.; Soni, N.; Trivedi, U.B.; Patel, K.C. An enhancement of red pigment production by submerged culture of *Monascus purpureus* MTCC 410 employing statistical methodology. *Biocatal. Agric. Biotechnol.* **2014**, 3,140-145. [CrossRef]

49. Kang, B.; Zhang, X.; Wu, Z.; Qi, H.; Wang, Z. Effect of pH and nonionic surfactant on profile of intracellular and extracellular Monascus pigments. *Process. Biochem.* **2013**, 48, 759-767. [CrossRef]

50. Orozco, S.F.B.; Kilikian, B.V. Effect of pH on citrinin and red pigments production by *Monascus purpureus* CCT3802. *World J. Microbiol. Biotechnol.* **2007**,24, 263-268. [CrossRef]

51. Lee, B.-K.; Park, N.-H.; Piao, H.Y.; Chung, W.-J. Production of red pigments by *Monascus purpureus* in submerged culture. *Biotechnol. Bioprocess Eng.* **2001**, 6, 341-346. [CrossRef]

52. Chen, M.-H.; Johns, M.R. Effect of carbon source on ethanol and pigment production by *Monascus purpureus. Enzym. Microb. Technol.* **1994**,16, 584-590. [CrossRef]

53. Liu,J.;Guo,T.;Luo,Y.;Chai,X.;Wu,J.;Zhao,W.;Jiao,P.;Luo,F.;Lin,Q.Enhancement of Monascus pigment productivity via a simultaneous fermentation process and separation system using immobilized-cell fermentation. *Bioresour. Technol.* **2019**, 272, 552-560. [CrossRef]

54. Wong, H.C.; Lin, Y.C.; Koehler, P.E. Regulation of growth and pigmentation of *Monascus purpureus* by carbon and nitrogen concentrations. *Mycologia* **1981**, 73, 649-654. [CrossRef]

55. Pastrana, L.; Blanc, P.; Santerre, A.; Loret, M.; Goma, G. Production of red pigments by Monascus ruber in synthetic media with a strictly controlled nitrogen source. *Process. Biochem.* **1995**, 30, 333-341. [CrossRef]

56. Meinicke, R.M.; Vendruscolo, F.; Moritz, D.E.; de Oliveira, D.; Schmidell, W.; Samohyl, R.W.; Ninow, J.L. Potential use of glycerol as substrate for the production of red pigments by Monascus ruber in submerged fermentation. *Biocatal. Agric. Biotechnol.* **2012**, *1*, 238-242. [CrossRef]

57. Zhang, X.-W.; Wang, J.-H.; Chen, M.-H.; Wang, C.-L. Effect of nitrogen sources on production and photostability of Monascus pigments in liquid fermentation. *IERI Procedia* **2013**, 5, 344-350. [CrossRef]

58. Perendeci, N.A.; Gokgol, S.; Orhon, D. Impact of alkaline H2O2 pretreatment on methane generation potential of greenhouse crop waste under anaerobic conditions. *Molecules* **2018**, 23,1794. [CrossRef]

CHAPTER 13

Application of a Biosurfactant Produced by Bacillus cereus UCP 1615 from Waste Frying Oil as an Emulsifier in a Cookie Formulation

1. INTRODUCTION

Globalization and the growth of the population have promoted the expansion of the production of, and investments in, complex food supply dynamics. However, food safety issues arise such as the origin and properties of products and components added to food. Most of these substances, including thickeners, stabilizers and emulsifiers, are important additives for the quality of food [1] because they help maintain or even improve their appearance, freshness, flavor, texture and safety [2].

The search for "green" ingredients has intensified in the food industry thanks to the progress of studies and the increase in competition in the sector as well as the growing interest of consumers for natural ingredients over synthetic additives [3]. This interest is mainly associated with the growing demand for natural, organic and vegan food [4].

Biosurfactants are promising products obtained from biological sources whose attractiveness is due to their biodegradability, specific bioactivity, sustainable production and low toxicity [5,6]. These features give biosurfactants considerable potential for practical applications particularly in the food, cosmetic, healthcare, biomedical and pharmaceutical sectors [5].

The literature describes improvements in the texture, volume and conservation of baked goods due to the addition of biosurfactants. Researchers reported improvements in the viscosity of food products when using microbial emulsifiers, the efficient emulsification of fat from meat products, enhanced solubilization of aromas and a greater stability of salad dressings [1]. Biosurfactants are also effective in solubilizing vegetable oils, stabilizing fats during cooking processes and improving the organoleptic properties of bread. Biomolecules can be used in ice cream formulations, muffins (as an ingredient to replace baking powder and eggs), cookies and salad dressings. The use of microbial emulsifiers was also shown to reduce the use of currently marketed emulsifiers in farinaceous food and to improve their rheology [1,2].

Among the different types of biosurfactants explored, lipopeptides and glycolipids stand out due to their desirable properties for application in the food industry such as antibacterial activity against a variety of species, antioxidant activity and low cytotoxicity. The lipopeptides produced by bacteria of the genus *Bacillus* are examples of microbial biosurfactants obtained by fermentation [7,8] whose main characteristics are an emulsification capacity and a reduction of surface tension along with antioxidant, antiadhesive, antibiofilm, antibacterial, antifungal, antitumor and antiviral properties [6,9].

However, the industrial production of biosurfactants faces great challenges due to the high costs of microbial cultivation and their recovery [10]. One of the solutions to make the industrial production of these biomolecules economically feasible consists of the use of agro-industrial co-products as substrates for the fermentation process, given that the substrate accounts for up to 50% of the final manufacturing cost [7]. The food industry generates waste products that often contain high concentrations of carbohydrates, lipids and proteins, which makes these co-products attractive candidates for fermentation processes [11]. Among such substances, waste cooking oil—which is produced in the kitchens of homes, restaurants and industries—is considered to be hazardous to the environment and human health; therefore, it should be collected to reduce the environmental impact of its improper disposal [12].

A 2019 report on the prospects for the commodities market by the Organization for Economic Cooperation and Development (OECD) and the Food and Agriculture Organization (FAO) of the United Nations states that approximately 210 million tons of vegetable oils are produced and consumed by humans every year. Therefore, the estimated annual global production of waste cooking oil is around 42 million tons [13]. The use of agroindustrial waste products or renewable raw materials in fermentation processes to produce biosurfactants is in line with green chemistry and is an important tool for sustainable innovation, which meets the demands of the current market [14].

Therefore, the aims of the present study were (a) to investigate the use of a biosurfactant produced by *Bacillus cereus* UCP 1615 as an additive in a cookie formulation, (b) to analyze the nutritional benefits of its addition, (c) to check its non-toxicity, (d) to determine its antioxidant potential and (e) to evaluate its effects on the physicochemical properties as well as the texture of the product.

2. MATERIALS AND METHODS

2.1. Microorganism and the Production of the Biosurfactant

The bacterium *Bacillus cereus* UCP 1615 obtained from the culture bank of the Catholic University of Pernambuco was used as the biosurfactant-producing microorganism. This strain was previously isolated from environmental samples of water contaminated with petroleum byproducts spilled from ships (port area) in the Atlantic Ocean in the state of Pernambuco, Brazil. As described by Durval et al. [15], the bacterium was cultured by adding a 2% cell suspension (optical density of 0.7 at a wavelength of 600 nm corresponding to a 24 h inoculum of 10^7 colony-forming units/L) to a 500 mL flask containing 100 mL of a mineral salt medium (0.087% K2HPO4, 0.65% trisaminomethane, 0.02% KCl, 0.06% $MgSO_4$-7H2O, 0.01% NaCl and 0.005% yeast extract) supplemented with 2% waste frying soybean oil and 0.12% peptone with the pH adjusted to 7.0 ± 0.2. After culturing for 48 h at 28 ^{0} C and 250 rpm, the samples were withdrawn to determine the concentration of the biosurfactant.

2.2. Isolation of the Biosurfactant

The biosurfactant was extracted after centrifugation of the fermentation broth at 5000 x *g* for 30 min to remove the cells. A 6.0 M HCl solution was added to the supernatant to adjust the pH to 2.0, followed by the addition of a 2:1 (v/v) solution of CHQ3/CH3OH. After vigorous shaking by hand for 15 min and a phase separation, the organic phase was removed and the operation was repeated two more times. The organic phases were pooled and put on a rotary evaporator under a vacuum at 40 °C until the complete evaporation of the solvents and the obtention of the isolated biosurfactant [15]. The extraction procedure allowed the isolation of the biomolecule from the fermented broth and, at the same time, the suppression of any live cells still present in it.

2.3. Evaluation of the Biosurfactant Cytotoxic Potential

The biosurfactant cytotoxicity was assessed using the 3-(4,5-dimethylthiazol-2-yl)- 2,5-diphenyltetrazolium bromide) (MTT) method [16,17]. The L929 (mouse fibroblast) cells and the Vero (renal epithelial) cells from the African green monkey were maintained in Dulbecco's Modified Eagle Medium supplemented with 10% inactivated fetal bovine serum and a 1% antibiotic (penicillin and streptomycin) solution. The cells were kept in a chamber at 37 °C in an atmosphere enriched with 5% CO2 and 95% humidity.

The cells were plated (10^5 cells/mL) in 96-well plates and incubated for 24 h. Next, 10 pL of the biosurfactant solutions was added to the wells at a final concentration of 3.12 to 400 pg/mL. After a further incubation for 72 h, 25 pL of the MTT solution (5 mg/mL) was added, followed by an incubation of 3 h. The culture with the MTT was then aspirated and 100 pL of dimethyl sulfoxide was added to each well. The absorbance was measured in a microplate reader at a wavelength of 560 nm. The experiments were conducted in triplicate and the percentage of inhibition was calculated with the aid of the demo version of GraphPad Prism 7.0 software (GraphPad Software, San Diego, CA, USA).

An intensity scale was used for the determination of toxicity. Samples with 95 to 100% inhibitory activity were considered to be highly toxic, those with 70 to 90% inhibitory activity were considered to be moderately toxic and those with inhibitory activity less than 50% were considered to be non-toxic [18].

2.4. Antioxidant Activity

2.4.1. Total Antioxidant Capacity

The total antioxidant capacity (TAC) was determined using the phosphomolybdenum method [19], which is based on the reduction of Mo^{6}+ to Mo^{5}+ by the sample and the subsequent formation of a greenish phosphate/Mo^{5}+ complex. Tubes containing the samples and reagents (0.6 M sulfuric acid, 28 mM sodium phosphate and 4 mM ammonium molybdate) were incubated at 100 ^{0} C for 90 min. The absorbance of each solution was measured at wavelength of 695 nm against a blank. The TAC was expressed in relation to a solution of ascorbic acid at a concentration of 1 mg/mL assumed to be 100%. All assays were carried out in triplicate.

2.4.2. DPPH Scavenging Activity

The antioxidant activity of the biosurfactant was also evaluated using the free radical sequestration method based on hydrogen donation using the stable 2,2-diphenyl-1- picrylhydrazyl radical (DPPH*) [20]. The measurements were performed in triplicate and the inhibition activity was calculated based on the percentage of DPPH* scavenged. The vitamin E analogue 6-hydroxy-2,5,7,8-tetramethylchroman-2-carboxylic acid (Trolox) and butylated hydroxytoluene (BTH) were used as standards. The percentage of inhibition (I%) was calculated using the equation:

$$I\% = \frac{A_c - A_s}{A_c} \times 100 \tag{1}$$

in which Ac is the absorbance of the control and As is the absorbance of the sample.

2.4.3. ABTS Scavenging Activity

The determination of the antioxidant activity of a biosurfactant using the 2,2'-azino-bis (3-ethylbenzothiazoline-6-sulfonic acid) (ABTS) assay is based on the generation of the cationic chromophore radical obtained from the oxidation of ABTS by potassium persulfate [20]. The measurements were performed in triplicate and the inhibition activity was calculated based on the percentage of ABTS removed. Trolox and BTH were used as standards. The percentage of inhibition (I%) was calculated as described above for the DPPH scavenging activity.

2.5. Emulsification Activity

The emulsification activity of the biosurfactant was determined following the method described by Gaur et al. [21]. Vegetable oils from soybeans, corn, canola and peanuts were added at a 1:1 proportion (*v*/*v*) to an aqueous biosurfactant solution at concentrations of 10, 5.0 and 2.5 mg/mL in glass tubes and the content was vortexed for 2 min at a frequency of 50 Hz. After 24 h, the emulsification index (E24) was determined accordi ion:

$$E_{24} = \frac{h_e}{h_t} \times 100 \qquad (2)$$

in which he is the height of the emulsion layer and ht is the total height of the mixture. All samples were stored at room temperature.

2.6. Cookie Formulation and Preparation

The biosurfactant was tested in the standard cookie formulation described by Ribeiro et al. [22] (Table 1). The extract containing the biosurfactant was used in this formulation as an additive at three different concentrations (1%, 0.5% and 0.25%) in relation to the total dough for the analysis of the physical and physicochemical properties.

Table 1. Composition of the cookie dough.

Ingredient	Standard Formulation (%)
White wheat flour	47.0
Margarine	20.0
Sugar	15.0
Vanilla extract	3.0
Baking powder	1.0
Pasteurized egg white	40.0
Pasteurized egg yolk	4.0
Biosurfactant	0.0

The dough was also prepared following the method described by Ribeiro et al. [22]. The ingredients were purchased from local establishments and blended in a mixer (Turbomix Duo MX21, Arno Ciranda, Florianopolis, SC, Brazil) for 7 min. The dough was then spread on a polyethylene cutting board and shaped into circular forms with the aid of stainless steel molds with a 50 mm diameter to standardize the cookies. The specimens were placed in an oven for 5 min at 150°C after which the temperature was increased to 180°C for an additional 15 min. The cookies were then cooled, weighed, packed in vacuum-sealed plastic bags and stored at room temperature for 24 h.

2.7. Physicochemical Properties and the Energy Value of the Cookies

The weight, diameter, thickness and spread factor of the cookies were determined with and without the biosurfactant [23,24]. The total diameter was determined by using four randomly selected specimens placed side by side. The cookies were then turned 90° and the diameter was measured again. The final diameter was expressed as the mean of the two measurements divided by four. The thickness was determined by stacking four cookies and expressed as the total height divided by four. The spread factor was determined by dividing the diameter by the height.

The physicochemical properties of the cookies were determined using the analysis methods of the Association of Official Analytical Chemists [25]. The moisture content was determined gravimetrically considering the mass loss from the specimens submitted to heating in an oven at 105°C until a constant weight was reached. The concentration of total protein was calculated using the Kjeldahl method, which consists of the acid digestion of organic matter followed by distillation, the quantitative determination of nitrogen by titration and multiplication of the obtained value by a factor of 6.5. The gravimetric method was used for the determination of the fixed mineral residue (ash) based on the loss of mass from the specimens submitted to incineration at 550°C. The Bligh and Dyer [26] extraction method was employed to quantify the lipids. The energy value was determined by the sum of the values of carbohydrates, lipids and proteins multiplied by 4, 9 and 4, respectively [27].

2.8. Texture Profile Analysis

The texture profile analysis involved the determination of the hardness, cohesiveness, adhesiveness and springiness of the cookie dough with and without the biosurfactant after baking. For this purpose, a CT3 texture analyzer (Brookfield, Middleboro, MA, USA) was used with a 245 N load cell. The specimens were compressed at a constant velocity of 1 mm/s using a 60 mm-wide polymethacrylate plate. A second compression was performed after a 5 s interval and the hardness was defined as the force of half of the height on the specimen during the first compression. The cohesion was defined as the ratio between the compression work in the second compression cycle and the compression work in the first cycle. The sponginess was calculated using the relative height of the remaining specimen when the initial force was registered during the second compression [28].

2.9. Statistical Analysis

The data were analyzed statistically using the one-way procedure in Statistica (version 7.0) (StatSoft. Inc, Tulsa, OK, USA), followed by a linear one-way analysis of variance (ANOVA). The results were expressed as mean ± standard deviation (SD) determined from the triplicate experiments. The differences were examined using the Tukey post hoc test with a 95% confidence level, i.e., a significant level (p) of 0.05.

3. RESULTS AND DISCUSSION

3.1. Cytotoxicity of the Biosurfactant

Cytotoxicity tests were among the first in vitro bioassays used to predict the toxicity of substances and are performed in laboratories throughout the world to classify compounds and evaluate safety [29]. The biosurfactant produced by *Bacillus cereus* UCP 1615 was submitted to the MTT assay to monitor the response of the cells in the cultures and determine the viability of the biosurfactant for human consumption.

The results of the cytotoxicity tests regarding the viability of the L929 (mouse fibroblast) cells and the Vero (renal epithelial) cells from the African green monkey exposed to different concentrations of the biosurfactant are displayed in Table 2.

Table 2. Viability (percentage) of L929 and Vero cell lines after contact with the biosurfactant from Bacillus cereus UCP 1615 at different concentrations (data expressed as mean ± SD of the triplicate determinations).

Concentration (pg/mL)	Cell Viability (%)	
	L929 Line	Vero Line
Control	99.99 ± 3.33	100.03 ± 2.80
3.12	111.41 ± 5.18	102.66 ± 5.79
6.25	103.00 ± 5.42	103.51 ± 5.28
12.5	103.80 ± 2.56	101.25 ± 4.96
25	92.89 ± 2.56	98.13 ± 3.56
50	92.17 ± 2.49	94.84 ± 5.60
100	98.61 ± 0.56	86.05 ± 5.21
200	104.30 ± 2.09	74.77 ± 6.40
400	53.89 ± 1.80	12.13 ± 0.55

The viability of the L929 cells was 54% when submitted to the highest concentration of the biosurfactant (400 pg/mL) but above 93% when submitted to concentrations between 3.12 and 200 pg/mL. The viability of the Vero cells was 12.13, 74.77 and 86.05% when exposed to concentrations of 400,200 and 100 pg/mL, respectively, but above 95% when exposed to concentrations between 3.12 and 50 pg/mL (Table 2). Substances that enable an 80% or higher cell viability rate are considered to be without cytotoxic activity [30].

The MTT results revealed that the biosurfactant may have a potential application in food as it did not exhibit cytotoxicity to either cell line at concentrations of up to 100 pg/mL, equivalent to 0.1 g/L. Moreover, the viability of the L929 cell line was 100% at the relatively high concentration of 200 pg/mL. These results are compatible with those reported by Ribeiro et al. [22] for the biosurfactant produced by *Saccharomyces cerevisiae* incorporated into a cookie formulation, which exhibited no toxicity to the L929 and RAW 264.7 (mouse macrophage) cell lines.

The use of the biosurfactant produced by *Candida bombicola* URM 3718 in a cupcake formulation was successful after the determination of its toxicity to the L929 and Vero cell lines at concentrations up to 50 pg/mL [31]. The survival rate of the BHK-21 cell line (kidney cells from hamster pups) was 63% after exposure to a biosurfactant produced by *Bacillus cereus* MMC at a concentration of 10^4 pg/mL [32]. The biosurfactant produced by *Lactobacillus helveticus* also exhibited no cytotoxicity to the L929 cell line at concentrations of up to 25 x 10^3 pg/mL [33].

3.2. Antioxidant Activity of the Biosurfactant

Oxidation can occur during the processing and/or storage of food, resulting in the deterioration of their nutritional value, color, flavor, texture and safety. The most effective, convenient and economical method employed to control oxidation is the use of antioxidants [34]. The food industry uses antioxidants to stabilize lipids in food, which are the most sensitive compounds to the oxidation process [35]. In addition to the preservation of food, antioxidants are also used in fields related to health and wellbeing due to their capacity to protect the body from oxidative damage.

The results of the total antioxidant capacity (TAC) as well as those of scavenging the DPPH' radical and the ABTS cation radical (ABTS*+) by the biosurfactant are presented in Table 3.

Table 3. Total antioxidant capacity (TAC) and oxidative inhibition based on the DPPH* and ABTS*+ scavenging capacity of the biosurfactant produced by Bacillus cereus UCP1615 at different concentrations.

Biosurfactant Concentration (mg/mL)	TAC (%)	DPPH' (I%)	ABTS'+ (%)
40.00	476.43 ± 12.34	28.45 ± 3.24	36.67 ± 4.23
20.00	353.46 ± 10.45	19.34 ± 5.34	25.62 ± 3.52
10.00	218.25 ± 14.37	11.23 ± 3.28	18.24 ± 4.23

Biosurfactant Concentration (mg/mL)	TAC (%)	DPPH' (I%)	ABTS'+ (%)
5.00	147.56 ± 17.45	4.34 ± 1.35	10.23 ± 3.60
2.50	98.35 ± 8.56	2.14 ± 1.11	7.24 ± 2.49
1.25	49.56 ± 4.03	0.35 ± 0.04	4.65 ± 0.97
0.62	27.45 ± 3.15	0.00 ± 0.00	3.35 ± 1.38
0.32	15.34 ± 7.36	0.00 ± 0.00	2.84 ± 1.87

The biosurfactant showed promising results in terms of reducing the phosphomolybdenum complex when comparing its percentage of the total antioxidant capacity (TAC) with that of ascorbic acid at a concentration of 1 mg/mL. The addition of the biosurfactant at the lowest concentration (0.32 mg/mL) led to a TAC of 15.34%; it exceeded 100% with concentrations above 2.5 mg/mL and reached 476.43% at the highest concentration (40 mg/mL). A linear increase in the TAC with the increase in biosurfactant concentration was observed.

One percent (equivalent to 10 mg/mL) was the maximum concentration of the biosurfactant tested in the present study for application in the cookie formulation. The TAC at this concentration was 218.25%, demonstrating that the biosurfactant had potential regarding protection from oxidation in food. The total antioxidant activity in this study was consistent with the indices reported for biosurfactants from *Candida bombicola* [31] and *Saccharomyces cerevisiae* [22], which were also evaluated for use in food.

In the assessment of the oxidative inhibition in terms of the DPPH' scavenging capacity of the biosurfactant, the results were low even at the highest concentration tested (40 mg/mL), preventing only 28.45% of oxidation (Table 3). In the assessment of the antioxidant activity based on ABTS'+ scavenging, the index achieved at the highest biosurfactant concentration was 36.67%. These results show that the biosurfactant under investigation did not have sufficient antioxidant potential to serve as the only antioxidant agent in a formulation.

3.3. Emulsification Activity

Stability is an important indicator when determining the commercial value of food products with water-in-oil emulsions. However, these emulsions are thermodynamically unstable due to the large interfacial area of the dispersed phase [36]. Their structural organization and amphiphilic nature make biosurfactants excellent emulsifiers acting at the oil-water interface, promoting the thermodynamic stability of unstable systems. Moreover, the characteristics of biosurfactants enable these natural compounds to interact with carbohydrates and proteins in food products [37].

The choice of vegetable oils in the emulsification tests was based on their importance and use in the food industry. Soybean oil stands out in terms of production and consumption whereas the other oils were selected due to their specific beneficial and functional properties for human consumption. In particular, peanut oil has a high vitamin E content, canola oil has a low content of saturated fatty acids and contains omega 3, and corn oil has essential acids and is considered to be of a high quality [38]. The emulsifying capacity of the biosurfactant produced by *B. cereus* at different concentrations against the selected vegetable oils is displayed in Table 4 in terms of the emulsification index (E24).

Table 4. Emulsification index (E24) of the biosurfactant produced by B. cereus UCP 1615 for different vegetable oils (data expressed as mean ± SD of the triplicate determinations). Data are expressed in %.

Biosurfactant Concentration (mg/mL)	Vegetable Oil			
	Corn	Soybean	Peanut	Canola
10.0	64.5 ± 1.1	56.0 ± 1.0	68.1 ± 0.0	65.8 ± 1.5
5.0	65.9 ± 1.4	54.0 ± 0.0	62.7 ± 1.7	64.4 ± 1.6
2.5	42.1 ± 1.1	47.7 ± 1.3	53.5 ± 1.5	36.0 ± 2.0

The results indicated that the biosurfactant was able to ensure a satisfactory emulsification of all the oils studied. As expected, the increase in the concentration of the biosurfactant led to an E24 increase. The best results were achieved at concentrations of 5.0 and 10.0 mg/mL, with E24 values ranging from 54 to 68%.

Few studies have investigated the capacity of biosurfactants produced by the genus *Bacillus* to emulsify vegetable oils or the application of these natural compounds as bioemulsifiers in food products. A study involving a biosurfactant produced by *Bacillus subtilis* ICA56 reported E24 > 50% for soybean oil [39]. Studies involving bioemulsifiers produced by *Candida albicans* reported E24 values around 50% for peanut, mustard, olive and soybean oils [21]. The bioemulsifier from *Candida utilis* showed indices around 30% for corn and sunflower oil under different conditions of pH and salinity [40]. The biosurfactant produced by *C. bombicola* achieved indices of 56% for corn oil, 51% for soybean oil, 69% for peanut oil and 50% for canola oil [31]. Thus, the present results were consistent with the findings described in the literature.

3.4. Characterization of the Cookies

The concentrations of the biosurfactant chosen for this study were defined based on the maximum concentrations recommended for most of the emulsifying additives authorized by both the Brazilian Health Vigilance Agency (Agencia Nacional de Vigilancia Sanitaria, ANVISA) and the US Food and Drug Administration [41,42].

Figure 1 illustrates the c ookies b efore and after baking, and the mean values of their physical properties (weight, diameter, thickness and spread factor) are gathered in Table 5.

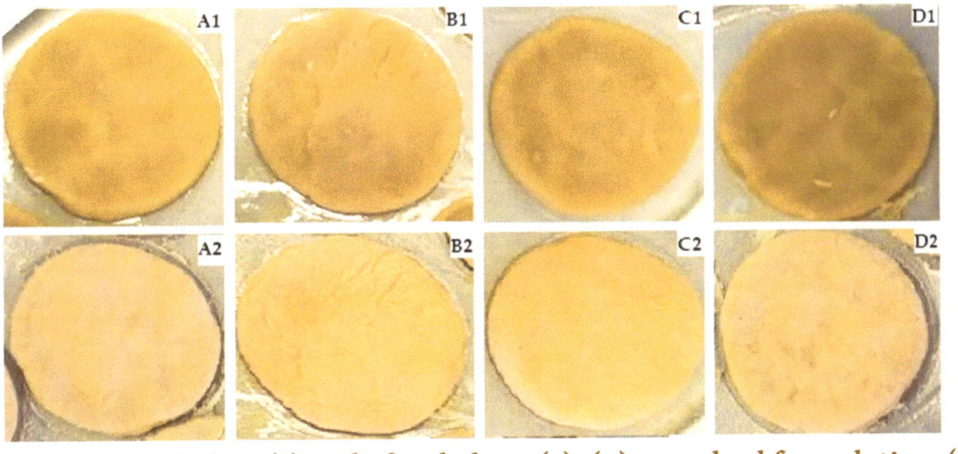

Figure 1. Cookies before (1) and after baking (2). (A): standard formulation; (13): formulation with 1% biosurfactant; (C): formulation with 0.5% biosurfactant; (D): bormulation with 0.25% biosubfactant.

Table 5. Physical properties of the cookies after baking for standard formulation, formulation A (1% biosurfactant), formulation B (0.5% biosurfactant) and formulation C (0.25% biosurfactant).

Formulation	Weight (g)	Diameter (mm)	Height (mm)	Spread Factor
Standard	5.85 ± 0.02 ^{a}	48.66 ± 0.31 ^{a}	6.42 ± 0.15 ^{a}	7.59 ± 0.19 ^{a}
A	4.88 ± 0.25 ^{b}	48.87 ± 0.40 ^{a}	5.79 ± 0.14 ^{b}	8.44 ± 0.19 ^{b}
B	5.87 ± 0.53 ^{a}	49.11 ± 0.16 ^{a}	6.08 ± 0.04 ^{c}	8.08 ± 0.06 ^{c}
C	5.37 ± 0.14 ^{ba}	50.18 ± 0.09 ^{b}	6.50 ± 0.16 ^{da}	7.72 ± 0.18 ^{da}

a,b, c, d. Different letters in same column denote statistically significant differences ($p < 0.05$, Tukey test). The values for each property were compared statistically taking the standard formulation as a reference.

To complement the ANOVA results, the Tukey test was used to evaluate statistically significant differences ($p < 0.05$) between the standard formulation and the formulations containing different concentrations of the biosurfactant with regard to weight, diameter, height and spread factor.

The increase in the concentration of the biosurfactant led to a linear increase in the spread factor. In addition to the benefits regarding the dough homogenization, the addition of the biosurfactant produced by *B. cereus* UCP 1615 promoted an increase in the quantity of lipids in the cookies due to the presence of fatty acids in its non-polar portion, which is a biochemical characteristic of the biomolecule previously described by Durval et al. [15]. This was reflected in the greater spread rate (increase in diameter) during cooking, which was likely related to the increase in the mobility of the system as the lipid fraction melted. It is a well-known fact that the spread rate exerts a direct influence on the diameter and height. Indeed, significant differences in height were found among the formulations as a higher spread factor led to a shorter height.

The physicochemical composition and energy value of the baked cookies are listed in Table 6. This study was necessary to determine whether the addition of the biosurfactant maintained the pre-established standards of identity and quality of the cookie.

Table 6. Physicochemical composition and energy value of cookies prepared with standard formulation, formulation A (1% biosurfactant), formulation B (0.5% biosurfactant) and formulation C (0.25% biosurfactant) (data expressed as mean ± SD of triplicate determinations).

Variable	Standard Formulation	Formulation A	Formulation B	Formulation C
Moisture (%)	6.35 ± 0.04 [a]	5.44 ± 0.06 [ba]	4.87 ± 0.06 [cb]	5.87 ± 1.24 [cba]
Ash (%)	1.60 ± 0.08 [a]	1.92 ± 0.07 [b]	1.74 ± 0.17 [cb]	1.66 ± 0.01 [ca]
Lipids (%)	11.10 ± 0.13 [a]	13.83 ± 0.82 [b]	11.69 ± 0.69 [a]	11.47 ± 0.31 [a]
Proteins (%)	1.07 ± 0.00 [a]	1.02 ± 0.06 [b]	1.07 ± 0.00 [a]	1.07 ± 0.00 [a]
Carbohydrates (%)	87.83 ± 0.13 [a]	85.14 ± 0.88 [b]	87.74 ± 0.69 [a]	87.47 ± 0.31 [a]
Energy Value (cal)	455.52 ± 0.67 [a]	469.17 ± 4.10 [b]	455.97 ± 3.46 [a]	457.33 ± 1.56 [a]

[a,b,c]: Different letters on same line denote statistically significant differences ($p < 0.05$, Tukey test). The values obtained for each physicochemical variable in the different formulations were compared statistically taking the standard formulation as a reference.

A 24% reduction in moisture was found in formulation B containing 0.5% biosurfactant compared with the standard formulation. Thus, the use of the biosurfactant was quite promising as the reduction in moisture minimizes the proliferation of microorganisms, thereby enhancing the durability of the product. On the other hand, the ash content only differed (around 20%) when the highest concentration of the biosurfactant was used.

Seventy-eight percent of the fatty acids in the composition of the biosurfactant were essential and unsaturated fatty acids [15], which suggests that its addition should not compromise the nutritional aspects of the cookie. The biochemical composition of the biosurfactant investigated in this study had a direct impact on the quantity of lipids in the cookies with a statistically significant increase in formulation A containing 1% biosurfactant compared with the other formulations, promoting a proportional increase in the caloric value of the cookies. A cupcake and cookie containing a glycolipid biosurfactant also presented a caloric value increase [22,31]. Conversely, the protein content was unaffected, which may be related to the low protein moiety of the biosurfactant.

3.5. Texture Profile

Studies on the texture profile of food emerged due to the need for a better understanding of human sensorial sensitivity in relation to food. Over time, such studies have acquired a greater relevance due to the need for the presence of functional ingredients and the emergence of innovative technologies that improve texture to ensure quality and satisfy the preferences of consumers [43].

Table 7 shows the results of the texture profile analysis of the cookies after the addition of the biosurfactant. The Tukey test revealed that the addition of the biosurfactant did not cause any significant change in most of the variables analyzed (hardness, cohesiveness and sponginess).

Table 7. Texture profile analysis of the dough 24 h and 45 days after baking for standard formulation, formulation A (1% biosurfactant), formulation B (0.5% biosurfactant) and formulation C (0.25% biosurfactant) (data expressed as mean ± SD of triplicate determinations).

Formulation	Hardness (N)		Cohesiveness (mm)		Sponginess (mm)	
	24 h	45 days	24 h	45 days	24 h	45 days
Standard	2133.70 ± 570.75 [a]	2440.30 ± 94.80 [a]	0.34 ± 0.03 [a]	0.49 ± 0.03 [a*]	0.40 ± 0.00 [a]	0.60 ± 0.00 [a]
A	2827.70 ± 82.59 [a]	1882.00 ± 63.32 [a*]	0.31 ± 0.04 [ba]	0.42 ± 0.06 [a]	0.63 ± 0.12 [b]	0.63 ± 0.06 [a]
B	4557.00 ± 566.77 [b]	2310.00 ± 763.42 [a*]	0.45 ± 0.03 [ca]	0.52 ± 0.02 [a*]	0.60 ± 0.00 [cb]	0.57 ± 0.06 [a]
C	3234.70 ± 236.90 [a]	1969.30 ± 278.44 [a*]	0.49 ± 0.07 [c]	0.42 ± 0.06 [a]	0.63 ± 0.06 [cb]	0.63 ± 0.06 [a]

[a,b,c]: *Different letters in same column denote statistically significant differences ($p < 0.05$, Tukey test). The values for each texture component in the different formulations were compared statistically taking the standard formulation as a reference. * Asterisk on same line denotes a significant difference after storage ($p < 0.05$, Tukey test).*

The similarities among the results for the different formulations are important for the maintenance of the typical characteristics of a cookie; the bioemulsifier proved to be effective in this sense, having led to an improvement in the properties of the cookies in comparison with those of the standard formulation. A considerable difference in hardness was found for the cookie made with formulation B containing 0.5% biosurfactant especially in comparison with the standard formulation, which suggested an increase in crispness when using this biosurfactant concentration. The significant increases in cohesiveness and sponginess can be expected to enhance the chewability.

Zouari et al. [28] reported that the addition of the biosurfactant produced by *B. subtilis* at a concentration of 0.1% to a cookie formulation promoted a significant improvement in the texture profile of the dough; moreover, the action of this bioemulsifier was more pronounced than that of a commercial emulsifier (glycerol monostearate). Kiran et al. [44], who incorporated a lipopeptide at a concentration of 0.75% to a muffin formulation, found an improvement in the final softness due to the increase in sponginess and cohesiveness as well as a reduction in hardness.

The literature offers other reports on the potential of microbial biosurfactants [1,3,45] but with few examples of applications in the formulation of products for human consumption. Such examples include the addition of microbial bioemulsifiers to the formulations of mayonnaises [46], cupcakes [31] and cookies [22].

The biosurfactant at the lowest concentration (formulation C) did not promote a significant improvement in the texture profile compared with the standard formulation. However, cohesiveness was maintained after 45 days of storage, which did not occur with the standard formulation. Campos et al. [40] found that the use of a bioemulsifier produced by *C. utilis* at a concentration of 0.7% offered a greater stability and hardness to a salad dressing formulation after 30 days of storage; the product was considered a good emulsifier compared with commercial products such as guar gum and carboxymethylcellulose.

In the production of food, the useful life of emulsions during long-term storage should be considered to ensure the consistent quality of the product. Studies have shown that the oily phase composition and the type of emulsifier exert significant effects on the long-term stability and sensorial properties such as spreadability, viscosity and appearance [36].

As the biosurfactant investigated herein had no significant negative effect from the statistical standpoint on the texture profile of the cookie, it could be considered to be a potential ingredient for the food industry. A simple assessment of aroma, flavor, color and texture revealed no significant differences between the formulations containing the biosurfactant and the standard formulation. However, sensory assessments are needed and the biosurfactant should be added to other formulations of flour-based products to determine whether it can be incorporated into other food without compromising the desired characteristics. Such investigations could expand the applications of this biosurfactant.

4. CONCLUSIONS

The present findings demonstrate that the biosurfactant produced by *Bacillus cereus* UCP 1615 grown in a medium containing waste frying oil has the potential for use as a bioemulsifier in food systems because it has been shown to be an effective emulsifying agent for various vegetable oils. The addition of the biosurfactant did not drastically affect the final product as the biosurfactant-containing formulations showed energetic and physical characteristics similar to those of the standard formulation, indicating the feasibility of applying this biomolecule in the formulation of cookies. The biosurfactant was non-toxic, which suggested its safe use, and had a considerable antioxidant activity. The biosurfactant demonstrated promising results as an ingredient for a flour-based product in terms of the physical, physicochemical and textural properties of the cookies formulated. The biosurfactant also ensured a good preservation of the cookies. Based on the results obtained in this study, the bacterial surfactant could be tested in other products as a green additive in the food industry. However, further studies are needed to enhance the economic viability of the production of this microbial surfactant on an industrial scale for its use as an emulsifier in food.

REFERENCES

1. Ribeiro, B.G.; Guerra, Sarubbo, L.A. Biosurfactants: Production and application prospects in the food industry. *Biotechnol. Prog.* **2020**, 36, e3030. [CrossRef]
2. Durval, I.J.B.; da Silva, I.A.; Sarubbo, L.A. Application of microbial biosurfactants in the food industry. In *Microbial Biosurfactants. Environmental and Microbial Biotechnology;* Inamuddin, M., Ahamed, M.I., Prasad, R., Eds.; Springer: Singapore, 2021; Volume 112, pp. 1-10. [CrossRef]
3. Salek, K.; Euston, S.R. Sustainable microbial biosurfactants and bioemulsifiers for commercial exploitation. *Process. Biochem.* **2019**, 85,143-155. [CrossRef]
4. Nitschke, M.; Silva, S.S.E. Recent food applications of microbial surfactants. *Crit. Rev. Food Sci. Nutr.* **2017**,58, 631-638. [CrossRef]
5. Liu, K.; Sun, Y.; Cao, M.; Wang, J.; Lu, J.R.; Xu, H. Rational design, properties, and applications of biosurfactants: A short review of recent advances. *Curr. Opin. Colloid Interface Sci.* **2020**, 45, 57-67. [CrossRef]
6. Drakontis, C.E.; Amin, S. Biosurfactants: Formulations, properties, and applications. *Curr. Opin. Colloid Interface Sci.* **2020**, 48, 77-90. [CrossRef]
7. Farias, C.B.B.; Almeida, F.C.G.; Silva, I.A.; Souza, T.C.; Meira, H.M.; da Silva, R.D.C.F.S.; Luna, J.M.; Santos, V.A.; Converti, A.; Banat, I.M.; etal. Production of green surfactants: Market prospects. *Electron. J. Biotechnol.* **2021**, 51, 28-39. [CrossRef]
8. Geissler, M.; Heravi, K.M.; Henkel, M.; Hausmann, R. Lipopeptide biosurfactants from *Bacillus* species. In *Biobased Surfactants,* 2nd ed.; Hayes, D., Solaiman, D., Richard, A., Eds.; Elsevier: Amsterdam, The Netherlands, 2019; pp. 205-240. [CrossRef]

9. Carolin, C.F.; Kumar, P.S.; Ngueagni, P.T. A review on new aspects of lipopeptide biosurfactant: Types, production, properties and its application in the bioremediation process. *J. Hazard. Mater.* **2021**, 407,124827. [CrossRef] [PubMed]

10. Markande, A.R.; Patel, D.; Varjani, S. A review on biosurfactants: Properties, applications and current developments. *Bioresour. Technol.* **2021**, 330,124963. [CrossRef]

11. Santos, F.F.; Freitas, K.M.L.; Pereira, A.S.; Fontes-Sant'Ana, G.C.; Rocha-Leao, M.H.M.; Amaral, P.F.F. Butter whey and corn steep liquor as sole raw materials to obtain a bioemulsifier from *Yarrowia lipolytica* for food oil-in-water emulsions. *Ciencia Rural* **2021**, *51*, 1-12. [CrossRef]

12. Foo, W.H.; Chia, W.Y.; Tang, D.Y.Y.; Koay, S.S.N.; Lim, S.S.; Chew, K.W. The conundrum of waste cooking oil: Transforming hazard into energy. *J. Hazard. Mater.* **2021**, 417,126129. [CrossRef]

13. OECD-FAO. *OECD-FAO Agricultural Outlook 2019-2028,*15th ed.; OECD Publishing: Paris, France, 2019; pp. 1-140. [CrossRef]

14. Ferreira, J.F.; Vieira, E.A.; Nitschke, M. The antibacterial activity of rhamnolipid biosurfactant is pH dependent. *Food Res. Int. 116*, 737-744. [CrossRef]

15. Durval, I.J.B.; Mendonca, A.H.R.; Rocha, I.V.; Luna, J.M.; Rufino, R.D.; Converti, A.; Sarubbo, L.A. Production, characterization, evaluation and toxicity assessment of a *Bacillus cereus* UCP 1615 biosurfactant for marine oil spills bioremediation. *Mar. Pollut. Bull.* **2020**, *157*, 111357. [CrossRef]

16. Alley, M.C.; Scudiero, D.A.; Monks, A.; Hursey, M.; Czerwinski, M.J.; Fine, D.L.; Abbott, B.J.; Mayo, J.G.; Shoemaker, R.; Boyd, M.R. Feasibility of drug screening with panels of human tumor cell lines using a microculture tetrazolium assay. *Cancer Res.* **1988**, *48*, 584-588.

17. Mosmann, T. Rapid colorimetric assay for cellular growth and survival: Application to proliferation and cytotoxicity assays. *J. Immunol. Methods* **1983**, 65, 55-63. [CrossRef]

18. Rodrigues, F.A.R.; Bomfim, I.S.; Cavalcanti, B.C.; Pessoa, C.; Goncalves, R.S.B.; Wardell, J.L.; Wardell, S.M.S.V.; de Souza, M.V.N. Mefloquine-oxazolidine derivatives: A new class of anticancer agents. *Chem. Biol. Drug Des.* **2014**, 83,126-131. [CrossRef]

19. Prazeres, L.D.K.T.; Aragao, T.P.; Brito, S.A.; Almeida, C.L.F.; Silva, A.D.; De Paula, M.M.F.; Farias, J.S.; Vieira, L.D.; Damasceno, B.P.G.L.; Rolim, L.A.; et al. Antioxidant and antiulcerogenic activity of the dry extract of pods of *Libidibia ferrea* Mart. ex Tul. (Fabaceae). *Oxid. Med. Cell. Longev.* **2019**, *2019,*1983137. [CrossRef]

20. Veras, B.O.; Oliveira, M.B.M.; Oliveira, F.G.S.; Santos, Y.Q.; Oliveira, J.R.S.; Lima, V.L.M.; Almeida, J.R.G.S.; Navarro, D.M.A.F.; Aguiar, J.C.R.O.F.; Aguiar, J.S.; et al. Chemical composition and evaluation of the antinociceptive, antioxidant and antimicrobial effects of essential oil from *Hymenaea cangaceira* (Pinto, Mansano & Azevedo) native to Brazil: A natural medicine. *J. Ethnopharmacol. 247*, 112265. [CrossRef]

21. Gaur, V.K.; Regar, R.K.; Dhiman, N.; Gautam, K.; Srivastava, J.K.; Patnaik, S.; Kamthan, M.; Manickam, N. Biosynthesis and characterization of sophorolipid biosurfactant by *Candida* spp.: Application as food emulsifier and antibacterial agent. *Bioresour. Technol.* **2019**, 285,121314. [CrossRef]

22. Ribeiro, B.G.; Guerra, J.M.C.; Sarubbo, L.A. Potential food application of a biosurfactant produced by *Saccharomyces cerevisiae* URM 6670. *Front. Bioeng. Biotechnol.* **2020**, 8, 434. [CrossRef]

23. Aziah, A.A.N.; Noor, A.Y.M.; Ho, L.H. Physicochemical and organoleptic properties of cookies incorporated with legume flour. *Int. Food Res. J.* **2012**,19,1539-1543. [CrossRef]

24. Mudgil, D.; Barak, S.; Khatkar, B.S. Cookie texture, spread ratio and sensory acceptability of cookies as a function of soluble dietary fiber, baking time and different water levels. *LWT-Food Sci. Technol.* **2017**, 80, 537-542. [CrossRef]

25. Feldsine, P.; Abeyta, C.; Andrews, W.H. AOAC International methods committee guidelines for validation of qualitative and quantitative food microbiological official methods of analysis. *J. AOAC Int.* **2002**, 85,1187-1200. [CrossRef]

26. Bligh, E.G.; Dyer, W.J. A rapid method of total lipid extraction and purification. *Can. J. Biochem. Physiol.* **1959**, 37, 911-917. [CrossRef]

27. Pires, T.C.S.P.; Dias, M.I.; Barros, L.; Ferreira, I.C.F.R. Nutritional and chemical characterization of edible petals and corresponding infusions: Valorization as new food ingredients. *Food Chem.* **2017**,220, 337-343. [CrossRef] [PubMed]

28. Zouari, R.; Besbes, S.; Ellouze-Chaabouni, S.; Ghribi-Aydi, D. Cookies from composite wheat-sesame peels flours: Dough quality and effect of *Bacillus subtilis* SPB1 biosurfactant addition. *Food Chem.* **2016**,*194,* 758-769. [CrossRef] [PubMed]

29. Tolosa, L.; Donato, M.T.; Gomez-Lechon, M.J. General cytotoxicity assessment by means of the MTT assay. In *Protocols in In Vitro Hepatocyte Research,* 1st ed.; Vinken, M., Rogiers, V., Eds.; Humana Press: New York, NY, USA, 2015; Volume 1250, pp. 333-348. [CrossRef]

30. Silva, J.A.G.; da Silva, G.C.; Silva, M.G.F.; da Silva, V.F.; Aguiar, J.S.; da Silva, T.G.; Leite, S.P. Physicochemical characteristics and cytotoxic effect of the

methanolic extract of *Croton heliotropiifolius* Kunth (Euphorbiaceae). *Afr. J. Pharm. Pharmacol.* **2017**,11, 321-326. [CrossRef]

31. Silva, I.A.; Veras, B.O.; Ribeiro, B.G.; Aguiar, J.S.; Guerra, J.M.C.; Luna, J.M.; Sarubbo, L.A. Production of cupcake-like dessert containing microbial biosurfactant as an emulsifier. *PeerJ* **2020**, 8, 9064. [CrossRef]
32. Basit, M.; Rasool, M.H.; Naqvi, S.A.R.; Waseem, M.; Aslam, B. Biosurfactants production potential of native strains of *Bacillus cereus* and their antimicrobial, cytotoxic and antioxidant activities. *Pak. J. Pharm. Sci.* **2018**, *31* (Suppl. 1), 251-256.
33. Sharma, D.; Saharan, B.S.; Chauhan, N.; Bansal, A.; Procha, S. Production and structural characterization of *Lactobacillus helveticus* derived biosurfactant. *Sci. World J.* **2014**,*2014*, 493548. [CrossRef]
34. Shahidi, F.; Zhong, Y. Measurement of antioxidant activity. *J. Funct. Foods* **2015**,18, 757-781. [CrossRef]
35. Comert, E.D.; Gokmen, V. Evolution of food antioxidants as a core topic of food science for a century. *Food Res. Int.* **2018**,105, 76-93. [CrossRef]
36. Zhu, Q.; Pan, Y.; Jia, X.; Li, J.; Zhang, M.; Yin, L. Review on the stability mechanism and application of water-in-oil emulsions encapsulating various additives. *Compr. Rev. Food Sci. Food Saf.* **2019**,18,1660-1675. [CrossRef]
37. Satpute, S.K.; Zinjarde, S.S.; Banat, I.M. Recent updates on biosurfactants in the food industry. In *Microbial Cell Factories;* CRC Press: Boca Raton, FL, USA; Taylor & Francis: Abingdon, UK, 2018; pp. 1-20. [CrossRef]
38. El-Hamidi, M.; Zaher, F.A. Production of vegetable oils in the world and in Egypt: An overview. *Bull. Natl. Res. Cent.* **2018**, 42,19. [CrossRef]
39. Fran a, I.W.L.; Lima, A.P.; Lemos, J.A.M.; Lemos, C.G.F.; Melo, V.M.M.; De Sant'ana, H.B.; Gonalves, L.R.B. Production of a biosurfactant by *Bacillus subtilis* ICA56 aiming bioremediation of impacted soils. *Catal. Today* **2015**,255,10-15. [CrossRef]
40. Campos, J.M.; Stamford, T.L.M.; Sarubbo, L.A. Characterization and application of a biosurfactant isolated from *Candida utilis* in salad dressings. *Biodegradation* **2019**, 30, 313-324. [CrossRef]
41. ANVISA. Diario Oficial da Uniao-Resolu^ao da Diretoria Colegiada-RDC No 239, DE 26 DE JULHO DE 2018. Available online: https://www.in.gov.br/materia/-/asset_publisher/Kujrw0TZC2Mb/content/id/34380515 (accessed on 25 June 2021).

42. U.S. Food and Drug Administration. FDA Food Additive Status List I FDA. 2019; pp. 1-95. Available online: https://www.fda. gov/food/food-ingredients-packaging/food-additive-listings (accessed on 14 August 2021).

43. Chen, J.; Rosenthal, A. Food texture and structure. In *Modifying Food Texture: Novel Ingredients and Processing Techniques;* Chen, J., Rosenthal, A., Eds.; Woodhead Publishing: Cambridge, UK; Elsevier Ltd.: Amsterdam, The Netherlands, 2015; Volume 1, pp. 3-24. [CrossRef]

44. Kiran, G.S.; Priyadharsini, S.; Sajayan, A.; Priyadharsini, G.B.; Poulose, N.; Selvin, J. Production of lipopeptide biosurfactant by a marine *Nesterenkonia* sp. and its application in food industry. *Front. Microbiol.* **2017**, 8,1138. [CrossRef]

45. Anjum, F.; Gautam, G.; Edgard, G.; Negi, S. Biosurfactant production through *Bacillus* sp. MTCC 5877 and its multifarious applications in food industry. *Bioresour. Technol.* **2016**,213, 262-269. [CrossRef] [PubMed]

46. Campos, J.M.; Stamford, T.L.M.; Rufino, R.D.; Luna, J.M.; Stamford, T.C.M.; Sarubbo, L.A. Formulation of mayonnaise with the addition of a bioemulsifier isolated from *Candida Utilis. Toxicol. Rep.* **2015**,2,1164-1170. [CrossRef]

CHAPTER 14

Production of Omega-3 Fatty Acids from the Microalga Crypthecodinium cohnii by Utilizing Both Pentose and Hexose Sugars from Agricultural Residues

1. INTRODUCTION

Lignocellulosic biomass has become the subject of a great deal of attention from researchers in order to mitigate the diminution of fossil reserves stemming from growing energy requirements worldwide. Biomass is considered one of the most low-cost and largest sources of carbon, rendering it the most important raw material with great potential to support the development of a bio-based economy. This renewable feedstock can contribute to the production of versatile and value-added products in either stand-alone processors or biorefineries [1]. Lignocellulose is encountered in hardwood, softwood, grasses, agricultural and forest residues, domestic and municipal solid wastes, as well as food/feed industry residues. The main components of these materials are cellulose, hemicellulose and lignin which form a compact and recalcitrant structure [2]. The present study focuses on agricultural residues and specifically on wheat straw from Greek wheat fields, aiming to exploit the maximum potential of lignocellulosic-derived sugar streams valorization. Towards this direction, the choice of integrated processes for the conversion of all sugar streams is considered to be of paramount importance. Hence, not only the pretreatment step of biomass but also the performance of a successful fractionation of biomass to its constituents holds a prominent role in the biorefinery concept.

Due to the recalcitrant structure of lignocellulosic biomass, pretreatment and hydrolysis are deemed necessary, in order not to rate-limit the subsequent fermentation [3]. Therefore, aiming for the adequate disruption of the compact structure of biomass, an initial pretreatment procedure is required in order to remove hemicellulose/lignin and expand the accessible surface area of cellulose for enzymes [4]. Among all pretreatment processes that have been thoroughly examined throughout the recent years, organosolv stands out due to the numerous advantages it offers such as the high purity of cellulose and the isolation of high-quality, sulfur-free lignin [5,6]. The organosolv treatment uses a liquid phase reaction medium of organic solvents or their aqueous solutions and allows for the separation and the simultaneous recovery of three streams, namely a cellulose-rich solid pulp, an aqueous hemicellulose liquid fraction and a dry solid lignin that can be obtained after evaporation of the organic solvent [7]. Moreover, organosolv pretreatment contributes to high lignin dissolution, thus resulting in high delignification rates [8] and preservation of *fl-O-4* linkages for downstream lignin utilization [9] while maintaining the formation of sugar degradation products at very low levels. It is worth noting that despite the costs of the chemicals that are used through the organosolv pretreatment, there are many reports in the literature that highlight the recyclability of these solvents, thus reducing the economic constraints as much as possible [7].

Microalgae include both photosynthetic and heterotrophic unicellular organisms that commonly live in freshwater or marine aquatic environments, and they can be of a high value or can be used as a green feedstock for many important products [10]. Nowadays, microalgae are already cultivated for commercial purposes, adding value to the market by producing microalgae-based products, such as health food supplements (nutraceuticals), pharmaceuticals, cosmetics, lubricants and feed for aquaculture hatcheries in agriculture and in many other applications [11]. In this context, the cultivation of heterotrophic microalgae on sustainable, abundant and available carbon sources, such as lignocellulosic residues, not only provides an option to take advantage of these streams but also significantly reduces the cost and carbon footprint compared to autotrophic or conventional microalgae production systems that use pure sugars in the growth medium. This is achieved by converting biomass-derived hydrolysates as low-value resources into value-added bioproducts, including compounds of nutritional value, such as omega-3 fatty acids (FA) and carotenoids [12,13]. Poly-unsaturated fatty acids (PUFAs), especially those with very long chains (LC-PUFAs) such as eicosapentaenoic acid (20:5n-3, EPA) and docosahexaenoic acid (22:6n-3, DHA), are bioactive compounds of great importance within the food and nutraceutical industry, where they are used as ingredients in functional foods, fulfilling the appropriate specifications of high value-added products [14]. *Crypthecodinium cohnii* is a heterotrophic microalga widely known for its ability to produce high percentages of DHA [15], a necessary provider for the proper cardiovascular and neural development of humans. This microorganism has been shown to be able to utilize a variety of different carbon sources apart from glucose, including volatile fatty acids from anaerobic digestion [16], ethanol [17], liquor produced after exhausted olive pomace pretreatment [18] as well as sugar-rich hydrolysates obtained after enzymatic hydrolysis of organosolv-pretreated beechwood [19,20]. In this study, we attempted to expand the range of substrates that can support the growth of *C. cohnii* and lead to accumulation of DHA by assessing wheat straw from Greek agricultural residues as a feedstock. Both pentose and hexose streams yielded after organosolv pretreatment were evaluated as carbon sources. The present work reports an efficient holistic approach for the integrated valorization of all sugar-containing fractions of biomass towards the production of this valuable product through fermentation.

2. MATERIALS AND METHODS

2.1. Material and OxiOrganosolv Pretreatment Process

The wheat straw feedstock used in this work was provided by Flourmills Thrakis S.A. It originated from soft white wheat and it was collected from wheat fields in Northern Greece in 2019. The wheat straw was analyzed according to the methods described in Section 2.2.

The fractionation of the wheat straw feedstock was carried out with an acid-free oxidative organosolv fractionation (OxiOrganosolv), as described previously [6]. An amount of 25 g of wheat straw and an organic solvent/water mixture (solid: liquid ratio 1:20, organic solvent: H2O ratio 1:1) were added into an autoclave reactor, which was pressurized with 100% O_2 gas to 16 bar and then heated to the desired temperature. The organic solvents investigated in this work were acetone (ACO) and ethanol (EtOH), which were chosen to comply with food industry regulations. Three reaction temperatures were studied, 150,160 and 175 ^{0} C, while the reaction time was kept constant at 120 min. The starting time of the reaction (t_0) was considered as the time when the system reached the desired temperature. After the reaction, the cellulose-rich solid fraction was separated from the liquid with vacuum filtration. The liquid fraction contained water, organic solvent, dissolved lignin and hydrolyzed hemicellulose fragments. The organic solvent was separated from the liquid fraction in a rotary evaporator under vacuum, which caused the lignin to precipitate. The lignin was then separated from the liquid fraction by vacuum filtration. The solid fraction, referred to as pulp, was washed first with 500 mL of the organic solvent and then with 250 mL of distilled water before it was finally air dried to a moisture content of around 5-8 wt%.

2.2. Analysis Methods

2.2.1. Moisture Content

The moisture content was determined by weighing ca. 2 g of as-received wheat straw in a pre-weighed crucible and placing it in a muffle furnace at 100 °C for 3 h. The crucible was then placed in a desiccator to cool down to room temperature, and then it was weighed. The moisture content was calculated from the weight difference of the sample before and after drying.

2.2.2. Ash Content

The ash content was determined by weighing 1 g of as-received or extractives-free wheat straw in a pre-weighed crucible and placing it in a muffle furnace at 575 ° C for 3 h, in the presence of ambient air. The crucible was then removed from the furnace and placed in a desiccator to cool down to room temperature, and then it was weighed. The ash content was calculated based on the final sample weight, which was considered to be the ash in the sample.

2.2.3. Elemental Analysis

The carbon and hydrogen content of the as-received wheat straw feedstock was determined by analysis in a CHN 628 elemental analyzer from Leco Corporation, according to ASTM D5291. The oxygen content was determined by difference (O% = 100% — C% — H% — Ash%).

2.2.4. Compositional Analysis of Lignocellulosic Samples

The extractives content of the wheat straw feedstock was determined according to the method described by the National Renewable Energy Laboratory [21]. The hemicellulose, a-cellulose and lignin content of the wheat straw feedstock and the wheat straw pulp were determined after hydrolysis with 4 wt% H2SO4 in an autoclave (121° C for 60 min) according to the procedures described by the National Renewable Energy Laboratory [22]. The hydrolyzed samples were filtered to remove any solids and analyzed by Ion Chromatography (IC) on an ICS5000 (Dionex, Sunnyvale, CA, USA) to quantify the content of sugars. The quantification was based on external calibration, using standard solutions of sugars (glucose, mannose, xylose, fructose, galactose, arabinose and rhamnose) and sugar alcohols (sorbitol and mannitol). The analysis was performed using a CarboPac PA1 (10 gm, 4 x 250 mm) column and guard column (10 gm, 4 x 30 mm) connected to a pulsed amperometric detector (PAD). The eluent was 20 mM NaOH at a 0.6 mL/min flow rate, and the total analysis time was 75 min. The compositional analyses were carried out twice, and the mean values are reported.

2.3. Enzymatic Hydrolysis of Solid Pulps

Both the cellulose-rich solid fraction and the hemicellulose-rich liquid fraction of the OxiOrganosolv pretreatment were exploited for the production of DHA with *C. cohnii*. An SHF approach (separated hydrolysis and fermentation) was employed because the conditions of these operations are different.

The solid fraction was enzymatically hydrolyzed to fermentable sugars following a procedure previously reported [19]. The reaction took place by employing a commercial enzyme cocktail, namely Cellic® CTec2 (Novozymes A/S, Bagsvwrd, Denmark), with a protein content of 95.6 mg/mL, as determined by Bradford assay [23]. Hydrolysis was performed in 250 mL glass flasks, at 50°C, under agitation (160 rpm) and at an initial dry matter (DM) of 9% (w/v), in order to obtain a hydrolysate with high sugar concentration, suitable for *C. cohnii* cultivation. An enzyme loading of 9 mg/g of biomass was used, and the pH levels were maintained constant at 5.5 upon addition of 80 mM MES (2-N- morpholino-ethanesulfonic acid) buffer solution. The final ratio of the reaction volume to shake flask volume was 1/10.

In order to track down the glucose and total reducing sugars (TRS) production, the glucose oxidase/peroxidase (GOD/POD) [24] and 3,5-dinitrosalicylic acid (DNS) method [25] were used, respectively. Samples were taken at different time intervals (8, 24, 48 and 72 h), centrifuged before each analysis, and the supernatant was filtered with 0.22 gM pore size filters. As a complement, in order to accurately determine the monosaccharide (glucose, xylose) and acetic acid concentration, a chromatographic analysis was performed (HPLC) by isocratic ion-exchange chromatography using an Aminex HPX-87H column with a micro-guard column, at 50°C (Bio-Rad Laboratories, Hercules, CA, USA), using 3 mM H2SO4 as a mobile phase at a flow rate of 0.6 mL/min. After 72 h of hydrolysis and analysis of released products, all supernatants containing fermentable sugars were collected after vacuum filtration and kept in the freezer (-20°C) for future use.

2.4. Detoxification and Enzymatic Hydrolysis of Liquid Fractions

Detoxification and enzymatic hydrolysis of liquid fractions were required before the evaluation of pretreatment aqueous liquors as carbon source for *C. cohnii*. Even after the initial removal of the OxiOrganosolv organic solvent, as described in Section 2.1, some residual organic solvent with potential to alter the microalga metabolism was still present (as detected by HPLC analysis), so an additional evaporation step was necessary. After the removal of the organic solvent, the liquid fraction was filled with water up to its original volume in order to maintain the same sugar concentration as the corresponding sample prior to evaporation. Moreover, the OxiOrganosolv-derived liquid fractions contained phenolic compounds that are toxic to the microalgae used. Therefore, 5% (w/v) active carbon was used for the removal of phenol compounds, whose levels were determined with the Folin–Ciocalteu method [26].

The liquid fractions are richer in hemicelluloses in comparison with the solid fractions. However, most of the compounds are found in the form of oligosaccharides, which cannot be metabolized by *C. cohnii*. As a result, hydrolysis with Cellic® HTec2 enzyme cocktail (Novozymes A/S, Bagsvwrd, Denmark) was carried out in order to produce fermentable monosaccharides. The protein content was 75 mg/mL, as calculated by the Bradford method [23]. Hydrolysis experiments were performed in 500 mL glass flasks at 45°C under agitation (100 rpm) with an enzyme loading of 10 mg/g of oligosaccharides at pH 5.5 (self pH, no addition of buffer). The total reaction volume to shake flask volume was 1/10. Samples were analyzed with HPLC by isocratic ion-exchange chromatography using an Aminex HPX-87P column with a micro-guard column at 85°C (Bio-Rad Laboratories, Hercules, CA, USA) and using water as a mobile phase at a flow rate of 0.6 mL/min. With this HPLC analysis, the levels of glucose, xylose, mannose, galactose and arabinose were determined. The hydrolysis yield was calculated based on the increase of monosaccharides after enzymatic reaction.

2.5. Microalga Cultures

The heterotrophic marine microalga *C. cohnii* ATCC® 30772™ (American Type Culture Collection) was used to study the production of omega-3 fatty acids. Stock cultures and growth procedure were performed according to ATCC guidelines and protocols, as previously described [16]. Initially, the microalgal cells were incubated in standing precultures at 27°C in the dark, on a medium containing 9 g/L glucose, 25 g/L sea salts and 2 g/L yeast extract. The pH of the medium was set at 6.5. After 4 days of growth, the standing cultures were used as inoculum for shaken cultures.

The medium for the shaken cultures contained 25 g/L sea salts and 2 g/L yeast extract, while hydrolysates after enzymatic treatment of the solid pulps (6 samples) or the liquid fractions (3 samples) were employed as a carbon source. In order to increase the total sugar content of the hydrolysates from solid pulps, enzymatic reactions took place at 12% (w/v) initial solids, while all other parameters (temperature, enzyme loading, pH, agitation) were the same as those mentioned in Section 2.3. The total culture volume was 30 mL in 100 mL shake flasks, 27 mL of either 2-times diluted enzymatic hydrolysate from solid fraction or undiluted enzymatically treated liquid fraction and 3 mL inoculum, as it had to consist of 10% (v/v) of the final culture volume. The pH of the culture was set at 6.5 with NaOH solution. The temperature of the culture was set at 27°C and agitation was essential for homogeneity, namely 160 rpm [27]. Each experiment was performed in duplicate. The growth of cells was supervised on a daily basis by measuring the optical density (OD) at 685 nm and the TRS concentration with the DNS method.

Moreover, to track carbon consumption more accurately, an HPLC analysis was performed with an Aminex HPX-87H column, as mentioned above. At the end of cultivation, cells were harvested by centrifugation, washed thoroughly with dH_2O contacting 25 g/L sea salts, lyophilized and weighed, in order to determine the biomass concentration.

2.6. Extraction of Lipids, Quantification and Evaluation of Fatty Acids Profile

For the extraction of lipids, the Folch method was utilized [28] with some alterations [19]. Namely, 50 mg of dried cells were mixed with 10 mL of chloroform: MeOH 2:1 mixture and left overnight at room temperature. Then, distilled H_2O was added at a volume equal to 20% of that of the organic mixture, and the mixture was centrifuged. After centrifugation, the lower phase was collected in glass tubes and washed with a MeOH: dH_2O 1:1 mixture. The organic solvents used were evaporated in a vacuum oven for a maximum of 24 h in order to minimize the risk of sample oxidation. After the evaporation of the solvent, the total lipids were weighted. Following this, the collected fatty lipids had to be transformed into fatty acid methyl esters (FAME). For this reason, they were diluted in 1 mL pure chloroform and mixed with 2.5 mL solution of MeOH: HCl 92:8. The samples at that point were incubated at 60°C for 15 min for the esterification to take place. The reaction was halted with the addition of 2.5 mL of CaCl2 5% (v/v). The methyl esters were extracted from the mixture using hexane, which proved to be a very good solvent, and analyzed by GC-MS, as described by Chalima et al. [19]. The quantification of fatty acids took place by employing a commercially accessible FAME combination (Supelco® 37, Sigma-Aldrich, St. Louis, MI, USA), whereas estimation of wt% DHA was considered out of the whole sum of five fatty acids that were primarily distinguished (C14:0, C16:0, C18:0, C18:1 and DHA).

3. RESULTS

3.1. OxiOrganosolv Fractionation of Wheat Straw with ACO and EtOH as Organic Solvents

Initially, wheat straw biomass was organosolv-pretreated by following the OxiOrganosolv process [6] with an H_2O/ACO 50:50 or an H_2O/EtOH 50:50 co-solvent system. The pretreatment time was kept constant at 120 min in all cases upon addition of 16 bar O_2 as a catalyst. The only parameter shifting was the temperature that was regulated at 150, 160 and 175°C for both solvent mixtures, resulting in six different experimental runs. Pretreatment conditions as well as compositional analysis, cellulose/hemicellulose recovery and lignin removal of solid fraction are shown in Table 1. The results showed that, with a constant residence time of 120 min, increasing the pretreatment temperature results in a corresponding increase in the % biomass solubilization in the case of both ACO and EtOH. This was attributed to the removal of the hemicellulose and lignin fractions from the wheat straw substrate. Delignification peaked at 175°C of pretreatment both in the case of ACO and EtOH, reaching 87.5 ± 0.6% and 86.9 ± 0.1% for sample S3 and S6, correspondingly. Notably, cellulose recovery in the solid fraction was almost 100% in almost all cases in both solvents, indicating only very limited hydrolysis of cellulose in the pre-treatment conditions that were studied. On the other hand, hemicellulose recovery in the solid fraction reached the highest values of 82.6 ± 0.8% in the case of ACO (sample S1) and 84.6 ± 0.1% in the case of EtOH (sample S4) in the lowest temperature of 150°C, while increasing temperature resulted in increased hemicellulose removal and also combined with increasing lignin removal. As such, both solvents achieved very good delignification of the wheat straw, while ACO was also more effective for the removal of hemicellulose, resulting in the most cellulose-rich pulps.

Table 1. Pretreatment conditions, compositional analysis of solid pulps, lignin removal and % recovery of cellulose and hemicellulose. Standard error is <2.5% in all measurements.

Sample	Temperature (°C)	Reaction Time (Min)	Solvent (50:50%)	% Solubilization	Composition of Solid Pulp					% Recovery in the Pulp	
					Lignin (wt%)	Cellulose (wt%)	Hemicellulose (wt%) [1]	Total Mass Closure	Total Lignin Removal	C6 Cellulose Recovery	C6/C5 Hemicellulose Recovery
S1	150	120	H2O/ACO	32.6	12.5	60.1	28.3	100.8	49.3	101.5	83.3
S2	160	120	H2O/ACO	44.6	7.8	71.0	23.2	102.0	73.8	98.6	56.1
S3	175	120	H2O/ACO	55.2	4.4	83.2	11.4	99.0	88.1	93.3	22.3
S4	150	120	H2O/EtOH	31.1	14	56.2	28.4 [2]	98.6	41.8	97.0	85.4
S5	160	120	H2O/EtOH	44.4	8.6	70.0	23.6	102.2	70.9	97.5	57.3
S6	175	120	H2O/EtOH	52.7	4.6	82.4	17.3	104.3	86.8	97.6	35.7
S7			Untreated		16.6	39.9	22.9 [3]	98.0 [4]			

[1] Unless stated, hemicellulose fraction consists only of xylose-containing structures. [2] Contains 28.1 wt% xylose and 0.3 wt% galactose.

[3] Contains 21.8 wt% xylose, 0.5 wt% mannose and 0.8 wt% galactose. [4] Includes also extractives 15.5 wt% and inorganic content (ash) 3.1 wt%.

The aqueous liquid fraction resulting from the organosolv pretreatment was examined for hemicellulose recovery after organic solvent evaporation and lignin removal by vacuum filtration. The results, presented in Figure 1 and in Table 2, demonstrate that the highest % of hemicellulose recovery in the form of monosaccharides and polysaccharides was achieved when the temperature was the highest for both organic solvents, reaching up to 32.7 ± 1.2% for ACO, 175°C, 120 min (sample L3) and 34.4 ± 0.9% for EtOH, 175°C, 120 min (sample L6). These values accounted for 119 and 88.3 mg/g of untreated wheat straw biomass after ACO or EtOH pretreatment, respectively. The aforementioned results are in agreement with the % hemicellulose recovery values in solid pulps, as depicted in Table 1, where the samples that underwent pretreatment at higher temperatures stood out for their high cellulose and low hemicellulose content. As such, the results in Table 2 demonstrated a dwindling trend in the oligosaccharides/monosaccharides ratio in the aqueous phase when the pretreatment conditions became more severe in the case of ACO. More specifically, the ratio was 10.5 for 150°C, 120 min (sample L1) noting a remarkable decline to 2.2 for 175°C, 120 min (sample L3), which was linked to the degradation of oligosaccharides in severe conditions. Contrariwise, the corresponding data in the case of EtOH did not follow the same trend, instead exhibiting a similar ratio regardless of the pretreatment temperature, and more specifically, 10.8 for 150°C, 120 min (sample L4) and 10.5 for 175°C, 120 min (sample L6).

Table 2. Compositional analysis of monosaccharides/oligosaccharides in the liquid fraction. Standard error is <2.5% in all measurements.

Sample	mg/g of Untreated Biomass			% Hemicellulose Recovery in Liquid Fraction			
	Mono-	Oligo-	Total	Mono-	Oligo-	Oligo-/ Mono- Ratio	Total
L1 (ACO 150°C)	3.9	40.3	44.2	1.5	15.7	10.5	17.2
L2 (ACO 160°C)	7.3	65.1	72.3	2.8	25.4	9.1	28.2
L3 (ACO 175°C)	37.2	81.7	119.0	14.5	31.9	2.2	32.7
L4 (EtOH 150°C)	3.0	33.0	36.0	1.2	12.9	10.8	14.0
L5 (EtOH 160°C)	2.6	14.2	16.9	1.0	5.6	5.6	6.6
L6 (EtOH 165°C)	7.7	80.6	88.3	3.0	31.4	10.5	34.4

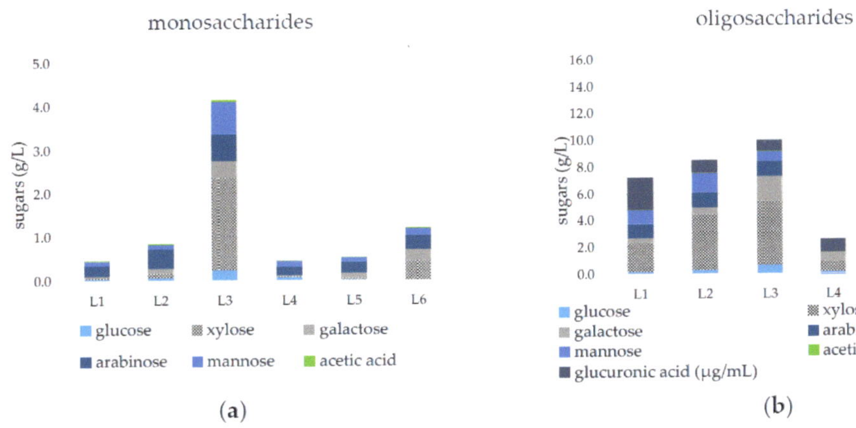

Figure 1. Compositional analysis and hemicellulose sugar profile of liquid fractions. The concentration of sugars in the form of (a) monosaccharides and (b) oligosaccharides is presented. Standard error is ≤ 2.5% in all measurements.

The optimal temperature for efficient % hemicellulose recovery in the liquid fraction was 160°C (sample L2) in the case of ACO, resulting in 25.4 ± 2.1% hemicellulose recovery in the form of oligosaccharides and only 2.8 ± 0.2% in the form of monosaccharides. In the case of EtOH, the optimal temperature for high hemicellulose recovery was 171°C (sample L6), achieving 31.4 ± 1.4% hemicellulose recovery in the form of oligosaccharides and only 3.0% in the form of monosaccharides. Furthermore, no presence of sugar degradation products was detected, while phenolic compounds were found only in low quantities, as identified by the Folin-Ciocalteu method.

3.2. Saccharification of Solid Fraction towards the Production of a C6 Sugar-Rich Hydrolysate

Cellulose-rich solid pulps obtained after pretreatment were enzymatically hydrolyzed with a commercial cellulase cocktail in order to produce a glucose-rich syrup to be utilized as carbon source for the subsequent fermentation. The time course data of hydrolysis and overall % cellulose conversion to glucose are presented in Figure 2.

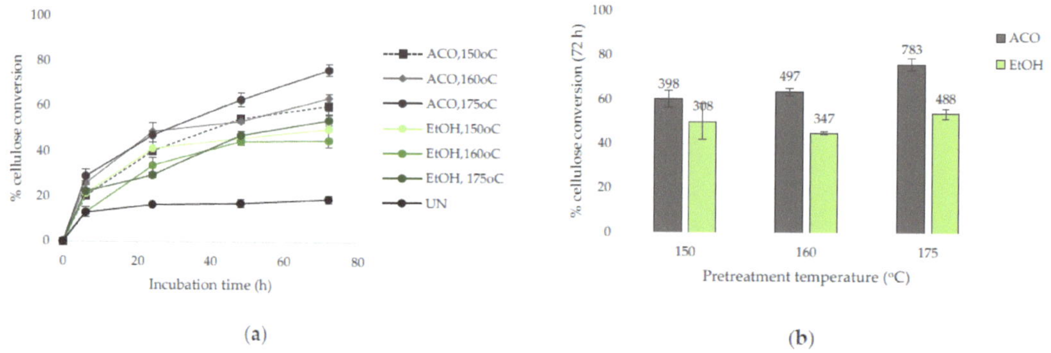

Figure 2. (a) Time course of hydrolysis (solid fraction) as function of incubation time and (b) effect of temperature and solvent on the % cellulose conversion. Labels in (b) represent the mg of glucose/g of pretreated biomass.

Even though the total number of pretreatment runs was relativelysmall, a correlation between pretreatment conditions and cellulose; conversion to glucose can be extracted. Broadly, wheat straw samples pretreated with ACO showed better hydrolysabHity that their EtOH counterp arts, while there was an upsurge in cellulose conversion as pretreatment temperature increased. More specificaliy, pretrcatment at 150°C with ACO yielded a solid pulps that exhibited 60.2 ± 3.7% cellulose conversion to glucose after 72 h of hydrolysis, while pretreatment with EtOH at 150°C resulted in 50.1 ± 8.0% conversion. The advantageous effect of ACO over EtOH was also profound at higher temperatures (160 and 175°C), while at 175°C, cellulose conversion reached the highest values, namely 76.2 ± 2.6% and,± 2.3% for ACO and EtOH pulps, respectively. These values corresponded to 783 ± 26 and 488 ± 21 mg of glucose/g of pretreated biomass. Regarding the amount of TRS, similar trends were observed, with the highest titer achieved at 175°C with ACO, reaching ± 1.1 mg/mL (721 ± 13 mg/g of pretreated biomass).

3.3. Enzymatic Hydrolysis of Liquid Fraction for the Enrichment in C5 Fermentable Sugars

Taking into account that a substantial amount of sugars is required for *C. cohnii* cultivation, L2, L3 and L6 samples were selected to be hydrolyzed and further evaluated as carbon sources, due to the higher concentration of total sugars compared to other samples and due to the presence of xylose. Cellic® HTec2 was used in order to hydrolyze hemicellulose-derived oligosaccharides to monomers that can be utilized by *C. cohnii*. The concentration of monosaccharides was increased after hydrolysis, and the results are presented in Table 3, while the sugar profile is described in Figure 3. The data show that the hydrolysis was efficient, resulting in more than a three-fold increase in the concentration of monosaccharides, especially for the L3 sample (ACO, 175°C) which reached a hydrolysis yield of 73.4 ± 1.8%. Moreover, all monosaccharides were increased, especially xylose, thus confirming the results from sugar analysis prior to hydrolysis that showed the presence of xylo-oligosaccharides (Figure 1). However, the sample population is rather low to extract a statistically significant correlation between hydrolysis yield and pretreatment conditions, both in terms of solvents and temperature. The enzymatically treated liquid fractions were further utilized in *C. cohnii* cultures, as described below.

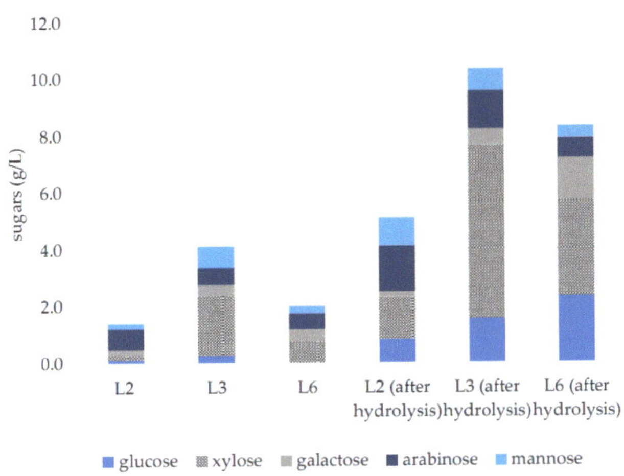

Figure 3. Sugar profile of monosaccharide fraction before and after enzymatic hydrolysis of liquid fraction. Standard error is <2.5% in all measurements.

Table 3. *Hydrolysis yield in terms of monosaccharide increase. The fraction of oligosaccharides refers to the sugar streams only (the presence of acetic or uronic acids was not considered). Standard errors are given in parenthesis.*

	Before Hydrolysis		After Hydrolysis	
Sample	**Mono- (mg/ mL)**	**Oligo- (mg/mL)**	**Mono- (mg/mL)**	**Hydrolysis Yield (%)**
L2	1.4 (0.04)	7.6 (0.2)	5.1 (0.1)	49.0 (1.i22))
L3	3.6 (0.09)	9.2 (0.3)	10.3 (0.2)	73.4 (1.8)
L6	2.0 (0.05)	12.6 (0.2(	8.3 (0.2)	50.2 (1.3)

3.4. Lipid A cumulation and DHA Production by C. cohnii Growing on Biomass-Derived Enzymatic Hydrolysates

Hydrolysates after enzymatic treatment of solid and liquid fractions were evaluated as carbon sources for *C. cohnii* cultures towards the production of a DHA-rich oil. The results are presented in Table 4, showing that both cellulose and hemicellulose-derived streams were able to support the growth of microalgal cells andaccumulatiou of fatty acids. What can be observed is that solid fractions showed a better performance than liquids in terms of growth, TFA accumulation and °% DI IA concentration. Regarding the solid fractions, EtOH pretreated samples showed higher TFA synthesis (60.5-70.3% of dried cell weight) than Fieri ACO pretreated counterparts (37-39°%), while; the cell biomass was also sligjhtly higher (Supplementary Figure S1). ACO pretruaeed samples reached the highest peak for cell biomass productivity at 150°C pretreatment temperature, where the cell biomass concentration was 6.72 ± 0.67 g/L. On the other hand, EtOH-pretreated solid fraction reached a peak at production at 160°C, namely 6.23 ± 0.25 g/L. The highest TFA concentration observed was 4.38° ±0.52 and 2.49 ±0.38 g/L of culture medium for EtOH and ACO pretreatment, respectively. No eignificant differences were observed in the % DHA contene of the oil extracted from *C. cohnii* cells in different pretreatment conditions; infact, only a slight increase could be observed in samples S1-S3 (ACO) compared to S4-S6 (EtOH). As far as the DI IA concentration is concerned, the highest values were observed at 150°C, 120 min with 0.97 ±0.19 g/L in ACO pretreated samples end at 160°C, 120 min with 1.41 ± 0.45 g/L % in EtOH pretreated samples, respectively. Noeorrelation was obseraed betweon the TEA accumulation, the % DHA and the cellulose cote tent of the initial solid pulps. Moreover, solril foactions pretreated with the same soleent have producedsimilar levels of fatty acids, indicating that pretreatment temperature was of little significance for %o TFA yields.

Table 4. *Cell biomass growth and accumulation of total fatty acid (TFA) by C. cohnii cell biomass. The carbon source used was hydrolysates from both solid and liquid fraction of organosolv process at wheat straw biomass. Standard errors are given in parenthesis. % TFA represents g of lipids per 100 g of cell biomass, while % DHA refers to the weight percentagewise concentration of DHA in the C. cohnii oil extracted. Pure glucose (4.5 wt%) was also used as a carbon source for comparison.*

Sample	Cell Biomass (g/L)	TFA (%)	TFA (g/L)	DHA (%)	DHA (g/L)
S1	6.72 (0.67)	37.04 (2.02)	2.49 (0.38)	38.82 (1.84)	0.97 (0.19)
S2	4.30 (0.11)	38.35 (1.2)	1.65 (0.10)	33.78 (2.60)	0.56 (0.08)
S3	3.07 (0.06)	39.02 (4.61)	1.20 (0.16)	33.50 (0.30)	0.40 (0.06)
S4	5.01(0.48)	64.06 (2.98)	3.21 (0.45)	22.28 (0.83)	0.72 (0.13)
S5	6.23 (0.25)	70.29 (5.6)	4.38(0.52)	32.20 (3.40)	1.41 (0.45)
S6	3.75 (0.97)	60.55 (3.34)	2.27(0.41)	29.19 (1.18)	0.66 (0.15)
Untreated	4.29 (0.11)	10.60 (2.73)	0.45 (0.13)	12.11 (1.83)	0.06 (0.01)
Glucose 4.5%	2.67(0.10)	40.61 (3.29)	1.08 (0.09)	47.20 (1.75)	0.51 (0.06)
L2	3.62 (0.34)	48.50 (4.76)	1.76 (0.34)	9.68 (0.89)	0.17(0.05)

Sample	Cell Biomass (g/L)	TFA (%)	TFA (g/L)	DHA (%)	DHA (g/L)
L3	3.75 (0.25)	22.45 (4.44)	0.84 (0.22)	2.21 (0.30)	0.02 (0.01)
L6	4.04 (0.11)	66.20 (4.73)	2.67 (0.27)	13.03 (1.41)	0.35 (0.07)

All enzymatic hydrolysates from pretreated solid pulps showed higher cell growth, TFA accumulation and % DHA compared to the untreated one, which is indicative of the efficiency of the OxiOrganosolv pretreatment in order to enhance lipid production from biomass. However, it is worth mentioning that *C. cohnii* cell growth on pure glucose resulted in lower biomass production, which was attributed to the high initial concentration of the carbon source (4.5 wt%), which hinders microalga growth [17]. Similar results regarding low cell growth were also observed in samples S3 and S6, where the initial sugar concentration was 42.3 ± 0.8 and 43.2 ± 0.5 g/L, respectively (Supplementary Table S3), which may be attributed to the high concentration of the carbon source. When pure glucose was used as the carbon source, cells accumulated 40.61 ± 3.29% of dry weight TFA, which was comparable to hydrolysates from ACO pretreated samples, resulting in a concentration of 1.08 g/L of culture medium. % DHA was significantly higher (47.2 ± 1.75% of total lipids), which corresponded to 0.51 ± 0.06 g/L. As far as the profile of fatty acids is concerned (Figure 4), one can observe that a low % DHA content is accompanied by a higher accumulation of C18:1, such as in samples S2 and S4, or C16:0 such as in sample S3.

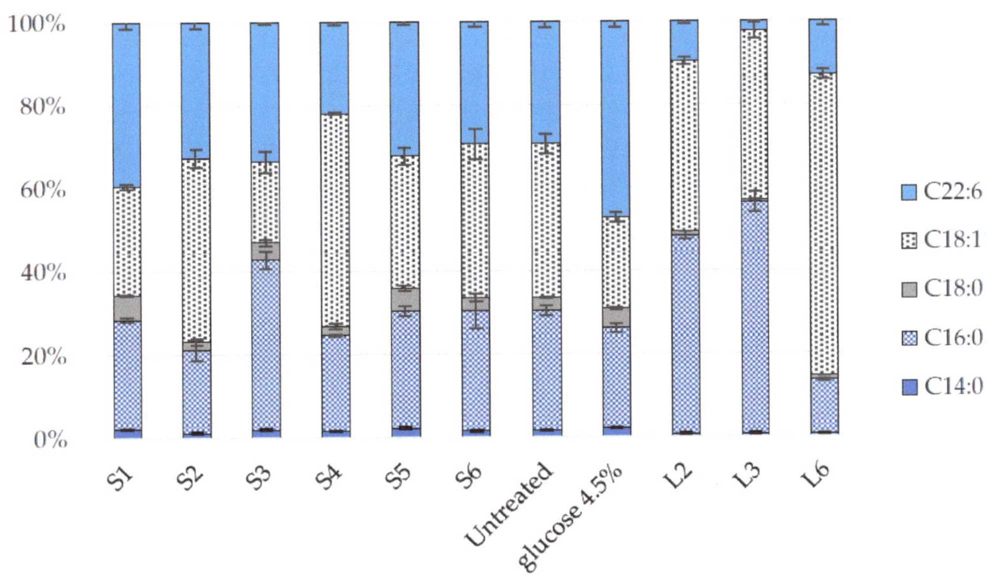

Figure 4. Fatty acid profile of oil extracted by *C. cohnii* cells after 120 h of cultivation on biomass- derived sugars.

C. cohnii cultivation using enzymatically treated liquid fractions resulted in lower yields compared to those from the solid fractions, regarding not only the TFA accumulation (with the exception of L2) but also % DHA content; the latter reached up to 13.03 ± 1.41% of total lipids in the L6 sample. One possible explanation would be the presence of sugar degradation products such as furans or phenolic compounds that inhibit the cells biomass; however, after detoxification with activated carbon, the concentration of phenolic compounds was negligible, while no presence of HMF or furfural was detected on HPLC, indicating that the low amounts of DHA could arise from the different sugar profile. The profile of fatty acids, as presented in Figure 4, reveals that hydrolysates from liquid fractions have produced far less DHA and more C18:1 and C16:0 than their solid counterparts. More specifically, C16:0 was 47.8 ± 0.9% and 55.6 ± 2.4% in L2 and L3, respectively, while C18:1 reached 72.7 ± 1.2% of total fatty acids in L6.

In an attempt to provide a clear view of the efficiency of the suggested process for the production of TFAs and, more specifically, DHA, the results are expressed in mg/g of pretreated and untreated biomass, as described in Table 5. Regarding the results from the solid fractions, EtOH pretreatment favored the production of TFA and DHA when considering the pulp recovery yield after pretreatment, the saccharification efficiency, *C. cohnii* cell growth, TFA accumulation and % DHA content. When hydrolysate from biomass pretreated with EtOH at 160°C was used as carbon source (sample S5), 20.3 ± 1.17 mg TFA/g of untreated biomass were produced, corresponding to 7.1 ± 1.2 mg DHA/g of untreated biomass, which was the highest achieved. The above results, when compared to those of the untreated sample (3.8 ± 1.09 mg TFA and 1.1 ± 0.05 mg DHA/g of untreated biomass), were significantly higher, thus demonstrating the efficiency of the pretreatment and the fractionation for the downstream process yields. The liquid fractions resulted in lower yields, with the highest being 10.71 ± 1.08 mg TFA and 1.41 ± 0.39 mg DHA/g of untreated biomass, providing the first documented evidence that hemicellulose-rich streams can be valorized as carbon sources for *C. cohnii.*

Table 5. Summary of total TFA and DHA yields per g of pretreated solid pulp or mL of aqueous liquid fraction and overall yields per g of untreated biomass. Standard errors are given in parenthesis.

Sample	TFA (mg/g of Pretreated Biomass or mg/mL of Liquid Fraction)	TFA (mg/g of Untreated Biomass)	DHA (mg/g of Pretreated Biomass or mg/mL of Liquid Fraction)	DHA (mg/g of Untreated Biomass)
S1	20.75 (3.19)	14.0 (1.09)	8.06 (1.62)	5.4 (1.09)
S2	13.76 (0.80)	7.6 (0.35)	4.65 (0.63)	2.6 (0.35)

Sample	TFA (mg/g of Pretreated Biomass or mg/mL of Liquid Fraction)	TFA (mg/g of Untreated Biomass)	DHA (mg/g of Pretreated Biomass or mg/mL of Liquid Fraction)	DHA (mg/g of Untreated Biomass)
S3	9.97(1.36)	4.5 (0.22)	3.34 (0.49)	1.5 (0.22)
S4	26.74 (3.79)	18.4 (0.74)	5.96 (1.07)	4.1 (0.74)
S5	36.48 (4.36)	20.3 (1.17)	11.75 (1.66)	7.1 (1.2)
S6	18.93 (5.95)	9.0 (0.93)	5.53 (1.96)	2.6 (0.93)
L2	1.95 (0.38)	9.23 (1.77)	0.19 (0.05)	0.90 (0.23)
L3	0.94 (0.25)	4.68 (1.23)	0.02 (0.01)	0.10 (0.05)
L6	2.97 (0.30)	10.71 (1.08)	0.39 (0.08)	1.41 (0.39)
Untreated	3.69 (1.06)	3.8 (1.09)	0.46(0.02)	1.1 (0.05)

4. DISCUSSION

Wheat straw was preferred in this case study since high amounts are made available in Greece as side streams from the agricultural sector. The term "agricultural residues" refers to any kind of by-products or agricultural derivatives without any economic value for the enterprise or their further management or any profitable utilization [29]. In Europe, agricultural waste consists mainly of wheat straw, while its production equals about 32% of the worldwide production [30]. In particular, according to the Centre for Renewable Energy Sources & Saving (CRES) in Greece for the year 2017, wheat straw residues were estimated to be 1,150,738 tonnes of dry matter, thus rendering the valorization of these feedstocks of pivotal importance.

The OxiOrganosolv pretreatment was applied in our previous studies and was proven optimal for the maximum delignification efficiency [6,19,20]. The use of ACO and EtOH as organic solvents in the pretreatment of various lignocellulosic biomass residues has been widely reported in the literature, since they are milder and green solvents that can be easily recovered and reused, making the whole process sustainable and applicable on an industrial scale [31-33]. Moreover, they have the ability not only to cleave the bonds between lignin and hemicellulose during the organosolv pretreatment but also to remove hemicellulose and solubilize the lignin, thereby causing an increase in surface area and pore volume of cellulose, which renders cellulose more accessible to enzymatic hydrolysis [34]. ACO is reported as the most favored ketone used for delignification [7]. However, the prevalence of EtOH is probably linked to its low cost and lack of toxicity, while also leading to the efficient deconstruction of lignocellulosic biomass [33]. The advantageous use of EtOH is supported by recent approaches, such as the integration of first- and second-generation ethanol, to enhance the commercialization of cellulose-derived EtOH

[34]. In addition, the concentration of EtOH in the reaction medium of organosolv processes has been mainly within 30-60% (v/v) [35-37], which is in concert with the concentration acquired after the first distillation at first-generation EtOH industries. Both ACO and EtOH have aroused research interest on account of their distinct properties: they dissolve lignin and enable recovery of solid lignin after solvent removal, since the latter is insoluble in aqueous solutions [35]; they enable fractionation of hemicellulose-derived oligosaccharides from cellulose because the former exhibit some solubility in the organic solvent: H_2O mixture [36]; they result in cellulose structure swelling, causing the crystal structure of cellulose fibers to be unfolded and thus rendering cellulose more eligible for hydrolysis [38]; and they can be both collected and recycled, thus reducing operating costs. This process seems to be promising for lignocellulose fractionation in order to obtain a solid and a liquid fraction that can be further processed separately, provided that a techno economical study including the solvent recycling is carried out prior to setting up larger scale units.

In our previous works, the advantages of OxiOrganosolv both for the efficient fractionation and increased enzymatic digestibility of hardwoods (beechwood) and softwoods (pine) has been demonstrated [6,20], thus highlighting the advantages of this process compared to the traditional organosolv pretreatment upon addition of sulfuric acid. In the present study, we attempted to further apply the process in crops, showing that it is also efficient for delignification of this type of substrate and the production of sugar-rich streams. Salapa et al. [39] reported the organosolv treatment of wheat straw in the presence of sulfuric acid as catalyst by evaluating five different organic solvents; the results showed that ACO was the most efficient in biomass delignification, achieving 76.4% lignin removal after pretreatment at 180°*C* for 40 min. However, enzymatic hydrolysis of the produced solid pulp reached only 49.97% cellulose conversion, while in the case of the present study, the yield was 76.2%. In another work, organosolv fractionation of wheat straw upon addition of acid catalyst resulted in 75.8% delignification at 190°C for 60 min, while in the absence of acid at 160 and 170°C (conditions comparable to those in this study), lignin removal was 4.7 and 14.4% respectively, and the enzymatic digestibility of the pulps was only 30.5% and 31.7% [40].

For the development of an efficient organosolv process taking advantage of all sugar streams, valorization of the pentose-rich liquid fraction together with the hexose-rich solid pulp is a prerequisite. Hemicellulose streams have been already studied as a potential substrate for the production of valuable chemicals, such as xylitol produced from corncob [41], biobutanol from birch kraft black liquor [42] and lactic acid from corn stover [43]. Moreover, the produced aqueous fraction contains a high number of xylo-oligosaccharides which have be studied as prebiotic compounds [20,44]. However, no work has yet been made towards the utilization of the liquid fraction for the production of fatty acids from microalga. The great advantage of C. *cohnii* is that this microorganism is able to utilize not only hexoses but also pentoses, accumulating high amounts of DHA-rich oil, as it has been originally reported from Karnaouri et al. using pure xylose and arabinose [19]. In this study, we further evaluated the ability of the microalga to grow on a true hydrolysate from wheat straw pretreatment. Pentose utilization is still halted because hemicellulose-rich fraction originating from biomass pretreatment may be contaminated with sugar degradation products that act as inhibitors [45]. However, the OxiOrganosolv pretreatment method yields an inhibitor-free solid fraction [6], as was also shown in the present study, which is the great advantage of this process. In the process that is suggested in this work, tuning the pretreatment for the removal of hemicellulose and the release of pentose sugars in the form of monosaccharides would be an option in order to eliminate the necessity for enzymatic treatment. Additionally, concerning wheat straw, no studies have taken place regarding fatty acids production, despite the fact that wheat straw is a widely available agricultural waste.

Regarding the growth of C. *cohnii* and the production of fatty acids, all enzymatic hydrolysates from solid pulps efficiently served as carbon sources in this work. Notably, higher biomass pretreatment temperatures with ACO or EtOH produced very cellulose-rich pulps that progressively lowered the cell biomass productivity when used as carbon source for cell growth. This was attributed to high initial concentration of glucose in the growth medium, which was confirmed with the pure glucose sample tests and is also supported by the literature [46]. Since high initial concentration of carbon source hinders C. *cohnii* growth, employing a fed-batch strategy could improve cell productivity yields and reduce incubation time. As far as the % DHA content is concerned, the results are in accordance with the previous reports in the literature; a variety of different substrates, such as carob pulp syrup, resulting in 48% DHA of total lipids (however with a much lower accumulation of TFA 9.2% of cell biomass) [47], rapeseed meal hydrolysate mixed with crude molasses, yielding 22-34% DHA [48], and cheese whey with corn steep liquor [49] have been used as carbon sources in media formulations for C. *cohnii* fatty acid production. Cultivation on liquid fractions in this work produced far less DHA and more C18:1 when compared to solid fraction hydrolysates, corroborating the idea that the initial sugar concentration and the type of carbon source affected not only the amount of accumulated TFAs but also the quality of the oil and the % DHA content. It has been already reported that when pure xylose and arabinose were used as carbon sources, the oil accumulated in *C. cohnii* cells contained more C18:1 than DHA [19]. Moreover, the initial concentration of sugars in liquid fractions was much lower than that of their solid counterparts, leading to a lower

C/N ratio in the culture medium, since the initial concentration of the nitrogen source was constant in all experiments. In addition, although strain-specific, it has been demonstrated that reduced degree of unsaturation and chain-length is related to lower specific growth rate and higher TFA accumulation yields [50].

Cellular stress and nutrient deprivation can affect not only the lipid accumulation but also the fatty acid profile. A lower cell growth has been correlated with a higher lipid content, and this has been observed for many microalgal strains [51]. The differences observed in this study regarding the TFA accumulation and DHA content can be attributed to the fact that *C. cohnii* responds differently to multi-stress factors that are related to the addition of hemicellulosic fraction as carbon source, even after detoxification. Moreover, the difference in the lipid profiles when using either the solid or the liquid fractions might be also correlated with the difference in their sugar profiles; the hydrolysates derived from solid fraction are richer in glucose, whereas the hydrolysis of liquid fractions produced more xylose. Despite the lower yields observed with liquid fractions, this is the first report of the cultivation of *C. cohnii* on the pentose-rich fraction and the valorization of this fraction for the production of DHA.

5. CONCLUSIONS

In the present work, hexose- and pentose-rich streams after organosolv pretreatment and fractionation of wheat straw with ACO and EtOH were used as carbon sources for *C. cohnii* cells for the production of fatty acids and, more specifically, DHA. Enzymatic hydrolysates from solid fractions have shown great potential, achieving up to 70.3 wt.%

TFA accumulation and 32.2% DHA of total lipids. Pentose-rich liquid fractions resulted in lower DHA yield (up to 13%), indicating the presence of compounds in the hydrolysate that may affect the lipid synthesis and alter the fatty acids profile. Moreover, the type of carbon source affects not only the amount of lipids but also the relative proportion of DHA that is accumulated in the cells. It is worth mentioning that this is the first report demonstrating, as a proof of concept, the valorization of all sugar streams towards the production of omega-3 fatty acids. Despite the challenges of the liquid fraction utilization, the suggested process is a promising approach towards the production of omega-3 fatty acids from non-edible sources and opens up new routes, considering the availability of wheat straw residues that are made available from the Greek agricultural sector, as well as worldwide.

REFERENCES

1. Solarte-Toro, J.C.; Chacon-Perez, Y.; Cardona-Alzate, C.A. Evaluation of biogas and syngas as energy vectors for heat and power generation using lignocellulosic biomass as raw material. *Electron. J. Biotechnol.* **2018**, 33, 52-62. [CrossRef]

2. da Costa Lopes, A.M.; Joao, K.G.; Rubik, D.F.; Bogel-Lukasik, E.; Duarte, L.C.; Andreaus, J.; Bogel-Lukasik, R. Pre-treatment of lignocellulosic biomass using ionic liquids: Wheat straw fractionation. *Bioresour. Technol.* **2013**,*142*,198-208. [CrossRef]

3. Zhang, H.; Zhang, J.; Xie, J.; Qin, Y. Effects of NaOH-catalyzed organosolv pretreatment and surfactant on the sugar production from sugarcane bagasse. *Bioresour. Technol.* **2020**, 312,123601. [CrossRef]

4. Alvira, P.; Tomas-Pejo, E.; Ballesteros, M.; Negro, M.J. Pretreatment technologies for an efficient bioethanol production process based on enzymatic hydrolysis: A review. *Bioresour. Technol.* **2010**,101, 4851-4861. [CrossRef]

5. Jargensen, H.; Kristensen, J.B.; Felby, C. Enzymatic conversion of lignocellulose into fermentable sugars: Challenges and opportunities. *Biofuel Bioprod. Biorefin.* **2007**,1,119-134. [CrossRef]

6. Kalogiannis, K.G.; Karnaouri, A.; Michailof, C.; Tzika, A.M.; Asimakopoulou, G.; Topakas, E.; Lappas, A.A. OxiOrganosolv: A novel acid free oxidative organosolv fractionation for lignocellulose fine sugar streams. *Bioresour. Technol.* **2020**, 313,123599. [CrossRef]

7. Zhao, X.; Cheng, K.; Liu, D. Organosolv pretreatment of lignocellulosic biomass for enzymatic hydrolysis. *Appl. Microbiol. Biotechnol.* **2009**, 82, 815-827. [CrossRef]

8. Zhao, X.; Li, S.; Wu, R.; Liu, D. Organosolv fractionating pre-treatment of lignocellulosic biomass for efficient enzymatic saccharification: Chemistry, kinetics, and substrate structures. *Biofuel Bioprod. Biorefin.* **2017**,*11*, 567-590. [CrossRef]

9. Dong, C.; Meng, X.; Yeung, C.S.; Ho-Yin, T.S.E.; Ragauskas, A.J.; Leu, S.Y. Diol pretreatment to fractionate a reactive lignin in lignocellulosic biomass biorefineries. *Green Chem.* **2019**, 21, 2788-2800. [CrossRef]

10. Tang, D.Y.Y.; Khoo, K.S.; Chew, K.W.; Tao, Y.; Ho, S.H.; Show, P.L. Potential utilization of bioproducts from microalgae for the quality enhancement of natural products. *Bioresour. Technol.* **2020**, 304,122997. [CrossRef]

11. Levasseur, W.; Perre, P.; Pozzobon, V. A review of high value-added molecules production by microalgae in light of the classification. *Biotechnol. Adv.* **2020**, 41,107545. [CrossRef]

12. Silambarasan, S.; Logeswari, P.; Sivaramakrishnan, R.; Incharoensakdi, A.; Cornejo, P.; Kamaraj, B.; Chi, N.T.L. Removal of nutrients from domestic wastewater by microalgae coupled to lipid augmentation for biodiesel production and influence of deoiled algal biomass as biofertilizer for Solanum lycopersicum cultivation. *Chemosphere* **2021**, 268,129323. [CrossRef]

13. Brasil, B.D.S.A.F.; de Siqueira, F.G.; Salum, T.F.C.; Zanette, C.M.; Spier, M.R. Microalgae and cyanobacteria as enzyme biofactories. *Algal Res.* **2017**, 25, 76-89. [CrossRef]

14. Khan, R.S.; Grigor, J.; Winger, R.; Win, A. Functional food product development-Opportunities and challenges for food manufacturers. *Trends Food Sci. Technol.* **2013**, 30, 27-37. [CrossRef]

15. Chalima, A.; Oliver, L.; Fernandez de Castro, L.; Karnaouri, A.; Dietrich, T.; Topakas, E. Utilization of volatile fatty acids from microalgae for the production of high added value compounds. *Fermentation* **2017**, 3, 54. [CrossRef]

16. Chalima, A.; Hatzidaki, A.; Karnaouri, A.; Topakas, E. Integration of a dark fermentation effluent in a microalgal-based biorefinery for the production of high-added value omega-3 fatty acids. *Appl. Energy* **2019**, 241,130-138. [CrossRef]

17. de Swaaf, M.E.; Pronk, J.T.; Sijtsma, L. Fed-batch cultivation of the docosahexaenoic-acid-producing marine alga *Crypthecodinium cohnii* on ethanol. *Appl. Microbiol. Biotechnol.* **2003**, 61, 40-43. [CrossRef]

18. Paz, A.; Karnaouri, A.; Templis, C.C.; Papayannakos, N.; Topakas, E. Valorization of exhausted olive pomace for the production of omega-3 fatty acids by *Crypthecodinium cohnii. Waste Manag.* **2020**,118, 435-444.

19. Karnaouri, A.; Chalima, A.; Kalogiannis, K.; Varamogianni-Mamatsi, D.; Lappas, A.; Topakas, E. Utilization of organosolv pretreated lignocellulosic biomass for the production of omega-3 fatty acids by the heterotrophic marine microalga *Crypthecodinium cohnii. Bioresour. Technol.* **2020**, 303,122899. [CrossRef]

20. Karnaouri, A.; Asimakopoulou, G.; Kalogiannis, K.G.; Lappas, A.A.; Topakas, E. Efficient production of nutraceuticals and lactic acid from lignocellulosic biomass by combining organosolv fractionation with enzymatic/fermentative routes. *Bioresour. Technol.* **2021**, *341*, 125846. [CrossRef]

21. Sluiter, A.; Ruiz, R.; Scarlata, C.; Sluiter, J.; Templeton, D. *Determination of Extractives in Biomass;* TP-510-42619; NREL: Golden, CO, USA, 2008.

22. Sluiter, A.; Hames, B.; Ruiz, R.; Scarlata, C.; Sluiter, J.; Templeton, D.; Crocker, D.L.A.P. Determination of structural carbohydrates and lignin in biomass. *Lab. Anal. Proced.* **2008**,*1617*, 1-16.

23. Bradford, M.M. A rapid and sensitive method for the quantitation of microgram quantities of protein utilizing the principle of protein-dye binding. *Anal. Biochem.* **1976**, 72, 248-254. [CrossRef]

24. Raba, J.; Mottola, H.A. Glucose Oxidase as an Analytical Reagent. *Crit. Rev. Anal. Chem.* **1995**,25,1-42. [CrossRef]

25. Miller, G.L. Use of dinitrosalicylic acid reagent for determination of reducing sugar. *Anal. Chem.* **1959**, 31, 426-428. [CrossRef] Singleton, V.L.; Orthofer, R.; Lamuela-Raventos, R.M. Analysis of total phenols and other oxidation substrates and antioxidants by means of folin-ciocalteu reagent. *Meth. Enzymol.* **1999**, 299,152-178.

26. de Swaaf, M.E.; de Rijk, T.C.; Eggink, G.; Sijtsma, L. Optimisation of docosahexaenoic acid production in batch cultivations by *Crypthecodinium cohnii. J. Biotechnol.* **1999**, 70,185-192. [CrossRef]

27. Folch, J.; Lees, M.; Stanley, G.S. A simple method for the isolation and purification of total lipids from animal tissues. *J. Biol. Chem.* **1957**, *226*, 497-509. [CrossRef]

28. Alatzas, S.; Moustakas, K.; Malamis, D.; Vakalis, S. Biomass potential from agricultural waste for energetic utilization in Greece. *Energies* **2019**,12,1095. [CrossRef]

29. Kim, S.; Dale, B.E. Global potential bioethanol production from wasted crops and crop residues. *Biomass Bioenerg.* **2004**, 26, 361-375. [CrossRef]

30. Zhang, K.; Pei, Z.; Wang, D. Organic solvent pretreatment of lignocellulosic biomass for biofuels and biochemicals: A review. *Bioresour. Technol.* **2016**,199, 21-33. [CrossRef]

31. Zhang, Y.P Reviving the carbohydrate economy via multi-product lignocellulose biorefineries. *J. Ind. Microbiol. Biotechnol.* **2008**, 35, 367-375. [CrossRef]

32. Ferreira, J.A.; Brancoli, P.; Agnihotri, S.; Bolton, K.; Taherzadeh, M.J. A review of integration strategies of lignocelluloses and other wastes in 1st generation bioethanol processes. *Process Biochem.* **2018**, 75,173-186. [CrossRef]

33. Li, X.; Luo, Y.; Daroch, M.; Hou, J.; Gui, W. Oxygen-assisted ethanol organosolv pretreatment of sugarcane bagasse for efficient removal of hemicellulose and lignin. *Cellulose* **2018**, 25, 5511-5522. [CrossRef]

34. Martin-Sampedro, R.; Eugenio, M.E.; Fillat, U.; Martin, J.A.; Aranda, P.; Ruiz-Hitzky, E.; Ibarra, D.; Wicklein, B. Biorefinery of lignocellulosic biomass from an elm clone: Production of fermentable sugars and lignin-derived biochar for energy and environmental applications. *Energy Technol.* **2019**, 7, 277-287. [CrossRef]

35. Park, Y.C.; Kim, J.S.; Kim, T.H. Pretreatment of corn stover using organosolv with hydrogen peroxide for effective enzymatic saccharification. *Energies* **2018**,*11*,1301. [CrossRef]

36. Alves, L.A.; Almeida e Silva, J.B.; Giulietti, M. Solubility of D-glucose in water and ethanol/water mixtures. *J. Chem. Eng. Data* **2007**, 52, 2166-2170. [CrossRef]

37. Del Rio, L.F.; Chandra, R.P.; Saddler, J.N. The effects of increasing swelling and anionic charges on the enzymatic hydrolysis of organosolv-pretreated softwoods at low enzyme loadings. *Biotechnol Bioeng.* **2011**,108,1549-1558. [CrossRef]

38. Salapa, I.; Katsimpouras, C.; Topakas, E.; Sidiras, D. Organosolv pretreatment of wheat straw for efficient ethanol production using various solvents. *Biomass Bioenerg.* **2017**,100,10-16. [CrossRef]

39. Wildschut, J.; Smit, A.T.; Reith, J.H.; Huijgen, W.J. Ethanol-based organosolv fractionation of wheat straw for the production of lignin and enzymatically digestible cellulose. *Bioresour. Technol.* **2013**,135, 58-66. [CrossRef]

40. Kumar, V.; Krishania, M.; Sandhu, P.P.; Ahluwalia, V.; Gnansounou, E.; Sangwan, R.S. Efficient detoxification of corn cob hydrolysate with ion-exchange resins for enhanced xylitol production by *Candida tropicalis* MTCC 6192. *Bioresour. Technol.* **2018**, 251, 416-419. [CrossRef]

41. Kudahettige-Nilsson, R.L.; Helmerius, J.; Nilsson, R.T.; Sjoblom, M.; Hodge, D.B.; Rova, U. Biobutanol production by *Clostridium acetobutylicum* using xylose recovered from birch Kraft black liquor. *Bioresour. Technol.* **2015**,176, 71-79. [CrossRef]

42. Hu, J.; Lin, Y.; Zhang, Z.; Xiang, T.; Mei, Y.; Zhao, S.; Liang, Y.; Peng, N. High-titer lactic acid production by *Lactobacillus pentosus* FL0421 from corn stover using fed-batch simultaneous saccharification and fermentation. *Bioresour. Technol.* **2016**, 214, 74-80. [CrossRef]

43. Boonchuay, P.; Techapun, C.; Leksawasdi, N.; Seesuriyachan, P.; Hanmoungjai, P.; Watanabe, M.; Takenaka, S.; Chaiyaso, T. An integrated process for xylooligosaccharide and bioethanol production from corncob. *Bioresour. Technol.* **2018**, 256, 399-407. [CrossRef] [PubMed]

44. Jonsson, L.J.; Martin, C. Pretreatment of lignocellulose: Formation of inhibitory by-products and strategies for minimizing their effects. *Bioresour. Technol.* **2016**,199,103-112. [CrossRef]

45. Jiang, Y.; Chen, F. Effects of medium glucose concentration and pH on docosahexaenoic acid content of heterotrophic *Cryptheco- dinium cohnii. Process Biochem.* **2000**, 35,1205-1209. [CrossRef]

46. Mendes, A.; Guerra, P; Madeira, V.; Ruano, F.; da Silva, T.L.; Reis, A. Study of docosahexaenoic acid production by the heterotrophic microalga *Crypthecodinium cohnii* CCMP 316 using carob pulp as a promising carbon source. *World J. Microbiol. Biotechnol.* **2007**,23,1209-1215. [CrossRef]

47. Gong, Y.; Liu, J.; Jiang, M.; Liang, Z.; Jin, H.; Hu, X.; Wan, X.; Hu, C. Improvement of omega-3 docosahexaenoic acid production by marine dinoflagellate *Crypthecodinium cohnii* using rapeseed meal hydrolysate and waste molasses as feedstock. *PLoS ONE* **2015**, *10*, e0125368. [CrossRef]

48. Isleten-Hosoglu, M.; Elibol, M. Bioutilization of cheese whey and corn steep liqour by heterotrophic microalgae *Crypthecodinium cohnii* for biomass and lipid production. *Acad. Food J. Akad. GIDA* **2017**,15, 233-241.

49. Gachelin, M.; Boutoute, M.; Carrier, G.; Talec, A.; Pruvost, E.; Guiheneuf, F.; Bernard, O.; Sciandra, A. Enhancing PUFA-rich polar lipids in *Tisochrysis lutea* using adaptive laboratory evolution (ALE) with oscillating thermal stress. *Appl. Microbiol. Biotechnol.* **2021**, *105*, 301-312. [CrossRef]

50. Roleda, M.Y.; Slocombe, S.P.; Leakey, R.J.G.; Day, J.G.; Bell, E.M.; Stanley, M.S. Effects of temperature and nutrient regimes on biomass and lipid production by six oleaginous microalgae in batch culture employing a two-phase cultivation strategy. *Bioresour. Technol.* **2013**,129, 439-449. [CrossRef]

CHAPTER 15

Production and Maturation of Soaps with Non-Edible Fermented Olive Oil and Comparison with Classic Olive Oil Soaps

1. INTRODUCTION

In ancient times, soap was produced with rancid oils and animal fat waste, involuntarily creating a circular economy with the requalification of products no longer used for human nutrition [1].

In recent times, the production of soaps, defined by law as cosmetic products, is regulated by the European Union [2]. The regulation determines and limits the substances to be used for the saponification processes (Regulation (EC) No. 1223/2009 article 14-17 [2]) and indirectly determines the chemical, physical, and mechanical treatments to obtain adequate raw materials for the cosmetic sector. Exhausted oils and fats, dangerous for human nutrition, are necessarily treated to eliminate all toxic and dangerous components such as heavy metals, microplastics, peroxides, alkyl esters, and suspended particles [3,4].

Companies that produce soaps buy large quantities of regenerated oils and fats, adding chemical compounds such as surfactants, parabens, silicones, and sulphates. These companies contribute to the recycling of regenerated oils and fats but add pollutants to improve certain products' characteristics [5,6].

Often, artisanal companies use edible oils, such as olive oil, for obtaining natural and high-quality products. These soaps are obtained from food and not from food waste [7-9].

Scientific studies report the recycling of olive production waste, such as vegetation water, pomace, and plant parts of the tree, but there are relatively few studies in the literature on the recycling of damaged and overripe non-edible olives [10-13].

The authors were focused on the use of non-edible drupes (food waste) for the production of high-quality soaps through natural fermentation processes.

Non-edible olives are drupes with defects that develop negative and irreversible chemical, physical, and organoleptic alterations. Producers of fine oils eliminate these olives to obtain an excellent quality of the final product oils [14]. Furthermore, these drupes cannot be marketed because they do not have the minimum requirements defined by the European Community [15].

Non-edible olives are therefore used as fodder, natural fertilizer, and for the production of energy biomass [11,13].

The aim of this study was to produce and evaluate the different ripening stages of soaps produced with non-edible fermented olive oil (NEFOO soap) considering three parameters: pH, color, and solubility. The results obtained were compared with those obtained from soaps produced with extra virgin olive oil (EVOO soap).

2. MATERIALS AND METHODS

2.1. Olives

The olives were harvested in the third week of October 2020, in Messina (Sicily, Italy). The soils had homogeneous composition and climatic conditions, and the olive cultivars were native of Sicily *(Nocellara Etnea, Nocellara Messinese,* and *Cerasuola)* [16,17]. The olives were harvested with mechanical systems that do not damage the plant or the fruit.

The olives were selected with machines and manually to remove overripe fruit with defects, the vegetative waste, and impurities.

Non-edible olives were collected in permeable bags and stored in a dark room with a temperature of 20 °C and constant humidity until the complete harvest of the olives.

2.2. Olives' Fermentation

Non-edible olives, 24 kg, were cleaned with cold water and inserted into a plastic vessel with a non-pre-acidified brine and 5% m/m NaCl, in a final working volume of 75%. No sterilization procedures were adopted on the non-edible olives before starting the natural fermentation, and no starter strains of yeast or bacteria were inoculated. The olive-brine mixture was aerated manually every 3 days to facilitate the growth of aerobic microorganisms. The fermentation was carried out for 60 days and the pH was monitored during all the process. Neither sodium chloride nor acids or bases were added to the brine after fermentation starting [18-21].

Brine samples were withdrawn for analysis at regular time intervals. The samples (1 mL) were aseptically transferred to 9 mL sterile 1/4 Ringer's solution. Decimal dilutions in the same Ringer's solution were prepared and used for the enumeration of lactic acid bacteria, yeasts and molds, and for enterobacteria.

The lactic acid bacteria amount was estimated on de Man-Rogosa-Sharpe agar (MRS Agar, Oxoid, Basingstoke, UK) added with 50 mg/L of Nystatin (Sigma, Saint-Quentin Fallavier, France) at 25 ° C for 72 h.

The population of yeasts and molds was estimated on Yeast Glucose Chloramphenicol Agar (YGC, bioMerieux, Lyon, France) at 25 °C for 48 h.

Violet Red Bile Glucose agar (VRBGA; Oxoid, Basingstoke, UK) was used for enterobacteria counts, incubated at 37 °C for 24 h.

The analyses were carried out in triplicate and the plates were subjected to microbiological numbering by CFU counting.

2.3. pH Determination

The pH of the brine solutions and the pH of the soap, after the maturation process, were determined with a digital benchtop pH meter (HANNA Instruments, Edge PH-HI2020, Smithfield, Rhode Island, USA) with glass immersion electrode.

The pH of the brine was determined by taking 50 mL of solution after mixing the fermentation mass.

The pH of the soaps was determined by solubilizing 1 g of soap, the inner part, in 50 mL of deionized water [7,8].

2.4. Oil Extraction

Non-edible fermented olive oil (NEFOO) was obtained with a cold mechanical extraction system. The extra virgin olive oil (EVOO) was kindly offered by the company that harvested the olives. It was obtained from the olives harvested but not discarded and stored in a dark room with a temperature of 20°C.

2.5. Acidity Value of Oils

The acidity value (AV) of the olive oils was determined with the titration method reported in the Execution Regulation (EU) 2019/1604 and in the EN ISO 660: 2020 regulation.

The oil (1 g) was weighed into a flask and 50 mL of ethyl ether/ethanol mixture (1:1 ratio) and two drops of phenolphthalein ethanolic solution were added. The mixture was titrated with a KOH solution until the indicator changed color. The acidity was expressed as the weight percentage of free oleic acid present in 1 g of analyzed sample. The analyses were carried out in triplicate, and the average of the values are reported in Table 1 [22,23].

Table 1. Saponification, acidity, and peroxide values of EVO and NEFO oils.

Oil	SV (mg NaOH/g Oil)	NaOH mg/g Oil Soap 0%	NaOH mg/g Oil Soap 5%	Acidity (% Oleic Acid)	Peroxide (meq O_2/kg Oil)
EVOO	135.41 ± 0.61	135.41	128.64	0.15 ± 0.01	6.24 ± 0.07
NEFOO	133.34 ± 0.53	133.34	126.67	38.26 ± 0.04	29.32 ± 0.09

2.6. Peroxide Value of Oils

The peroxide value (PV) in olive oils was determined with iodometric titration, described in the Implementing Regulation (EU) 2019/1604 and in the ISO 3960: 2017 regulation. The oil (1 g) was weighed into a flask and 25 mL of a mixture of acetic acid and chloroform (ratio 3:2) with 0.5 mL of a saturated aqueous solution of KI were added. The solution was diluted with 25 mL of distilled water and stirred. The iodine formed into the solution was titrated with a sodium thiosulfate solution, and the starch was used as indicator [22,24].

The titrations were carried out in triplicate, and the average of the values, expressed in milliequivalents of oxygen per kg of sample, are reported in Table 1.

2.7. Saponification Value of Oils

The saponification value (SV) represents the quantity of sodium hydroxide (as an alternative to potassium hydroxide), expressed in milligrams, necessary to neutralize all the fatty acids present in one g of fat sample. The analytical determination of SV was carried out according with Wakita et al. [25].

The oil sample (2 g) was weighed into a 250 mL flask and 25 mL of NaOH 0.50 N ethanolic solution were added. The mixture was heated up to 50°C, under constant magnetic stirring (300 rpm) for 90 min. Subsequently, the mixture was cooled to 30°C and 0.5 mL of phenolphthalein solution was added.

The solution was titrated with HCl 0.50 N until the color of the indicator changed, the titration was also carried out without the sample, only with solution. The analyses were carried out in triplicate, obtaining an average of the volumes of HCl consumed; the values are reported in Table 1.

The saponification values were expressed in milligrams of sodium hydroxide needed to saponify 1 g of oil.

2.8. Oils' Saponification

The fatty acids of the EVO and NEFO oils were totally (0% of non-neutralized fatty acids) and partially (5% of non-neutralized fatty acids) neutralized and four different types of soaps were obtained:

- EVOO soap 0%, obtained from EVOO with 0% of non-neutralized fatty acids

- EVOO soap 5%, obtained from EVOO with 5% of non-neutralized fatty acids
- NEFOO soap 0%, obtained from NEFOO with 0% of non-neutralized fatty acids
- NEFOO soap 5%, obtained from NEFOO with 5% of non-neutralized fatty acids.

The quantity of NaOH used for the production of the described soaps was determined by SV. The quantities of NaOH used and the SV values are shown in Table 1.

A classic saponification procedure was applied for soap produced with EVO oil, while for the saponification of NEFO oil it was necessary to make changes with respect to the standard procedures reported by Spitz [7,8].

In 30 mL of deionized water 13.541 g and 13.334 g of NaOH were dissolved to obtain, respectively, the soaps with 0% and 5% of non-neutralized fatty acids (Table 1). Each alkaline solution was mixed with 100 g of oil at a temperature of 35°C. The oil-water mixture was shaken strongly to obtain a semi-liquid consistency. The dough was placed in a mold for the subsequent maturation process [7,8].

The saponification of NEFO oil was carried out by varying the "cold" method. In 30 mL of deionized water 12.864 g and 12.667 g of NaOH were dissolved to obtain, respectively, the soaps with 0% and 5% of non-neutralized fatty acids (Table 1).

Each NaOH solution was added with slow dripping to 100 g of oil in a thermostated system at 20°C with continuous stirring. A warm, semi-solid soap paste formed quickly due to the exothermic acid-base reaction between NaOH and fatty acids. The soap paste was placed in a mold for the subsequent maturation process.

The saponification for EVO oil and NEFO oil was carried out considering the complete neutralization of the fatty acids and the non-neutralization of 5% of the fatty acids using the SV of the oils. The quantity of NaOH used for the production of the different soaps and the SV are shown in Table 1. The soaps were stored in a dark room with low humidity on filter paper to prevent rancidity.

The soaps considered (0% and 5% of non-neutralized fatty acids) are those most produced by industries and most consumed [7,8].

2.9. Colorimetric Determination

A colorimetric sensor system (LISUN GROUP, CD-200, Kowloon, Hong Kong) was used to determine the color variations of the soap samples at different maturation times. The instrument consists of a D65 light source, a blue LEDs light source and a photoelectric diode detector, and an electronic system for determining the electrical impulses.

The color determination was carried out with the RGB model (red, green, and blue) and the CIELAB color space, which considers the parameters l*, a*, and b* that represent the brightness and the four unique colors of human vision (red, yellow, green, and blue) [26,27].

2.10. Solubility Determination

The solubility of the soaps was determined by solubilizing 1 g of soap, not in contact with air, in 50 mL of deionized water. Each solution, after the complete dissolution of the soap, was stirred constantly for 5 min and, subsequently, its transparency was observed.

3. RESULTS

3.1. Olives' Fermentation

Initial brine pH was 8.21, after 5 days it showed a pH of 7.22, representing an optimal condition for the proliferation of initial lactic bacteria but also enterobacteria.

In fact, enterobacterif counts showed a rapid increase within the first 10 days, followed bya sharp de cline theneafter. No viable counts were enumerated after 25 days of fermentation, according with Sanchez et al. [28] (Figure 1).

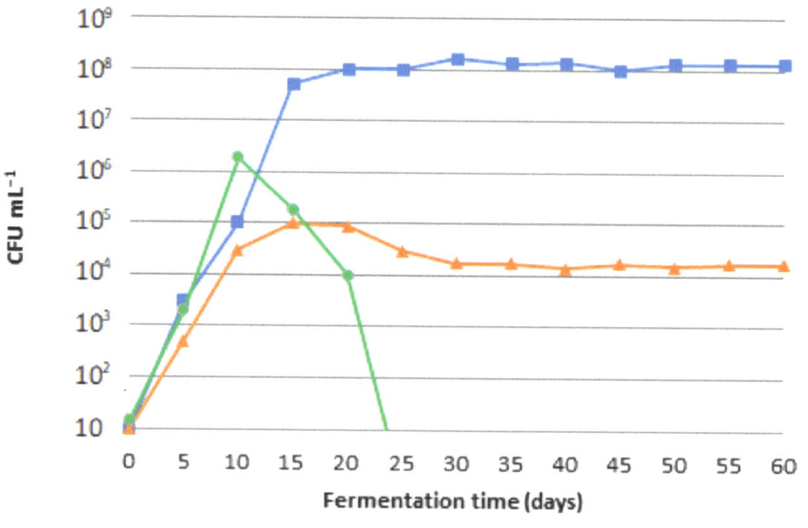

Figure 1. *Lactic bacteria (square), yeast and molds (triangle), enterobacteria (circle) concentration, reported as CFU per mL.*

The pH decreasing was due to lactic bacteria, which ferment a wide variety of watersoluble carbo hydrates and produce lactic acid [20,29,30]. In this ph aae the otives lost their classic hardness;: this indicates the presence of yeasts and molds that degrade pectin and cellulosic substances, important for the compactness of the olives [31,32].

Lactic acid bacteria (LAB) and yeasts increased steadily and became the dominant members of the microflora during fermentation [33]. I.aciic acid bacteria reached their maxima within the first 30 days (1.6 x 10^8 CFU mL^{-1}), and their enumeration was stable until the end of the fermentation process. Yeasts coexisted with lactic acid bacteria throughout the whole fermentation period. Their counts were lower than those of the LAB by approximately 3-4.5 log cyclee. They grew in similar populations with lactic acid bacteria during the first week, and their presence was stable until the end of fermentation reaching 1.7 x 10^4 CFU mL^{-1} (Figure 1).

Subsequently, the pH of the brane after 25 days was 5.87. In this phase, no viable counts of enterobacteria were enumeaated.

At the end of the fermentation process, after 60 days, the olives were very soft and the epicarp was easily detached from the pulp, and the pH was 4.52.

The pH, throughout the fermentation phase, decreased slowly compared to the common fermentation processes of table olives (Spanish and Greek method) [19,34]; moreover, anomalous phenomena such as the loss of hardness of the olives were observed [21,35,36].

3.2. Oil Analysis

The comparison of the values of acidity and peroxides of the NEFO and EVO oils are shown in Table 1. NEFO oil showed a low number of peroxides (produced by oxidative rancidity processes) but a high amount of free fatty^ oeids (produced by hydrolytic rancidity processes) [37b

The comparison of the results obtained is interesting; the acidity of NEFO oil is 255 times higher than the EVO oil, while the quantity of peroxides of the first oil is only 4.7 times higher than the second.

The high acidity of NEFO oil is caused by the lipases present in the pulp of the drupes but also by lipolytic enzymes produced by bacteria, yeasts, and molds developed in natural fermentation [38,39]

The oxidative rancidity, which forms peroxides, was probably limited by the brine because it blocked the direct contact of the drupes with oxygen. Furthermore, some species of lactobacilli degrade hydrogen peroxides and prevent the formation of peroxide radicals and hydroperoxide radicals, thus limiting the formation of peroxides and stabilizing the drupes in brine [38,40,41].

Natural fermentation with aerobic conditions determines the high AV and low PV of non-edible olive oil, these results are in agreement with the results of Girgis A. Y. and Alajtal A. I. et al. [40,41].

3.3. Soap Analysis

Saponification with EVOO developed normally, as described in Spitz L. [8]; the soap paste, initially semi-liquid, slowly solidified.

NEFOO soaps have different characteristics from EVOO soaps. The sodium salts of fatty acids are formed very quickly in NEFO oil due to the high acidity, so the soap paste hardens very quickly. The saponification process was developed at low temperature and with slow addition of the basic solution to obtain a homogeneous product.

The maturation times of the soaps were determined considering three parameters: pH, color, and solubility.

Initially, the pH of NEFOO soaps was lower than EVOO soaps (Table 2), probably due to the high amount of free fatty acids that neutralize part of the sodium hydroxide very quickly [41].

Table 2. *pH of the soap solutions produced for different maturation times.*

Day of Maturation	pH EVOO Soap 0%	pH EVOO Soap 5%	pH NEFOO Soap 0%	pH NEFOO Soap 5%
5	9.86	9.81	9.51	9.52
15	9.60	9.62	9.30	9.27
30	9.43	9.42	9.26	9.21
90	9.25	9.24	9.25	9.19
180	9.23	9.21	9.27	9.20

The pH of NEFOO soaps was constant from the thirtieth day of ripening, while the pH of EVOO soaps was constant after the ninetieth day of ripening. The pH values shown in Table 2 are the average of the results obtained in triplicate.

EVO oil (Figure 2A) showed a more intense color than NEFO oil (Figure 2B) due to the greater quantity of chlorophylls, carotenoids, polyphenols, and flavonoids, which determine the initial color of the soaps [16,17,42].

Soaps whiten in the ripening phase due to oxidative processes that act on the pigments. The color of the soaps depends on the type of oil initially used and the maturation time, generally, the soaps marketed have a natural color between soft yellow and white.

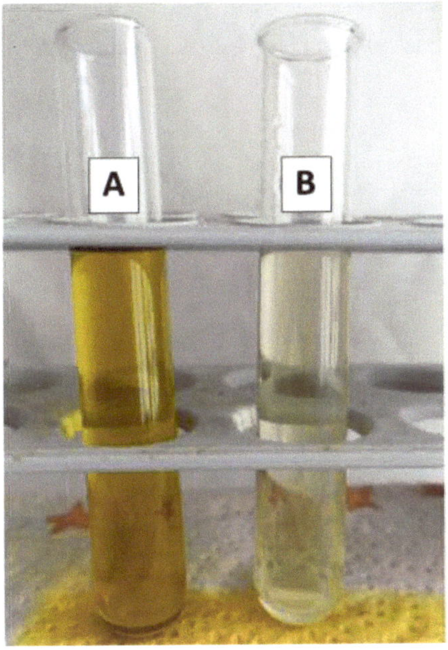

Figure 2. *Comparison of EVO oil (**A**) and NEFO oil(**B**) used for the production of soaps.*

The colors of the soaps produced are shown in Tablx 3 with the models RGB (model that uses red, green, and blue as base colors) and CIELAB (color space l* a* b*).

NEFOO soaps after 30 days of maturation were white and this color was maintained up to 180 days of maturation, while EVOO soaps tended to soft yellow after 90 days of maturation and to white at 180 days (Table 3).

Considering the CIELAB parameters, EVOO soap with 0% free fatty acids had a more intense yellow-green color than EVOO soap with 5% free fatty acids, this difference was not observable for NEFOO soaps.

Soaps' solub ility was determined considering the degree of clarity of the solutions with complete dissolution of the soap.

EVOO soaps, after 5 days of maturation, showed different turbidity (Figure 3A). / After 15 rnd 30 days of maturation (Figure 3B,C) the turbidity rapidly decreased for soap with 0% of non-neutralized fatty acids, while it slowly decreased for soap with 5% of non-neutralized fatty acids.

Table 3. Report colors of soaps with RGB and CIELAB models (l* brightness and a* b* mixture of red, yellow, green, and blue).

Days of Maturation	EVOO Soap 0%				EVOO Soap 5%				NEFOO Soap 0%				NEFOO Soap 5%			
	RGB	l*	a*	b*	RGB	l*	a*	b*	RGB	l*	a*	b*	RGB	1*	a*	b*
5		90.4	-24.8	87.2		93.3	-25.1	73.8		98.4	-9.3	29.1		98.8	-6.9	21.6
15		93.7	-22.5	74.1		95.4	-21.3	78.9		98.2	-2.62	8.5		97.7	-3.3	11.4
30		97.6	-14.4	48.7		95.8	-13.1	42.9		98.9	-0.3	1.0		96.9	- 0.2	1.3
90		97.0	-3.6	9.2		97.0	-3.7	10.6		97.1	-0.4	0.9		96.1	-0.2	0.8
180		99.3	-0.3	0.7		98.1	- 0.4	0.8		97.1	-0.3	0.8		97.4	-0.3	1.2

Figure 3. Images of the solutions of EVOO (A-C) and NEFOO (D-F) soaps in different ripening days 5 days (A,D) 15 days (B,E) and 30 days (C,F).

NEFOO soaps, after 5 days of maturation, had very similar turbidity (Figure 3D). After 15 and 30 days of maturation (Figure 3E,F) the turbidity of both soaps decreased until clear solutions were obtained.

After 90 and 180 days, all soaps showed a slight turbidity: this was more present in EVOO soaps (Figure 4A,B) than in NEFOO soaps (Figure 4C,D). This low turbidity for very long maturation times could be due to the; initial rancidity of the soaps.

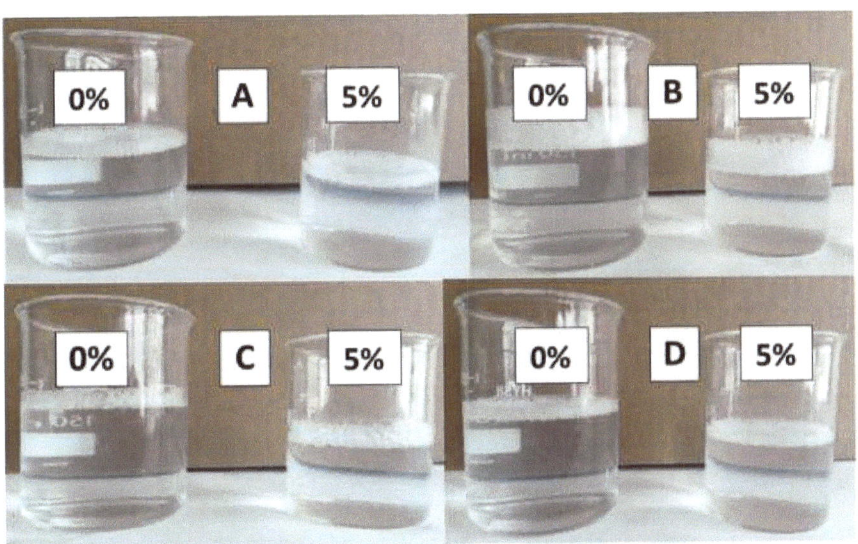

Figure 4. Images of the solutions of EVOO (A,B) and NEFOO (C,D) soaps after different ripening days: 90 days (A,C) and 180 days (B,D).

It is interesting to observe the hardness of the soaps: NEFOO soaps tended to be harder and more compact than EVOO soaps, especially in the early stages of maturation, with a consequent difficulty in their solubilization in wafer.

Considering the pH, color, and solubilify, NEFOO soaps showed a correct maturation after 30 days from their production, while EVOO soaps had an optimal maturation after 90-180 days. Soaps with maturation aiter 30 and 1810 days are shown in Figure 5.

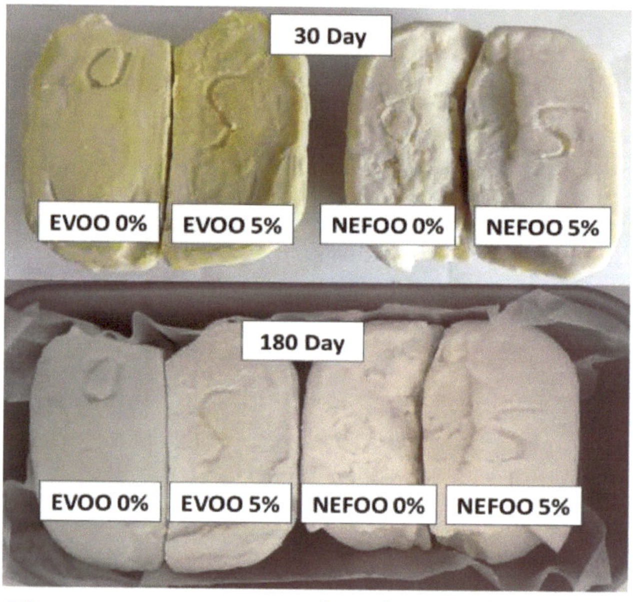

Figure 5. Soaps matured 30 days and 180 days.

There were no substantial differences, for the parameters considered in ttiis study, between soaps with 0% and 5% of free fatty acids produced with the same oil.

The stage o[a] maturation is not strictly linked to the use of the product but represents an impoatant parameter for its marketing.

4. CONCLUSIONS

This; study highlighted that non-edible fermented olives can be used for soaps' production.

The high acidity of NEFOO determtnes different and shorter sapontfication processes then the? normal grocedures reported by Spitz L. [7,8].

Considering the pH, color, and solubility data, NEFOO soaps ripen faster than EVOO soaps, while there are no substantial differences between soaps with $0\%_r$ and $5^0\%$ free fatty acids obtained with the same oil.

The fermentation of non-edible olives makes it possible to easily and quickly recycle an undervalued food waste to produce a widely used product, such as olive oil soaps, with lower costs, production times, and maturation.

The study described could be used to prod use "alternative" olive oils on a large scale, exploiding non-edible drupss currently used to produce fodder, naturai fertilizer, and energy biomass.

REFERENCES

1. Konkol, K.L.; Rasmussen, S.C. An Ancient Cleanser: Soap Production and Use in Antiquity. *ACS J.* **2015**, *9*, 245-266. [CrossRef]
2. European Commission. *Regulation (EC) No 1223/2009 of the European Parliament and of the Council of 30 November 2009 on Cosmetic Products;* European Commission: Brussels, Belgium, 2009.
3. Mannu, A.; Garroni, S.; Porras, J.I.; Mele, A. Available Technologies and Materials for Waste Cooking Oil Recycling. *Processes* **2020**, 8, 366. [CrossRef]
4. Kajdas, C. Major pathways for used oil disposal and recycling. Part 1. *Lubr. Sci.* **2006**, 7, 61-74. [CrossRef]
5. Abdel-Gawad, S.; Abdel-Shafy, M. Pollution control of industrial wastewater from soap and oil industries: A case study. *Water Sci. Technol.* **2002**, 46, 77-82. [CrossRef] [PubMed]
6. Bettley, F.R. Some Effects of Soap on the Skin. *Br. Med. J.* **1960**,1,1675-1679. [CrossRef]
7. Spitz, L. *Sodeopec: Soaps, Detergents, Oleochemicals, and Personal Care Products,* 3rd ed.; American Oil Chemists' Society: Champaign, IL, USA, 2004. [CrossRef]
8. Spitz, L. *Soap Manufacturing Technology,* 2nd ed.; Academic Press: Cambridge, MA, USA; AOCS Press: Urbana, IL, USA, 2016.
9. Felix, S.; Araujo, J.; Piresa, A.M.; Sousa, C. Soap production: A green prospective. *Waste Manag.* **2017**, 66,190-195. [CrossRef] [PubMed]
10. Azbar, N.; Bayram, A.; Filibeli, A.; Muezzinoglu, A.; Sengul, F.; Ozer, A. A Review of Waste Management Options in Olive Oil Production. *Crit. Rev. Environ. Sci. Technol.* **2004**, 34, 209-247. [CrossRef]
11. Charis, M.; Galanakis, C.M. *Olive Mill Waste: Recent Advances for Sustainable Management,* 1st ed.; Academic Press: Cambridge, MA, USA, 2017.
12. Garcia Martin, J.F.; Cuevas, M.; Feng, C.H.; Mateos, P.A.; Garcia, M.T.; Sanchez, S. Energetic valorisation of olive biomass: Olive-tree pruning, olive stones and pomaces. *Processes* **2020**, 8, 511. [CrossRef]
13. Garcia Martin, J.F.; Cuevas, M.; Bravo, V.; Sanchez, S. Ethanol production from olive prunings by autohydrolysis and fermentation with Candida tropicalis. *Renew. Energy* **2010**, 35,1602-1608. [CrossRef]

14. Cabrera, E.R.; Arriaza, M.; Rodriguez-Entrena, M. Is the extra virgin olive oil market facing a process of differentiation? A hedonic approach to disentangle the effect of quality attributes. *Grasas Aceites* **2015**, 66, 4. [CrossRef]

15. European Comission. *Commission Regulation: Commission Implementing Regulation (EU) 2017/2470 of 20 December 2017 Establishing the Union List of Novel Foods in Accordance with Regulation (EU) 2015/2283 of the European Parliament and of the Council on Novel Foods*; European Commission: Brussels, Belgium, 2017.

16. Rozanska, A.; Russo, M.; Cacciola, F.; Salafia, F.; Polkowska, Z.; Dugo, P.; Mondello, L. Concentration of Potentially Bioactive Compounds in Italian Extra Virgin Olive Oils from Various Sources by Using LC-MS and Multivariate Data Analysis. *Foods* **2020**, 9,1120. [CrossRef] [PubMed]

17. Dugo, L.; Russo, M.; Cacciola, F.; Mandolfino, F.; Salafia, F.; Vilmercati1, A.; Fanali, C.; Casale, M.; De Gara, L.; Dugo, P.; et al. Determination of the Phenol and Tocopherol Content in Italian High-Quality Extra-Virgin Olive Oils by Using LC-MS and Multivariate Data Analysis. *Food Anal. Methods* **2020**,13,1027-1041. [CrossRef]

18. Poiana, M.; Romeo, F.V. Changes in chemical and microbiological parameters of some varieties of Sicily olives during natural fermentation. *Grasas Aceites* **2006**, 57, 402-408. [CrossRef]

19. Panagou, E.Z.; Katsaboxakis, C.Z. Effect of different brining treatments on the fermentation of cv. Conservolea green olives processed by the Spanish-method. *Food Microbiol.* **2006**,23,199-204. [CrossRef] [PubMed]

20. Lanza, B. Abnormal fermentations in table-olive processing: Microbial origin and sensory evaluation. *Front. Biol.* **2013**, 4, 91. [CrossRef] [PubMed]

21. Kiain, H.; Hafidi, A. Chemical composition changes in four green olive cultivars during spontaneous fermentation. *LWT-Food Sci. Technol.* **2014**, 57, 663-670. [CrossRef]

22. Commission Regulation. *Commission Implementing Regulation (EU) 2019/1604 of 27 September 2019 Amending Regulation (EEC) No 2568/91 on the Characteristics of Olive Oil and Olive-Residue Oil and on the Relevant Methods of Analysis;* European Commission: Brussels, Belgium, 2019.

23. International Organization for Standardization. *ISO 660:2020-Animal and Vegetable Fats and Oils-Determination of Acid Value and Acidity*; ISO: Geneva, Switzerland, 2020.

24. International Organization for Standardization. *ISO 3960:2017-Animal and Vegetable Fats and Oils-Determination of Peroxide Value-Iodometric (Visual) Endpoint Determination;* ISO: Geneva, Switzerland, 2017.

25. Wakita, K.; Kuwabara, H.; Furusho, N.; Tatebe, C.; Sato, K.; Akiyama, H. A Comparative Study of the Hydroxyl and Saponification Values of Polysorbate 60 in International Food Additive Specifications. *Am. J. Anal. Chem.* **2014**, 5,199-204. [CrossRef]

26. Conesa, A.; Manera, F.C.; Brotons, J.M.; Fernandez-Zapata, J.C.; Simon, I.; Simon-Grao, S.; Alfosea-Simon, M.; Martinez Nicolas, J.J.; Valverde, J.M.; Garcia-Sanchez, F. Changes in the content of chlorophylls and carotenoids in the rind of Fino 49 lemons during maturation and their relationship with parameters from the CIELAB color space. *Sci. Hortic.* **2019**, 243, 252-260. [CrossRef]

27. Gatti, E.; Bordegoni, M.; Spence, C. Investigating the influence of colour, weight, and fragrance intensity on the perception of liquid bath soap: An experimental study. *Food Qual. Prefer.* **2014**, 31, 56-64. [CrossRef]

28. Sanchez, A.H.; de Castro, A.; Rejano, L.; Montano, A. Comparative study on chemical changes in olive juice and brine during green olive fermentation. *J. Agric. Food Chem.* **2000**, 48, 5975-5980. [CrossRef]

29. Tofalo, R.; Schiron, M.; Perpetuini, G.; Angelozzi, G.; Suzzi, G.; Corsetti, A. Microbiological and chemical profiles of naturally fermented table olives and brines from different Italian cultivars. *Antonie Van Leeuwenhoek* **2012**, 102, 121-131. [CrossRef] [PubMed]

30. Perpetuini, G.; Prete, R.; Garcia-Gonzalez, N.; Khairul Alam, M.; Corsetti, A. Table Olives More than a Fermented Food. *Foods* **2020**, 9,178. [CrossRef] [PubMed]

31. Arroyo-Lopez, F.N.; Querol, A.; Bautista-Gallego, J.; Garrido-Fernandezc, A. Role of yeasts in table olive production. *Int. J. Food Microbiol.* **2008**,128,189-196. [CrossRef] [PubMed]

32. Giannoutsou, E.P.; Meintanis, C.; Karagouni, A.D. Identification of yeast strains isolated from a two-phase decanter system olive oil waste and investigation of their ability for its fermentation. *Bioresour. Technol.* **2004**, 93, 301-306. [CrossRef]

33. Hurtado, A.; Reguant, C.; Bordons, A.; Rozes, N. Lactic acid bacteria from fermented table olives. *Food Microbiol.* **2012**, 31,1-8. [CrossRef] [PubMed]

34. Alexandraki, V.; Georgalaki, M.; Papadimitriou, K.; Anastasiou, R.; Zoumpopoulou, G.; Chatzipavlidis, I.; Papadelli, M.; Vallis, N.; Moschochoritis, K.; Tsakalido, E. Determination of triterpenic acids in natural and alkaline-treated Greek table olives throughout the fermentation process. *LWT-Food Sci. Technol.* **2014**, 58, 609-613. [CrossRef]

35. Sherahi, M.H.A.; Shahidi, F.; Yazdi, F.T.; Hashemi, S.M.B. Effect of Lactobacillus plantarum on olive and olive oil quality during fermentation process. *LWT-Food Sci. Technol.* **2018**, 89, 572-580. [CrossRef]

36. Borras, E.; Mestres, M.; Acena, L.; Busto, O.; Ferre, J.; Boque, R.; Calvo, A. Identification of olive oil sensory defects by multivariate analysis of mid infrared spectra. *Food Chem.* **2015**,187,197-203. [CrossRef]

37. Harzalli, U.; Rodrigues, N.; Veloso, A.C.A.; Dias, L.G.; Pereira, A.J.; Oueslati, S.; Peres, A.M. A taste sensor device for unmasking admixing of rancid or winey-vinegary olive oil to extra virgin olive oil. *Comput. Electron. Agric.* **2018**,144, 222-231. [CrossRef]

38. Singh, A.K.; Mukhopadhyay, M. Overview of Fungal Lipase: A Review. *Biotechnol. Appl. Biochem.* **2012**,166, 486-520. [CrossRef]

39. Zhu, H.; Wang, S.C.; Shoemakera, C.F. Volatile constituents in sensory defective virgin olive oils. *Flavour Fragr. J.* **2016**,31, 22-30. [CrossRef]

40. Alajtal, A.I.; Sherami, F.E.; Elbagermi, M.A. Acid, Peroxide, Ester and Saponification Values for Some Vegetable Oils Before and After Frying. *AASCIT J. Mater.* **2018**, 4, 43-47. [CrossRef]

41. Girgis, A.Y. Production of high quality castile soap from high rancid olive oil. *Grasas Aceites* **2003**, 54, 226-233. [CrossRef]

42. Kachouri, F.; Hamdi, M. Enhancement of polyphenols in olive oil by contact with fermented olive mill wastewater by Lactobacillus plantarum. *Process Biochem.* **2004**, 39, 841-845. [CrossRef]

CHAPTER

16

Green Husk of Walnuts (*Juglans regia* L.) from Southern Italy as a Valuable Source for the Recovery of Glucans and Pectins

1. INTRODUCTION

The walnut husk is the fleshy part of the fruit of the walnut fruit (Juglans regia L.) that covers the shell enclosing the edible kernel. When ripe, walnut husk has a green color that darkens over time. Shell and husk are discarded, causing environmental pollution. The green husks of the walnut fruit are the basic material for the traditional walnut liqueur, a wholesome alcoholic drink, which is rich in phenolic compounds and vitamins [1,2]. Additionally, the husks have been used since ancient times in traditional medicine and for the treatment of various diseases such as microbial infections, stomachache, thyroid dysfunctions, heart diseases and sinusitis [3,4]. Recently, scientific attention towards the valuable active constituents of walnut husks has been increasing. The potential of this lowcost natural material as a source of phenolic compounds with antiradical and antimicrobial activities has been demonstrated by several studies [5]. The influence of solvents of varying polarity on the extraction yields of phenolic extracts from walnut husks has been described in the literature [6] and the effects of geographical and climatic conditions on their phenolic and flavonoids contents were studied [7]. The walnut husk contains natural dyes, such as juglone which is a brown pigment [8], but also a potent antimicrobial and anticancer agent [9-11]. Several agricultural food wastes, such as pomegranate Akko peels [12], Viciafaba L. [13], banana peels [14], coffee pulp waste [15], mango peel and cocoa pod husk [16], were found to be potential sources of glucans and pectins. To the best of our knowledge, there are no reports on the recovery of nutraceuticals such as glucans and pectins from walnut husks for the valorization of these agricultural waste products. Therefore, with the aim of providing additional added value to these residual sources and basic support for their functional uses, the present study investigated glucan and pectin contents from the green husks of walnuts grown in two different soil and climate areas of Southern Italy (Montalto Uffugo e Zumpano). Further spectroscopic, morphological and thermal characterizations of the extracted high-value compounds were performed in order to determine if the effect of the different pedoclimatic conditions of the two areas could lead to significant differences both in the content of glucans and pectins and their characteristics, which could affect their functional uses.

2. MATERIALS AND METHODS

2.1. Standards and Chemicals

Analytical grade solvents were purchased from Carlo Erba. Sodium carbonate, hydrochloric acid and citric acid were from Sigma-Aldrich (Milan, Italy). The glucan contents were determined by the method published by McCleary and Codd [17] using a mushroom and yeast β-glucan kit (Cat. No. K-YBGL). The kit was obtained from Megazyme International (Bray, County Wicklow, Ireland).

2.2. Agri-Wastes

Walnut fruits with green hulls (*Juglans regia* L., Figure 1) were harvested from trees grown in two different agroclimatic localities of Calabria, in the south of Italy, which were Montalto Uffugo (latitude: 39°24′20”88 N, longitude: 16°9′31”68 E) and Zumpano (latitude: 39°18′42”84 N, longitude: 16° 17′34”44 E). Altitudes, geographical and climatic conditions of gathering areas are summarized in Table 1.

Figure 1. Walnut fruits with green hulls (Juglans regia L.).

Collection	Latitude	Longitude	Altitude (m)	Annual Raining Average (mm)	Daily Temperature Average (°C)	Humidity Average (%)
Montalto	39° 24'20"88 N	16°9'31"68E	490	20-110	10-20	85
Zumpano	39° 18'42"84 N	16°17'34"44 E	429	12.7-101.6	3.9-27.8	75

The whole fruits were collected in the period of October 2017, when the walnut fruits were mature while the husks (mesocarp) were green and firm. The husks were manually separated from the walnut, freeze-dried (Telstar freeze-dryer, mod. Cryodos) after freezing at -20°C and then reduced to powder using a pestle and mortar. The powder was sieved through a 60-mesh (250 pm) screen and then kept in screw-cap vials under nitrogen at -20°C, before any extraction. Three samples per area were used, and all the extractions were carried out in triplicate. The data were expressed on a dry matter basis as mean ± standard deviation.

2.3. Extraction of Glucans

The employed extraction procedure was a conventional method involving two stages to sequentially remove starch and proteins from the powder, namely an aqueous alkali extraction at pH 10 and 55°C and acidic precipitation, respectively [12,18-20].

Preliminary treatments of this method involve sequential extractions with three solvents in order of increasing polarity, to remove lipids and soluble material such as vitamins, polyphenols, monosaccharides, disaccharides, and others [21]. The flour (3 g) was first extracted with acetone (3 x 30 mL each), then with methanol (3 x 30 mL each) and finally with 70% aqueous ethanol (3 x 30 mL each). All the extractions were performed at room temperature (t = 2 h). For starch precipitation, residue solid was added to a solution of 20%, *w/v* sodium carbonate in distilled water (pH 10,30 mL), and the obtained suspension was warmed at 55°C in a water bath with stirring for 30 min and then was centrifuged (Model J2-21, Beckman Instrument Co., Mississauga, ON, Canada) at 5000 x *g* for 30 min. The solid was removed, and the pH of the supernatant was added to HCl 2 M until a pH of 4.5 was reached (isoelectric point of proteins), and centrifuged at 5000 x *g* for 30 min to separate proteins [22,23], which were discarded. An equal volume of absolute ethanol was added to the supernatant in order to precipitate glucans. After 12 h at 4°C, the solution was centrifuged at 5000 x *g* for 30 min. The precipitate was resuspended in ethanol, filtered, rinsed with ethanol and freeze-dried. The extractions were performed in triplicate on three samples of husk powder from each agroclimate locality and the results were expressed as mean ± standard deviation. The crude glucans were measured and characterized by FT-IR, SEM and DSC.

2.4. α- and β-Glucan Content

β-glucan content of husks was determined in triplicate using β-glucan assay kit from Megazyme Ltd., Bray, Co. Wicklow, Ireland [17]. The dry sample (100 mg) was weighed into a 25 mL flask and 12 M ice-cold sulphuric acid solution (2 mL) was added. The solution was vortexed and left to incubate for 2 h in an ice-water bath, and then it was diluted with distilled water (12 mL) and left for 2 h in a boiling-water bath (T = 100°C). After cooling the temperature, 10 M KOH solution (6 mL) and 200 mM sodium acetate buffer (pH 5) were added. After centrifugation (1500x g, t = 10 min), an aliquot (0.1 mL) of the supernatant was mixed with 0.1 mL of a mixture of *exo*-1,3-β-glucanase (20 U/mL) plus β-glucosidase (4 U/mL) at 40°C for 60 min. Finally, the content of glucose in the solutions was determined by incubating the solution with glucose-oxidase/peroxidase (GOPOD, 3.0 mL) at 40°C for 20 min.

For obtaining total glucan content, the solution was analysed by the spectrophotometer at A = 510 nm (Model UV-vis, JASCO, V-550) against the blank and the D-glucose standard solution (1 mg/mL), incubated with GOPOD reagent. The blank was prepared by adding 0.2 mL of 200 mM sodium acetate buffer at pH 5.0 to 3 mL of GOPOD. For obtaining ∞-glucan content, the husk flour (0.1 g) was added with 2 M KOH (2 mL), 1.2 M sodium acetate buffer (pH 3.8,8 mL) and incubated with 0.2 mL of amyloglucosidase plus invertase for 30 min at 40°C. After centrifugation (1500x g, t = 10 min), the supernatant (0.1 mL) was mixed with sodium acetate buffer (200 mM, pH 5.0,0.1 mL) and GOPOD (3 mL). The difference between total and ∞-glucan contents gave the β-glucan content.

2.5. Extraction of Pectins

Pectins from husk flour were extracted according to the experimental conditions reported by Fazio et al. [24]. The employed procedure was the conventional heating extractive method with citric acid [25,26].

Next, 10% *(w/v)* citric acid solution was added to husk flour (2 g) until pH of 2 was reached, creating a mixture which was magnetically stirred for 1 h at 90°C. After cooling and centrifugation (5000 x g, t = 30 min), the supernatant was added with an equal volume of absolute ethanol and then kept for 16 h at 4°C in order to allow for pectin flotation. After centrifugation (Universal 320, Hettich Zentrifugen, Merck, Italy), the floating pectins were rinsed with absolute ethanol, solubilized in distilled water, added to an equal volume of acetone to decolourise the pectins [18], and left to stand at 4°C for 12 h. Afterwards, the supernatant was separated by centrifugation at 5000x *g*, t = 15 min, and the gelatinous residue was dried under a vacuum. The extraction procedure was repeated three times and data expressed as mean ± standard deviation. All samples were characterized by FTIR, and SEM and DSC.

2.6. Determination of Esterification Degree

The esterification degree of pectins, defined as the ratio of esterified carboxy groups to the total number of carboxy groups was determined by the potentiometric titration method (DE)] and confirmed by the instrumental FT-IR method (DM) [13].

2.6.1. Potentiometric Titration Method

The esterification degree of pectins was determined by titrimetric method, as previously described by Fazio et al. [24]. The dried pectin (20 mg) sample was weighed in a beaker, wetted with ethanol and dissolved in 5 mL of distilled water. For complete dissolution of pectins, the suspension was heated at 45°C under magnetic stirring for 20 min. The resulting solution was titrated with 0.01 N NaOH in the presence of three drops of phenolphthalein. The initial titration volume (V1) was recorded once a pale pink colour appeared, and it indicated the number of free carboxy groups. Then, 3 mL of 0.01 N NaOH was added to neutralize polygalacturonic acid, and the solution was stirred at room temperature for 2 h in order to saponify the esterified carboxy groups of the polymer. Then, 3 mL of 0.01 N hydrochloric acid (HCl) was added to neutralize the sodium hydroxide NaOH, followed by titration of excess HCl with 0.01 N NaOH in the presence of phenolphthalein. The volume of titration required for pale pink colouring of the sample was recorded as final titration volume (V2, representing the esterified group number). The DE was determined as follows:

$$\%DE = [V2\ (mL)/V1\ (mL) + V2\ (mL)] \times 100$$

2.6.2. Instrumental FT-IR Method

In this methodology, the FT-IR spectra recorded for the characterization of pectins were used to determine the degree of methoxylation (DM), taking into account the band areas at 1747-1746 cm^{-1} and 1633-1621 cm^{-1} arising from methyl-esterified and carboxylate groups, respectively. The degree of methoxylation was obtained using the equation:

$$\%DM = [A\ COOCH_3/(A\ COOCH_3 + ACOO^-)] \times 100$$

2.7. FT-IR Spectroscopic Analysis

Glucans and pectins were analysed by FT-IR using a Bruker ALPHA FT-IR spectrometer equipped with a A241/D reflection module in 4000-400 cm^{-1} range, by following the standard KBr method. The sample was ground with KBr in the 1:40 ratio, and the resulting powder was hard-pressed into tablets. All experiments were carried out in triplicate at a spectral resolution of 4 cm^{-1}.

2.8. Scanning Electron Microscopy (SEM)

The surface morphological investigations were carried out by a scanning electron microscope (SEM) (Field Emission SEM FEI Quanta 200, Thermo Fisher Scientific, Hillsboro, OR, USA) and Electron Probe Micro Analyzer (EPMA)-JEOL-JXA 8230t (Kyoto, Japan). The analyses were performed under conditions reported by Fazio et al. [24].

2.9. Differential Scanning Calorimetry (DSC)

The calorimetric behaviour of the samples was analysed using DSC SETARAM 131 instrument. The calibration operations were carried out using a sample of indium with a known weight. Each sample (15-30 mg) was weighed in a crucible, and then the container was capped and sealed by mechanical press. All the tests were carried out under nitrogen flow at a temperature scan rate of 20 °C/min.

2.10. Statistical Analysis

DE and DM were analysed in triplicate and the results were expressed as mean ± standard deviation (SD). One-way ANOVA method and a Sidak comparison method via GraphPad Prism 8 were used. Significance was established at p values < 0.05 (*), $p < 0.01$ (**), $p < 0.001$ (***), and $p < 0.0001$ (****).

3. RESULTS

3.1. Extraction of Glucans

Extraction solvents (acetone, methanol and 70% aqueous ethanol) removed lipids and soluble material to facilitate the complete separation of fibre [27]. During this process, the resulting residue was slurred in alkaline aqueous solution to solubilize glucans and proteins, followed by precipitation of proteins under acidic conditions (pH = 4.5) and by recovery of crude glucans after precipitation from ethanol. More advanced technologies to obtain glucans have been used recently, such as accelerated solvent extraction [28] and microwave-assisted extractions [29], plus pressurized solvent extraction [30]. Although these procedures have been shown to be faster than traditional methodologies, they require a more accurate purification procedure for the isolation of nutraceuticals of interest, and they can be performed through the use of more specific devices, which increases the costs [30]. The experimental results are given in Table 1. Data reported in the table show that the content of glucans was slightly higher in walnut husks from Montalto (4.6 ± 0.2 g/100 g dm) than in those from Zumpano (3.74 ± 0.3 g/100 g_{DM}).

3.2. α- and β-Glucan Content

The enzymatic method for β-glucan measurement is widely used due to its accuracy and reliability. In this approach, complete hydrolysis of glucans requires a controlled acid measured with GOPOD reagent. α-glucans were determined using specific enzymes for α-glycosidic bond hydrolysis under alkaline conditions, and β-glucans were calculated by the difference between total and α-glucan content. This method aimed to achieve complete hydrolysis of both α- and β-glucans to glucose, minimizing the loss of glucose through secondary reactions (Table 2).

Table 2. Recovered glucans (g/100 g dm) and total α- and β-glucans (%).

Samples	Recovered Glucans (g/100 g dm)	Total Glucans (%)	α-Glucans (%)	β-glucans (%)
Montalto husk	4.6 ± 0.2	11.3 ±0.7	5.0 ± 0.2	6.3 ± 0.4
Zumpano husk	3.74 ± 0.3	7.2 ± 0.4	3.6 ± 0.3	3.6 ± 0.5

As can be seen from the data reported in the table, the percentage of total glucans determined by the enzymatic method confirmed that walnut husks from Montalto (11.3 ± 0.7%) had a higher content than walnut husks from Zumpano (7.2 ± 0.4%). Additionally, Montalto husks were richer in β-glucans (6.3 ± 0.4%) than in α-glucans (5.0 ± 0.2%), while Zumpano husks, compared to the Montalto husks, contained α- and β-glucans in equal percentages (3.65 ± 0.4%).

3.3. FT-IR Spectroscopic Analysis of Glucans

Infrared spectroscopic analysis of glucans provided information on the fundamental molecular vibrations of covalent bonds in the 4000-400 cm^{-1} IR region, showing the characteristic bands of the major functional groups (Figure 2). The comparison of glucan spectra fromMontalto and Zumpano husks highlighted that the qualitative profile of the structures does not significantly change. The IR band centred around 3437 cm^{-1} was generated by the symmetrical and asymmetric stretching of the OH groups present on the glucan backbone. The bands at 2920-2918 cm^{-1} were due to CH_2 stretching of CH_2OH groups. The presence of proteins in the sample linked to the glucan backbone by amide bonding provided the stretching of the CN and NH groups [31,32] generating the band at 1641 cm^{-1} (the amide I). OH and CH bending from in-plane ring deformation generated two IR peaks at 1322 cm^{-1} and 1261 cm^{-1}, respectively. COC and CC stretching vibrations of the glueosides ring originated two bands in the region of 1090-1025 cm^{-1}, indicating the presence of cyclic structures of monosaccharides [33]. The 950-780 cm^{-1} region is called the "anomeric region", where it was possible to distinguish between the two anomeric glycosidic bood types of the glucopyranose rings which generated absorption bands at 860-830 cm^{-1} for α-linkage and at 920-890 cm^{-1} for β-linkage [19].

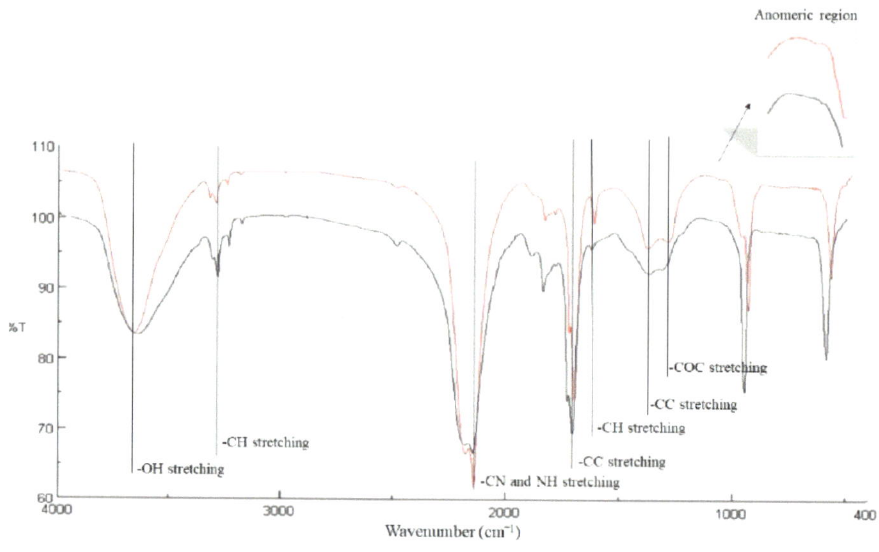

Figure 2. FTIR spectra in the 400 to 4000 cm 1 region of walnut husk glucans from Montalto (red line) and from Zumpano (black line).

3.4. Scanning Electron Microscopy (SEM) of Glucans

The morphological investigation was carried out to highlight any different surface structures arising in the same type of fibre (glucans) from different soil and climate conditions. All the sample were analysed under the same conditions: four magnifications equal to 100x, 2500x, 5000x and 10,000 with a scale equal to 200,10,4 and 2 gm, respectively. The surface of crude glucans from Zumpano husks (GZ) showed the presence of aggregates in the form of microcrystals, whose dimensions were variable as it was more evident at high magnifications (5000 x and 10,000 x) (Figure 3a-d). The size distribution of the crystals was uneven: some crystals were significantly larger than others. The microcrystals observed were rectangular and needle-like in nature. The presence of particles probably glassy in nature which were incorporated into the matrix were also observed. Microcrystals were also observed in the glucans from Montalto hulls (GM), but unlike those present on the surface of glucans from Zumpano husks, they were larger compared to GZ, not aggregated and had a cubic form (Figure 3A-D).

3.5. Differential Scanning Calorimetry (DSC) of Glucans

The calorimetric curves recorded for the glucan samples are very similar. Both mainly presented an endothermic peak (1) near 125 ° C and a glass transition (TG) between 150 and 220 °C (Figure 4).

Peak one of the GZ samples had an enthalpy value (E = 115.8 J/g) significantly higher than that recorded for the same peak of the GM samples (E = 52.3 J/g). The enthalpy of fusion was closely related to the degree of crystallinity. A higher value corresponded to a greater interaction between the molecules, and there was more interaction between the molecules in the GM samples than in the GZ samples.

Inflection points typical of glass transition (TG) [34] were present in both samples, but GZ samples had a higher TG (246.0 °C) than the GM samples (TG = 180.5 °C).

3.6. Extraction of Pectins

Recently, advanced technologies have been developed to improve the yield and to decrease the time of pectin extraction. The "so-called" green methodologies are nonthermal processes that apply, in place of the heat, the acoustic energy of ultrasounds (ultrasound- assisted extraction), or microwave radiation (microwave-assisted extraction) to increase the release of the target material [35,36]. Nevertheless, the employed process for the pectin extraction from walnut husks was the traditional method which involved hydrolysis of protopectin with citric acid and extraction from the primary cell wall of plant tissue using water under reflux, because preliminary extraction tests using ultrasounds resulted in lower yields. Citric acid, compared with mineral acids, had a lower environmental impact and it was cheaper [37].

The obtained results showed that the pectin content of husks from Montalto (PM, 69.8 ± 0.1 g/100 g dm) was about three times higher than that extracted from Zumpano hulls (PZ, 27.1 ± 0.3 g/100 g $_{DM}$), which indicated that different pedoclimatic areas influenced fibre content.

3.7. FT-IR Spectroscopic Analysis of Pectins

The FT-IR spectra of walnut husk pectins from Montalto and Zumpano, scanned at wave numbers ranging from 4000 to 400 cm^{-1} and corrected against the background spectrum of air, are reported in Figure 5. The qualitative profile of the spectra of the samples from different pedoclimatic areas only showed differences in the absorption intensity of the characterizing bands. The 3600-2500 cm^{-1} region presented two major peaks: the first one, centred around 3437 cm^{-1}, generated by OH stretching, due to inter- and intra-molecular hydrogen bonding of the galacturonic acid backbone, and the second one, at 2915 cm^{-1}, due to C-H absorption stretching. The 1800-1500 cm^{-1} region revealed the presence of a first band at 1747-1746 cm^{-1}, generated from ester carbonyl groups stretching, and a second one at 1633-1621 cm^{-1} due to the carboxylate ion stretching [38]. These bands were very important since it was possible to assess the degree of methoxylation of pectin by considering the corresponding intensities. Moreover, analysis of the spectra showed that in PM pectins (red line), the intensity of the band relating to the ester groups was greater than that relating to the carboxyiate groups, whereas in PZ (black line), the intensities of the two bands are reversed. In the "fingerprint region" (1450to 900 cm t) [39], Ramse ban ds were evident at 1019-1017 ent^{-1}, generated Prom COH deformation, at 1103-1106 cm^{-1} due to CC stretching and 1041-1040 cm^{-1} arising from asymmetric COC streching vibration. The presence of a-glycosidic linkage was highlighted by a weak band at 840 cm^{-1}, generated by the ring vrbration of [40].

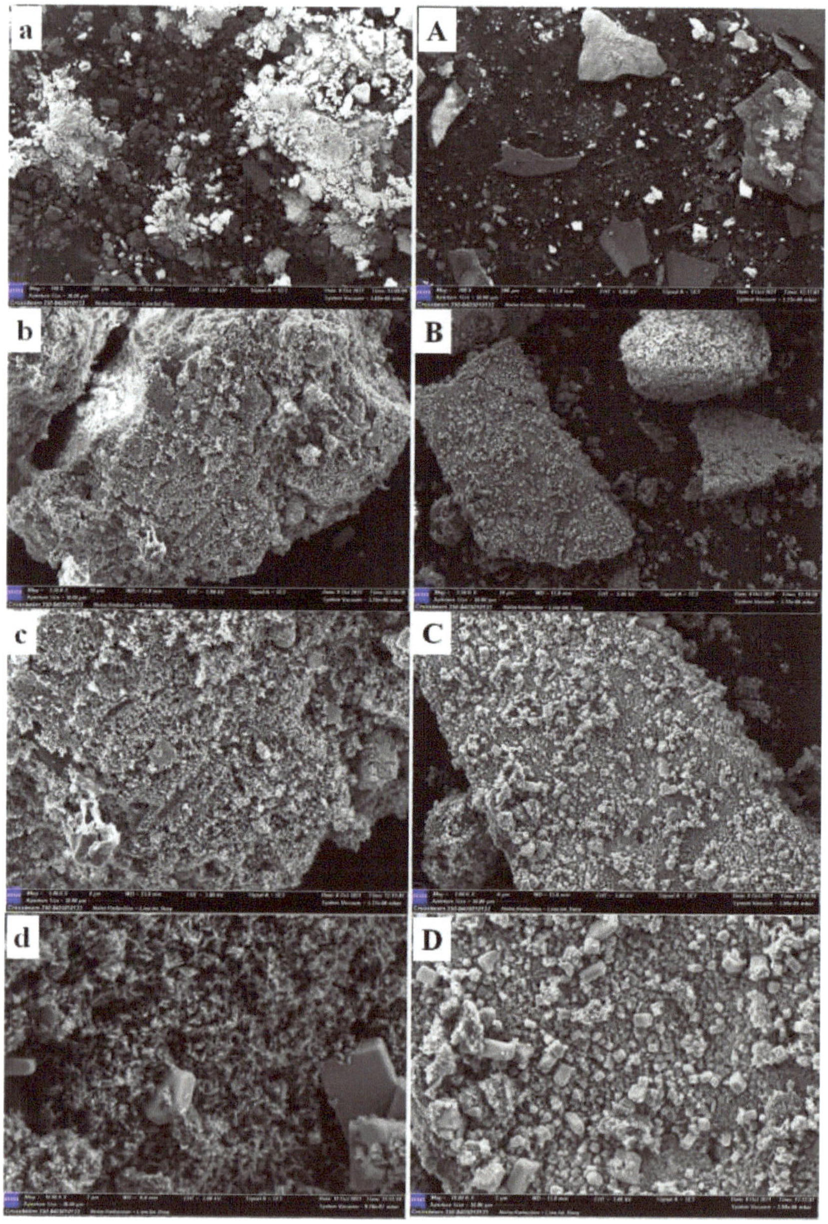

Figure 3. Scanning electron images of glucans from Zumjoano (GZ, a-d) and Montalto (GM, (A-D)) (at four different magnifications and srales. (a,A) (100x, 200 pm); (b,B) (2500x, 10 pm); (c,C) (5000x , 4 pm); (d,D) (00,000 x, 2 pm).

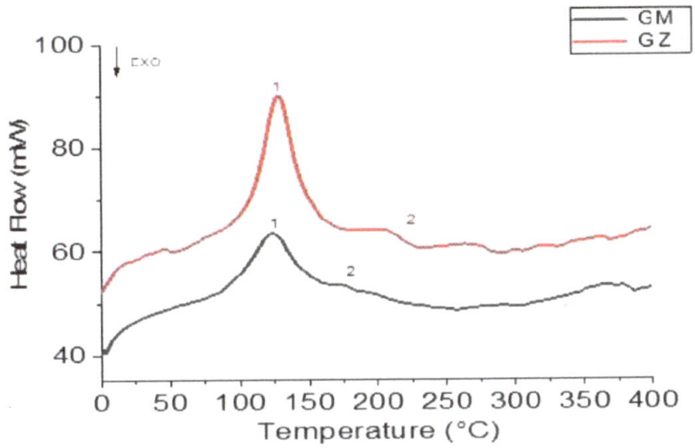

Figure 4. DSC records of walnut husks' glucans from Montalto (red line) and from Zumpano (blackline).

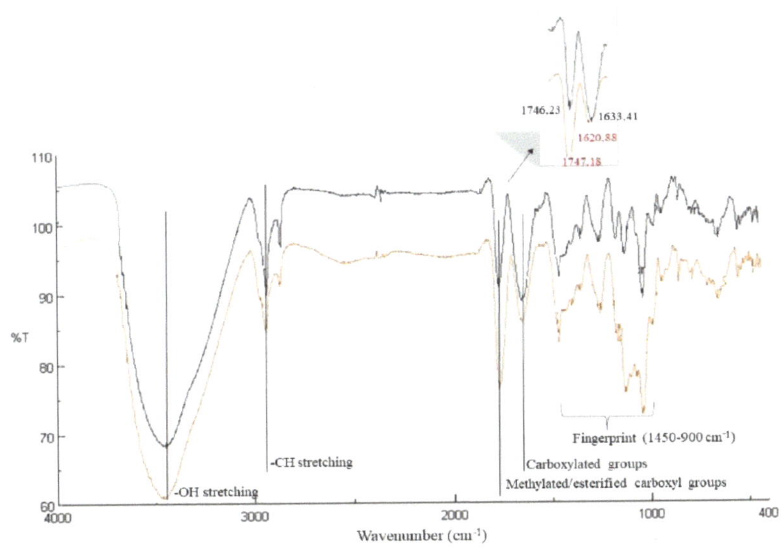

Figure 5. FTIR spectra in the 400 to 4000 cm 1 region of walnut husk pectins from Montalto (red line) and from Zumpano (black line).

3.8. Determination of Esterification Degree

Pectin is a complex polysaccharide composed of at least five different sugar moieties, with 80-90% of its dry weight being galacturonic acid (GalA). A major percentage of the GalA is present in homogalacturonan (HG) regions of pectin as unbranched chains in -which a variable proportion of the GalA contains a methyl ester at the C6 position [41]. The functional properties of pectins in foods, such as gelling capacity, and their reactivity towards calcium and other cations, are largely dependent on the amount of methylated GaSd subunits. Thus, degree of methylation (DM) is an important parameter for characterisation of food pectins [42]. DE was used to classify the pectins into high-methoxyl (HM) form, when the esterified group content was higher than 50%, and low-methoxyl (LM) form, if the content was lower than 50%. The esterification degree of PM and PZ was evaluated by titrimetric method and indicated as DE, while DM was the instrumental FTIIt value (Figure 6). The analyses on each sample were performed in triplicate by both titrimetric and instrumental methods and the results were reported as means ± standard deviation. The results showed that titrimetric values were slightly higher than those ontained by the instrumental method, and that the esterification degree, regardless of the method used, of pectin from Montalto was significantly higher than that of pectins from Zumpano. The titrimetric percentage values for DE of pectins from Montalto and Zumpano were 65.9 ± 0.8 and 39.4 ± 0.6, respectively, while the corresponding %DM were 56.3 ± 1.1 and 37.2 ± 0.5.

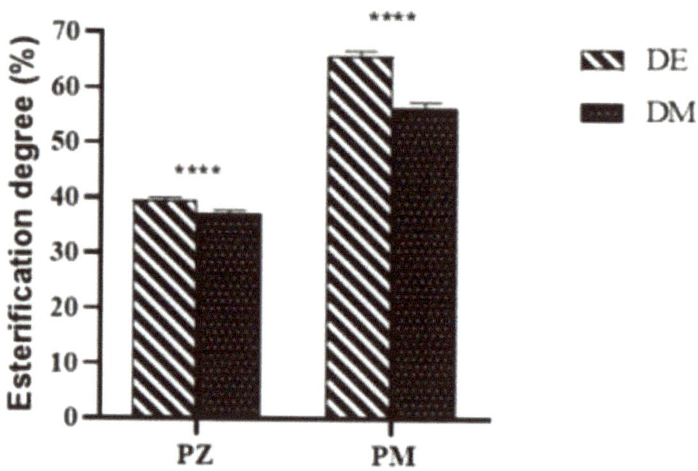

Figure 6. Effects of the two pedoclimatic areas (IM and Z) on esterification degree of walnut husks' pectins (P) evaluated by titrimetric method (DE) and instrumental methods (DM). Error bars indicated standard deviation (n = 3). Asterisks on the dashes indicate significant differences among the two methods used for esterification degree evaluation of pectins from tire same agroclimatic area (** p< 0.0001).**

3.9. Scanning Electron Microscopy (SEM) of Pectins

All the pectins were studied for morphological investigation under the same conditions: three magnifications equal to 100 x, 2P00x, and 5000 x with a scale equal to 200,10, and 4 gm, respectively. SEM images showed different surface structures depending on the pedoclimatic area of provenance of the walnut husks. The surface of PZ showed a leafy appearance and the tresence of crystals which were incorporated in a spherical matrix. Filaments with large vesicles were also present (Figure 7a-c).

The surface of PM exhibited a uniform structure, characterized by a very smooth lamellar appearance. This lamella also showed microcrystals incorporated in a matrix (Figure 7A-C).

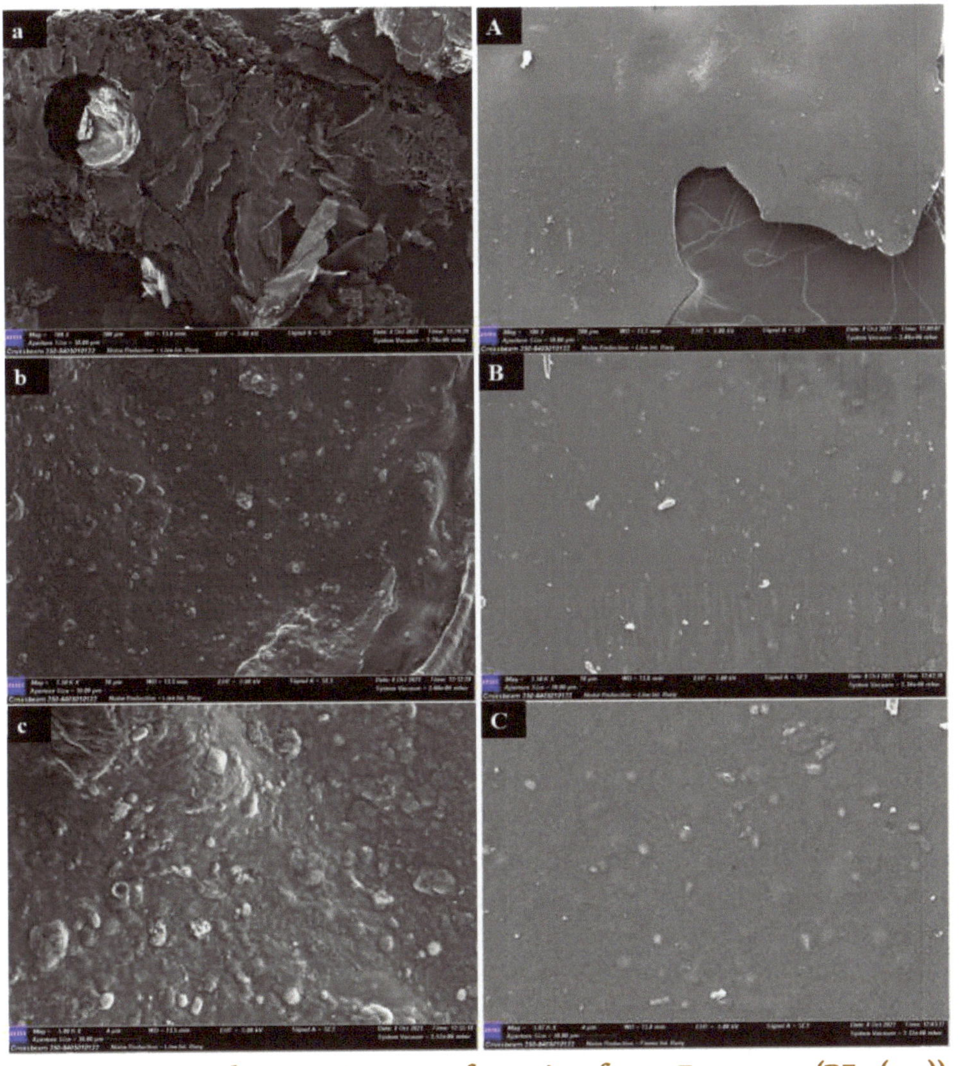

Figure 7. Scanning electron images of pectins from Zumpano (PZ, (a-c)) and Montalto (I'M, (A-C)) at three different magnifications and scales. (a,A) (100 x, 200 pm); (b,B) (2500 x, 10 pm); (c,C) (5000 x, 4 pm).

3.10. Differential Scanning Calorimetry (DSC) of Pectins

The calorimetric curves recorded for the PM (red line) and PZ (black line) were similar, but unlike the calorimetric curves for glucans, they showed two peaks, one of which was endothermic and the other exothermic (Figure 8). Peak one, which was very similar to that observed for glucan calorimetric curves, was endothermic and was located between 130 and 135°C. It had a higher enthalpy for the PM samples (E = 326.8 J/g) than for the PZ samples (E = 212.4 J/g). Peak two, between 240 and 250°C, was exothermic, and it showfd similat enthalpy values forboth samples (E = -90.6 J/g; E = — 77.1 J/g).

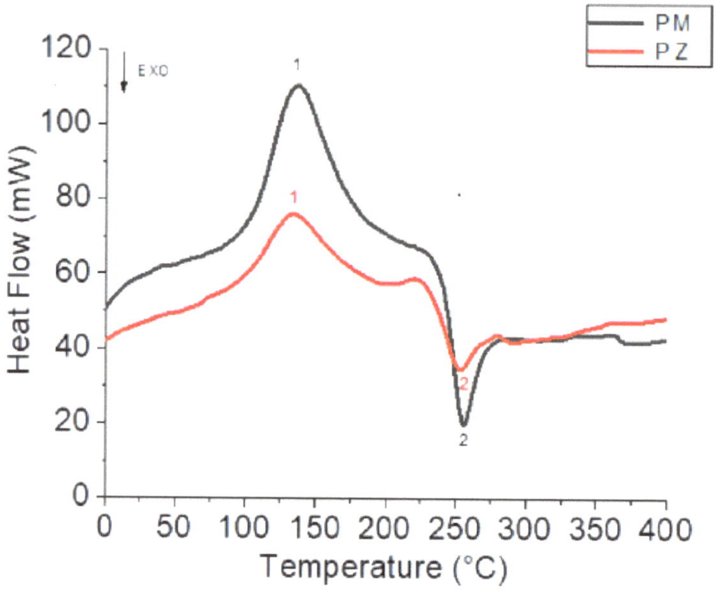

Figure 8. DSC records of walnut husks' pectins from Zumpano (red line) and from Montalto (black line).

4. DISCUSSION

Glucans and pectins are nonstarch polysaccharides and fibre components, which possess a broad spectrum of biological activities. Glucans have hypocholesterolemic and hypoglycemic effects but also improve the defence of the immune system and induce defence mechanisms to respond to wounding [20,43,44]. Pectins are applied in the pharmaceutical field to treat human pathologies, such as cancers, liver damage, and inflammations [45]. Husk glucans, occurring in the bran of cereal grains (barley and oats and to a much lesser degree in rye and wheat, in amounti of about 7⁰%, 5%, 2⁰% and less than 1%, respectively) and many kinds of mushrooms, have been researched in order to valorise the waste from nut production. In this context, walnuts with green hulls were harvested from two locations of southern Italy, where walnut trees were widespread. Although the two areas were characterized by similar climatic conditions (annual raining, daily temperature and humidity average), the recovery of glucans was different for the hulls from the two areas. In fatt, the glucan content from Monialto was 1.2 times higher than tha0 of GZ, most likely due to the different soil characteristics. The glucan conteni from both Montalto and Zumpano husks is nevertheless significant compared to that of some cereals such as oat samples (0.71-5.06%) [46], broad bean pods (3.0-4.5%) and pomegranate *Akko* peels (2.5⁰%) [12,13] The qualitative profile of trite IR spectra of botle GM and GZ glucans was similar. Furthermore, the observation of the anomeric region did not allow us to distinguish easily between the two glycosidic linkage types of the aldopyranooes. Despite this criticality, the use o0 the |3-glucan assay kit allowed us to determine the α- and β -glucan content in both samples, showing that Montalto husks were richer in β-glucans while Zumpano husks contained α- and β-glucans in equal percentages. However, the β-glucan content in the Montalto husks was 1.75 times of that in Zumpano husks. The morphological investigation highlighted different surface structures for GM and GZ. The surface of GZ showed the presence o0 aggregates of microcrystals whose dimensions were variable, while the GM surface showed not aggregated microcoyitals with a cubic form. The surface morphology of both GM and GZ was different from SEM images of barley and oat glucans [19,-47]. Glucans, recovered from other typeof biomesses such as *Vida faba* L. pods using the same extraction method, showed different morphological surfaces, which exhibited agglomerates with a spongy appearance [13].

Thermal analysis (DSC) confirmed the higher degree of crystallinity of GZ, since the corresponding melting enthalpy value was twice that recorded for the GM sample. Pectins have also been investigated in walnut husks from Montalto and Zumpano for their valorisation, carrying out the extraction under acidic conditions at reflux temperature. The pectin content, as well as that of glucans, was found to be different in the two matrices: recovery of PM was 2.6 times that of PZ, confirming that soil and climate conditions influenced fibre content. It has been reported that extraction of pectins from other types of wastes or by-products, carried out using similar chemical conditions, yielded 5.2-12.2% from banana peels [48], 3.7-7.7% from passionfruit peel [49], 3.9-11.2% from pomegranate peels [50], and 7.3-19.1% pistachio green hull [51]. The percentage yield in PM was higher than that of other food wastes extracted under alkaline conditions such as leek leaves (12 ± 7%), endive roots (22 ± 8%), onion hulls (14 ± 0%), endive leaves (36 ± 8%), pumpkin kernel cake (29 ± 2.13%), tomato skins (29 ± 9.15%), and grape pomace (15 ± 3.07%) [52]. The comparison of our results with these data highlighted that the walnut husks from Montalto were a good source of pectins. Similarly to glucans, pectins were characterized by FTIR spectroscopy, surface morphological analysis (SEM) and thermal characteristics (DSC). In addition, the degree of methoxylation was determined, since it is an important structural factor influencing the functional properties of pectins. The FT-IR spectra of both PM and PZ did not show any significant differences in the characteristic bands of the structure, except for the absorption intensities of the peaks in the 1800-1500 cm^{-1} region arising from ester carbonyl groups and carboxylate ion stretching. In fact, the IR spectra of PM showed that the intensity of the ester groups was greater than that relating to the carboxylate groups, whereas in PZ spectra the carboxylate ion band was stronger than the ester group band. These spectroscopic differences were reflected in the different degree of methoxylation (DM), calculated on the basis of the areas of the two bands between PM and PZ. The DM of PM was 1.5 times higher than PZ's. This was confirmed by the titrimetric method (DE), although the values obtained for DE are slightly higher than those obtained for DM. However, regardless of the method used, the esterification degree of PM was significantly higher than that of PZ, and considering that its percentage value (both DM and DE) was higher than 50, it is possible to classify PM as a high-methoxyl pectin, similarly to apple pectin, with a DM of 65.88% [53]. Additionally, pectins extracted from biomass (rind and peels) obtained from fruits with citric acid solution at high temperature, such as melon, kiwifruit, pomegranate and orange, were high-methoxyl pectins, having esterification degrees (DM) of 71.98%, 84.72%, 56.74% and 69.67%, respectively [54]. Morphological analysis of PM and PZ suggested different surface morphologies, exhibiting a very smooth lamellar appearance for PM with little pellets on it, comparable to the morphological characteristics of passion fruit pectin [25]. In contrast, the morphological structures of pectins from melon rind, kiwifruit, pomegranate and orange peels had some microfractures and hollow openings [54].

Calorimetric curves of PM and PZ were similar, showing two main peaks during the thermal analysis, one of which was endothermic and the other exothermic. The parameters associated with the two peaks were melting temperature and enthalpy (T_m and A_m, respectively), and degradation temperature and enthalpy (T_d and Ad, respectively). The first endothermic peak between 130 and 150°C is ascribed to water evaporation: PM and PZ showed little differences for Tm but Δm of PM samples was higher than the PZ samples, which indicated that more energy was needed to absolutely remove water from PM, likely due to higher esterification degree. The second exothermic peak, between 240 and 250°C, was caused by the degradation of pectin: PM and PZ showed little differences for T_d and Δm [55,56]. The above results indicated that geographical and climatic conditions in different regions could lead to significant differences both in the content of bioactive compounds and their morphological and thermal characteristics.

5. CONCLUSIONS

The results of the present study suggested that *J. regia* L. green husks, which represent a waste material, could be a source of glucans and pectins, with a potential use in food, cosmetics and pharmaceutical fields. Walnuts with green husks were collected from two different soil areas of Southern Italy (Montalto and Zumpano) which influenced the content, morphological and thermal characteristics of the extracted glucans and pectins, showing that the higher yields were obtained from the Montalto green husks.

In fact, the glucan content from the Montalto walnut husks was about 1.2 times that of the glucans from the Zumpano husks. In addition, GM were characterized by a prevalence of β-glucans over α-glucans. Morphological analysis of Montalto glucans showed a surface covered by cubic microcrystals, which differed from that of Zumpano ones, characterized by the presence of agglomerates of varying sizes of microcrystals. The morphological characteristic of GZ was confirmed by thermal analysis, which showed a higher degree of crystallinity than GM. Pectin recovery was 2.6 times higher from Montalto husks than from Zumpano ones. Additionally, PM, compared with PZ, showed a higher esterification degree that resulted in a higher Δm needed to remove retained water from the structure. The surface of the PM is lamellar in contrast to that of the PZ, which is leaf-shaped. Both glucans and pectins from Montalto and Zumpano walnut husks will be further studied to investigate how the differences in their chemical and physical characteristics are reflected in their biological activity.

REFERENCES

1. Stampar, F.; Solar, A.; Hudina, M.; Veberic, R.; Colaric, M. Traditional walnut liqueur—Cocktail of phenolics. Food *Chem.* **2006**, *95*, 627-631. [CrossRef]
2. Stampar, F.; Solar, A.; Hudina, M.; Veberic, R.; Colaric, M.; Fabcic, J. Phenolics in walnut liqueur. *Acta Hortic.* **2007**, *744*, 451-454. [CrossRef]
3. Wichtl, M.; Anton, R. Tradition, pratique officinale, science et therapeutique. In *Plantes Therapeutiques,* 2nd ed.; Tec & Doc-Cachan, Editions Technique et Documentation-Editions Medicales Internationales, Ed.; Librairie Eyrolles: Paris, France, 1999; Volume 1, p. XCVI-692.
4. Croitoru, A.; Ficai, D.; Craciun, L.; Ficai, A.; Andronescu, E. Evaluation and Exploitation of Bioactive Compounds of Walnut, *Juglans regia. Curr. Pharm. Des.* **2019**, 25,119-131. [CrossRef]
5. Oliveira, I.; Sousa, A.; Ferreira, I.C.F.R.; Bento, A.; Estevinho, L.; Pereira, J.A. Total phenols antioxidant potential and antimicrobial activity of walnut *(Juglans regia* L.) green husk. *Food Chem. Toxicol.* **2008**, 46, 2326-2331. [CrossRef] [PubMed]
6. Fernandez-Agullo, A.; Pereira, E.; Freire, M.S.; Valentao, P.; Andrade, P.B.; Gonzalez-Alvarez, J.; Pereira, J.A. Influence of solvent on the antioxidant and antimicrobial properties of walnut *(Juglans regia* L.) green husk extracts. *Ind. Crop. Prod.* **2013**, 42,126-132. [CrossRef]

7. Ghasemi, K.; Ghasemi, Y.; Ehteshamnia, A.; Nabavi, S.M.; Nabavi, S.F.; Ebrahimzadeh, M.A.; Pourmorad, F. Influence of environmental factors on antioxidante activity, phenol and flavonoids contents of Walnut (*Juglans regia* L.) green husks. *J. Med. Plants Res.* **2011**, 5,1128-1133.

8. Beiki, T.; Najafpour, G.D.; Hosseini, M. Evaluation of antimicrobial and dyeing properties of walnut (*Juglans regia* L.) green husk extract for cosmetics. *Color. Technol.* **2018**,134, 71-81. [CrossRef]

9. Kiran Aithal, B.; Sunil Kumar, M.; Nageshwar Rao, B.; Udupa, N.; Satish Rao, B. Juglone, a naphthoquinone from walnut, exerts cytotoxic and genotoxic effects against cultured melanoma tumor cells. *Cell Biol. Int.* **2009**, 33,1039-1049. [CrossRef]

10. Ji, Y.-B.; Qu, Z.Y.; Zou, X. Juglone-induced apoptosis in human gastric cancer SGC-7901 cells via the mitochondrial pathway. *Exp. Toxicol. Pathol.* **2011**, 63, 69-78. [CrossRef]

11. Xu, H.L.; Yu, X.F.; Qu, S.C.; Qu, X.R.; Jiang, Y.F.; Sui, D.Y. Juglone, from Juglans mandshruica Maxim, inhibits growth and induces apoptosis in human leukemia cell HL-60 through a reactive oxygen species-dependent mechanism. *Food Chem. Toxicol.* **2012**, 50, 590-596. [CrossRef]

12. Fazio, A.; Iacopetta, D.; La Torre, C.; Ceramella, J.; Muia, N.; Catalano, A.; Carocci, A.; Sinicropi, M.S. Finding solutions for agricultural wastes: Antioxidant and antitumor properties of pomegranate Akko peel extracts and beta-glucan recovery. *Food Funct.* **2018**, *9*, 6619-6632. [CrossRef] [PubMed]

13. Fazio, A.; La Torre, C.; Dalena, F.; Plastina, P. Screening of glucan and pectin contents in broad bean (*Viciafaba* L.) pods during maturation. *Eur. Food Res. Technol.* **2020**, 246, 333-347. [CrossRef]

14. Arias, D.; Rodriguez, J.; Lopez, B.; Mendez, P. Evaluation of the physicochemical properties of pectin extracted from Musa paradisiaca banana peels at different pH conditions in the formation of nanoparticles. *Heliyon* **2021**, 7, E06059. [CrossRef]

15. Manasa, V.; Padmanabhan, A.; Appaiah, K.A.A. Utilization of coffee pulp waste for rapid recovery of pectin and polyphenols for sustainable material recycle. *Waste Manag.* **2021**,120,762-771. [CrossRef]

16. Mellinas, C.; Ramos, M.; Jimenez, A.; Garrigos, M.C. Recent Trends in the Use of Pectin from Agro-Waste Residues as a Natural-Based Biopolymer for Food Packaging Applications. *Materials* **2020**,13, 673. [CrossRef] [PubMed]

17. McCleary, B.V.; Draga, A. Measurement of B-glucan in mushrooms and mycelial products. *J. AOAC Int.* **2016**, 99, 364-373. [CrossRef]

18. Temelli, F. Extraction and functional properties of barley B-glucan as affected by temperature and pH. *J. Food Sci.* **1997**, 62, 1194-1201. [CrossRef]

19. Limberger-Bayer, V.M.; de Francisco, A.; Chan, A.; Oro, T.; Ogliaru, P.J.; Barreto, P.L.M. Barley β-glucans extraction and partial characterization. *Food Chem.* **2014**,154, 84-89. [CrossRef]

20. Fazio, A.; La Torre, C.; Caroleo, M.C.; Caputo, P.; Plastina, P.; Cione, E. Isolation and Purification of Glucans from an Italian Cultivar of Ziziphus jujuba Mill. and In Vitro Effect on Skin Repair. *Molecules* **2020**, 25, 968. [CrossRef]

21. Nie, S.; Cui, S.W.; Xie, M.; Phillips, A.O.; Phillips, G.O. Bioactive polysaccharides from Cordyceps sinensis: Isolation, structure features and bioactivitie. *Bioact. Carbohydr. Diet. Fibre* **2013**,1, 38-52. [CrossRef]

22. Yalcin, E.; Celik, S. Solubility properties of barley flour, protein isolates and hydrolysates. *Food Chem.* **2007**, 104, 1641-1647. [CrossRef]

23. Bilgi, B.; Qelik, S. Solubility and emulsifying properties of barley protein concentrate. *Eur. Food Res. Technol.* **2004**, 218, 437-441. [CrossRef]

24. Fazio, A.; La Torre, C.; Caroleo, M.C.; Caputo, P.; Cannataro, R.; Plastina, P.; Cione, E. Effect of addition of pectins from jujubes *(Ziziphus jujuba* Mill.) on vitamin C production during heterolactic fermentation. *Molecules* **2020**,25, 2706. [CrossRef]

25. Rajia, Z.; Khodaiyana, F.; Rezaei, K.; Kiania, H.; Hosseini, S.S. Extraction optimization and physicochemical properties of pectin from melon peel. *Int. J. Biol. Macromol.* **2017**, *98*, 709-716. [CrossRef] [PubMed]

26. Liew, S.Q.; Chin, N.L.; Yusof, Y.A. Extraction and characterization of pectin from passion fruit peels. *Agric. Agric. Sci. Procedia* **2014**, *2*, 231-236. [CrossRef]

27. Bashir, K.M.I.; Choi, J.S. Clinical and physiological perspectives of betα-glucans: The past, present, and future. *Int. J. Mol. Sci.* **2017**, *18*, 1906. [CrossRef]

28. Du, B.; Zhu, F.; Xu, B. B-glucan extraction from bran of hull-less barley by accelerated solvent extraction combined with response surface methodology. *J. Cereal Sci.* **2014**, 59, 95-100. [CrossRef]

29. Yang, L.; Sun, X.W.; Yang, F.J.; Zhao, C.J.; Zhang, L.; Zu, Y.G. Application of ionic liquids in the microwave-assisted extraction of proanthocyanidins from Larix gmelini Bark. *Int. J. Mol. Sci.* **2012**,13, 5163-5178. [CrossRef]

30. Palanisamy, M.; Aldars-Garcia, L.; Gil-Ramirez, A.; Ruiz-Rodriguez, A.; Marin, F.R.; Reglero, G.; Soler Rivas, C. Pressurized water extraction of B-glucan

enriched fractions with bile acids-binding capacities obtained from edible mushrooms. *Biotechnol. Prog.* **2014**, *30*, 391-400. [CrossRef]

31. Wang, Y.; Ahmed, Z.; Feng, W.; Li, C.; Song, S. Physicochemical properties of exopolysaccharide produced by *Lactobacillus kefiranofaciens* ZW3 isolated from Tibet kefir. *Int. J. Biol. Macromol.* **2008**, 43, 283-288. [CrossRef]
32. Ahmad, A.; Anjum, F.M.; Zahoor, T.; Nawaz, H.; Ahmed, Z. Extraction and characterization of β-D-glucan from oat for industrial utilization. *Int. J. Biol. Macromol.* **2010**, 46, 304-309. [CrossRef]
33. Kacurakova, M.; Capeka, P.; Sasinkova, V.; Wellnerb, N.; Ebringerova, A. FT-IR study of plant cell wall model compounds: Pectic polysaccharides and hemicelluloses. *Carbohydr. Polym.* **2000**, 43,195-203. [CrossRef]
34. Hutchinson, J.M. Studying the Glass Transition by DSC and TMDSC. *J. Therm. Anal. Calorim.* **2003**, 72, 619-629. [CrossRef] Cui, J.; Zhao, C.; Feng, L.; Han, Y.; Du, H.; Xiao, H.; Zheng, J. Pectins from fruits: Relationships between extraction methods, structural characteristics, and functional properties. *Trends Food Sci. Technol.* **2021**,110, 39-54. [CrossRef]
35. Liew, S.Q.; Teoh, W.H.; Yusoff, R.; Ngoh, G. Comparisons of process intensifying methods in the extraction of pectin from pomelo peel. *Chem. Eng. Process. -Process. Intensif.* **2019**,143,107586. [CrossRef]
36. Pinheiro, E.R.; Silva, I.M.D.A.; Gonzaga, L.V.; Amante, E.R.; Teofilo, R.F.; Ferreira, M.M.C.; Amboni, R.D.M.C. Optimization of extraction of high-ester pectin from passion fruit peel *(Passiflora edulis Flavicarpa)* with citric acid by using response surface methodology. *Bioresour. Technol.* **2008**, 99, 5561-5566. [CrossRef] [PubMed]
37. Szymanska-Chargot, M.; Zdunek, A. Use of FT-IR spectra and PCA to the bulk characterization of cell wall residues of fruits and vegetables along a fraction process. *Food Biophys.* **2013**, 8, 29-42. [CrossRef] [PubMed]
38. Li, Q.; Xu, R.; Fang, Q.; Yuan, Y.; Cao, J.; Jiang, W. Analyses of microstructure and cell wall polysaccharides of flesh tissues provide insights into cultivar difference in mealy patterns developed in apple fruit. *Food Chem.* **2020**, 321,126707. [CrossRef]
39. Chylinska, M.; Szymanska-Chargot, M.; Zdunek, A. FT-IR and FT-Raman characterization of non-cellulosic polysaccharides fractions isolated from plant cell wall. *Carbohydr. Polym.* **2016**,*154*, 48-54. [CrossRef]
40. Ridley, B.L.; O'Neill, M.A.; Mohnen, D. Pectins: Structure, biosynthesis, and oligogalacturonide-related signaling. *PhytoChem.* **2001**, *57*, 929-967. [CrossRef]

41. Fissore, E.N.; Rojas, A.M.; Gerschenson, L.N.; Williams, P.A. Butternut and beetroot pectins: Characterization and functional properties. *Food Hydrocoll.* **2013**, 31,172-182. [CrossRef]

42. Rondanelli, M.; Opizzi, A.; Monteferrario, F. The biological activity of betα-glucans. *Minerva Med.* **2009**,100, 237-245. [PubMed]

43. Shahrahmania, N.; Akbarib, S.A.A.; Mojabc, F.; Mirzaid, M.; Shahrahmania, H. The effect of *Zizyphus Jujube* Fruit Lotion on Breast Fissure in Breastfeeding Women. *Iran. J. Pharm. Res.* **2018**,17,101-109.

44. Naqash, F.; Masoodi, F.A.; Rather, S.A.; Wani, S.M.; Gani, A. Emerging concepts in the nutraceutical and functional properties of pectin-A Review. *Carbohydr. Polym.* **2017**,168, 227-239. [CrossRef]

45. Shen, R.L.; Liu, X.Y.; Dong, J.L.; Si, J.L.; Li, H. The gel properties and microstructure of the mixture of oat |3-glucan/soy protein isolates. *Food Hydroll.* **2015**, 47,108-114. [CrossRef]

46. Wang, Y.X.; Li, L.Y.; Zhang, T.; Wang, J.Q.; Huang, X.J.; Hu, J.L.; Yin, J.Y.; Nie, S.P. Fractionation, physicochemical and structural characterization of polysaccharides from barley water-soluble fiber. *Food Hydrocoll.* **2021**,113,106539. [CrossRef]

47. Oliveira, T.I.S.; Rosa, M.F.; Cavalcante, F.L.; Pereira, P.H.F.; Moates, G.K.; Wellner, N.; Mazzetto, S.E.; Waldron, K.W.; Azeredo, H.M.C. Optimization of pectin extraction from banana peels with citric acid by using response surface methodology. *Food Chem.* **2016**,198,113-118. [CrossRef]

48. Liew, S.Q.; Chin, N.L.; Yusof, Y.A.; Sowndhararajan, K. Comparison of acidic and enzymatic pectin extraction from passion fruit peels and its gel properties. *J. Food Process. Eng.* **2016**, 39, 501-511. [CrossRef]

49. Pereira, P.H.F.; Oliveira, T.I.S.; Rosa, M.F.; Cavalcante, F.L.; Moates, G.K.; Wellner, N.; Moates, G.K.; Wellner, N.; Walder, K.W.; Azeredo, H.M.C. Pectin extraction from pomegranate peels with citric acid. *Int. J. Biol. Macromol.* **2016**, 88, 373-379. [CrossRef]

50. Chaharbaghi, E.; Khodaiyan, F.; Hosseini, S.S. Optimization of pectin extraction from pistachio green hull as a new source. *Carbohydr. Polym.* **2017**,173,107-113. [CrossRef] [PubMed]

51. Muller-Maatsch, J.; Bencivenni, M.; Caligiani, A.; Tedeschi, T.; Bruggeman, G.; Bosch, M.; Petrusan, J.; Van Droogenbroeck, B.; Elst, K.; Sforza, S. Pectin content and composition from different food waste streams. *Food Chem.* **2016**, 201, 37-45. [CrossRef] [PubMed]

52. Naqash, F.; Masoodi, F.A.; Gani, A.; Nazir, S.; Jhan, F. Pectin recovery from apple pomace: Physico-chemical and functional variation based on methyl-esterification. *Int. J. Food Sci. Technol.* **2021**, 9, 669-4679. [CrossRef]
53. Guzela, M; Akpinar, O. Valorisation of fruit by-products: Production characterization of pectins from fruit peels. *Food Bioprod. Process.* **2019**,115,126-133. [CrossRef]
54. Einhorn-Stoll, U.; Kunzek, H.; Dongowski, G. Thermal analysis of chemically and mechanically modified pectins. *Food Hydrocoll.* **2007**, *21*, 1101-1112. [CrossRef]
55. Wang, X.; Chen, Q.; Lu, X. Pectin extracted from apple pomace and citrus peel by subcritical water. *Food Hydrocoll.* **2014**, 38, 129-137. [CrossRef]

CHAPTER 17

Aspergillus oryzae Grown on Rice Hulls Used as an Additive for Pretreatment of Starch-Containing Wastewater from the Pulp and Paper Industry

1. INTRODUCTION

Climate change, together with the growing population expected over the coming years, makes food production a crucial issue. In this context, prevention and minimization of food waste are recognized as key actions [1]. In addition, food waste is highly polluting as it leads to the misuse of resources and significant greenhouse gas emission levels. To address these issues, the adoption of the biorefinery concept (circular economy approach), particularly strengthening the agri-food waste biorefinery, is a strategic point not only to make the cost of the process economical but also to reduce the pressure on natural resources [2,3]. There is no explicit mention or definition of the valorization of food supply chain waste in the Waste Framework Directive. However, the objective of using waste for value-added production comes within the spirit of the directive [4,5].

A circular economy provides a different flow model, where no resources are wasted; on contrary, they are considered as feedstocks. In particular, open-loop material flow patterns bring new supplies of secondary materials into the raw material pool that can be reclaimed by other industries [6,7]. Rice hulls are the largest by-product of rice milling in producing countries and most of them are thrown away as a waste byproduct, which will undoubtedly have too many negative influences on the global environment. Worldwide production amounts to approximately 100 million tonnes per year [8]. Rice hulls, the lignocellulosic outer coats of rice, have been only considered as combustible to recover energy or animal bedding, because of their low nutritive value as animal feed [9]. Although rice hulls have long been identified as a source for energy production, experiences from large-scale rice husk firing are quite limited, because of the high quantity of ash (about 20%) [10].

Many efforts have been made to use rice hulls as feedstock to produce fermentable sugars, followed by ethanol fermentation, but the application on large scale is still affected by the out-of-market costs of pretreatments and saccharification [9]. The pretreatment process is considered the most expensive step in the valorization of lignocellulosic byproducts, where it can contribute to about 30% of the total cost. In fact, even those experiences have demonstrated that rice hulls can be more conveniently re-used as untreated material [11]. For example, promising research has been carried out on the use of rice hulls, as well as other agricultural lignocellulosic waste, as inexpensive and efficient biosorbent for heavy metals removal from contaminated wastewater [12]. From this perspective, rice hulls have also been used as solid support for the production of various fermented products and enzymes by solid-state fermentation (SSF) [13]. SSF is defined as the cultivation of microorganisms on inert carriers or on insoluble substrates that can, in addition, be used as carbon and energy source. The fermentation takes place in the absence or near absence of free water, thus being close to the natural environment to which microorganisms, especially fungi, are adapted [14].

SSF aims to bring the cultivated fungi or bacteria into tight contact with the insoluble substrate and thus to achieve the highest substrate concentrations for fermentation [15]. This technology results, although so far only on a small scale, in several processing advantages of significant potential economic and ecological importance as compared with submerged fermentation. SSF holds tremendous potential to produce enzymes, as amylases, proteases, or extracellular lipases by fungal strains belonging to the genus *Aspergillus*. Several experiences have been reported on the production of amylase and glucosidase by *Aspergillus niger* from sugarcane bagasse, corn cobs and rice hulls using SSF [16], production of amylase by *Aspergillus oryzae* on spent brewing grain as solid substrate in SSF [17].

In this work, rice hulls have been used as support for SSF of the amylase-producing *Aspergillus oryzae* to verify the feasibility to add an amount of dried powder of *A. oryzae* adherent on rice hulls, as an additive in starch-containing wastewater treatments from pulp-and-paper mill [18]. Starch is presently the third most prevalent component by weight in papermaking, only surpassed by cellulose fiber and mineral pigments. It is used as a flocculant and retention aid, as a bonding agent, as a surface size, as a binder for coatings, and as an adhesive in corrugated board, laminated grades, writing paper, and other products. The starch-containing effluents generated by the papermaking industry are usually destined for the anaerobic digestion process or degraded by aerobes microorganisms in fluidized bed bioreactors. In this context, simple and low-cost strategies for accelerating starch digestion are desirable before both anaerobic and aerobic effluent treatments.

In a wider perspective, this application could represent a promising example of industrial symbiosis, where agri-food waste valorization can become the point of connection between two different supply chains, in which one company's waste is used as raw material by another company.

2. MATERIALS AND METHODS

2.1. Microorganism and Its Maintenance

Aspergillus oryzae DSM 1862 belongs to the collection of microorganisms of the Life Sciences and Biotechnology Department of the University of Ferrara and was purchased from the DSMZ (Leibniz Institute DSMZ-German Collection of Microorganisms and Cell Cultures GmbH, Braunschweig, Germany) company. It was propagated in potato dextrose agar medium (DIFCO, Wuerzburg, Germany), containing dextrose, 20 g/L, potato extract, 4 g/L, agar, 15 g/L added with chloramphenicol (Merck, Berlin, Germany), 50 mg/L for 72 h at 30°C and stored in Petri dishes at 4°C [19]. Amylase (CAS: 9000-90-2) lyophilized powder, >100 units/mg protein, was purchased from Merck, Berlin, Germany.

2.2. Inoculum Preparation

The liquid medium flasks (20 mL final volume) were inoculated with fungal cultures grown on PDA plates supplemented with chloramphenicol by aseptically transferring a block of mycelium and spores (5 x 5 mm^2 area) of the plate culture into the flasks [20]. Three different growth media were tested: PDB medium (DIFCO, Wuerzburg, Germany) having the same composition as the PDA but without agar; malt extract medium (DIFCO, Wuerzburg, Germany) containing malt extract, 6 g/L, maltose, 1.8 g/L, dextrose, 6 g/L, and yeast extract, 1.2 g/L; paper mill wastewater, supplied by a local company. The cultures were incubated at 28°C, 120 rpm for up to 96 h. All tests were carried out in triplicate for statistical significance.

2.3. Evaluation of Enzymatic Activity

From 20 mL inoculated flasks, 1 mL of the supernatant was taken immediately after inoculum and every 24 h until 72 h of growth. (NHj^SO.! sodium (657 mg) was added to each sample, once solubilized, the suspension was centrifuged (5600 RCF, 10 min). The precipitates were immediately resuspended in 10 mL of 100 mM acetate buffer containing 0.25% of starch at pH 5. Reaction samples (1 mL) were taken starting from T0 (immediately after adding the enzyme), after 6 h, and every 24 h up to 72 h. The same enzymatic reaction protocol was applied to 10 mL of paper mill water instead of the solution containing starch. All tests were carried out in triplicate for statistical significance. All the collected samples were analyzed by the DNS assay for the quantification of reducing sugars.

2.3.1. Dinitrosalicylic Acid Method (DNS)

DNS reagent was prepared by dissolving dinitrosalicylic acid (0.2 g, 0.88 mmol), phenol (0.04 g, 0.42 mmol), sodium thiosulfate (0.01 g, 0.04 mmol) and sodium-potassium tartrate (4.0 g, 14.2 mmol) in 10 mL of NaOH 2% (W/V). Distilled water was added to this solution to a final volume of 20 mL.

For determination of reducing sugars, 2 mL of DNS reagent and 500 gL of distilled water were added to 500 gL of enzymatic reaction sample. The mixture was brought to a boil for 5 min and left to cool at RT. The absorbance was measured at 540 nm in a Shimadzu UV-1601 spectrophotometer.

The concentration of reducing sugars produced was calculated by comparison with the previously constructed calibration curve.

2.3.2. Solid-State Fermentation (SSF): Substrate Preparation

Fifty mililiter Erlenmeyer flasks containing 1 g of rice hulls were autoclaved (121°C for 20 min). After cooling, the moisture content of rice hulls was brought up to 60% by the addition of 0.6 mL of a sterile water solution of KH_2PO_4 2 g/L, NaCl 1 g/L, $MgSO_4$, and $7H_2O$ 1 g/L.

2.3.3. Solid-State Fermentation (SSF): Inoculum Preparation

Two different types of inocula were used: the first consisted of 1 mL of spore suspension (S) obtained by spraying 10 mL of sterile 0.1% Tween-80 solution on 7-day-old PDA Petri dishes containing *A. oryzae*. One milliliter of spore suspension was used to inoculate 50 mL Erlenmeyer flasks containing 1 g of rice hulls. The number of spores was quantified by carrying out serial dilutions and seeding 0.1 mL of each one on the PDA plate. Petri dishes were incubated for 4 days at 28°C and then counted to establish the starting load.

The second type of inoculum consists of the supernatant of submerged liquid fermentation (SLF) and was obtained as follows: 6-day-old Petri dishes grown in PDA medium supplemented with chloramphenicol were used to inoculate a 50 mL Erlenmeyer flask containing 20 mL of PDB medium. After 4 days of growth, 1 mL of the medium was directly added to the rice hulls. The count of starting colonies was made following the same protocol used for spore suspension.

2.3.4. Solid-State Fermentation (SSF): Fermentation Conditions

For both inocula, the cultures were grown for 3,5, 7, and 10 days. After fermentation, each SSF was resuspended in 10 mL of previously sterilized physiological solution and left to stir at 100 rpm, RT for 2 h. One milliliter of suspension was then withdrawn, and the final microbial load (expressed in CFU/mL) was quantified by carrying out serial dilutions and seeding 0.1 mL of each one on PDA plates. Petri dishes were incubated for 4 days at 28°C and then counted to establish the final microbial load. For the fermentation that provided the best microbial load, a subsequent drying step was carried out bringing the temperature to 45°C for 48 h. All tests were carried out in triplicate for statistical significance. The relative standard deviation value for statistical analysis was also reported.

2.3.5. Solid-State Fermentation (SSF): Enzymatic Reaction on Paper Mill Water

After 6 days of *Aspergillus oryzae* growth of Petri dishes in PDA medium, a block of mycelium and spores (5 x 5 mm^2 area) of the plate culture with the fungus was used to inoculate a 50 mL Erlenmeyer flask containing 20 mL of PDB medium. The cultures were incubated at 28°C, 120 rpm for up to 96 h. One milliliter of the supernatant was used to inoculate 50 mL Erlenmeyer flasks containing 1 g of sterile rice hulls added with 3 mL of a sterile water solution of KH2PO4 2 g/L, NaCl 1 g/L, and MgSO4 x 7 H2O 1 g/L. The SSF was maintained in static conditions, at 28°C for 10 days, and was then inoculated in 100 mL of paper mill water and kept under stirring at 100 rpm, 28°C for 5 days. One milliliter of the supernatant was withdrawn and 657 mg of (NH4)2SO4 sodium was added. Once solubilized, the suspension was centrifuged (5600 RCF, 10 min). The precipitates were immediately resuspended in 10 mL of paper mill water. In the case of the addition of amylase, 100 U was added to the reaction. Reaction samples were taken starting from T_0, every 24 h up to 72 h. All tests were carried out in triplicate for statistical significance. The relative standard deviation value for statistical analysis was also reported. All the collected samples were analyzed by the DNS assay for the quantification of reducing sugars.

3. RESULTS AND DISCUSSION

Aspergillus oryzae was selected as a-amylase-producer fungal strain for bioremediation pretreatment of starch from the paper mill wastewater. The wastewater used was derived from several processes of the pulp and paper industry. Thus, its characteristics depend on the type of process, type of wood materials, process technology involved, management practices, internal recirculation of the effluent for recovery, and the amount of water to be used in the specific process [21]. Mandal et al. reported the pH, TS, SS, BOD5, COD, and color characteristics of wastewater at various pulp and paper processes [22]. In this work, we aimed to exploit the amylase activity of *A. oryzae* for the pretreatment of starch-containing wastewater. The pretreatment of the wastewater (in this case proposed through a biotechnological process) is necessary as a preliminary step that precedes the generally employed "fluidized bed reactors" bioremediation [23,24]. In particular, the purpose of this work was to evaluate the feasibility of using *Aspergillus oryzae* to eliminate starch from paper mill wastewater, proposing a bioremediation approach as pre-treatment (elimination of starch) of the wastewater given its subsequent further remediation steps. In the experimental design, the stringent operational needs—typical of the industry— were considered. To this end, attention was given to the amylase effect of the fungus in the wastewater, more than on the characteristics of the wastewater itself, or the fungal growth (which is, from an industrial point of view, assumed because of the observations of amylase activity). As mentioned, the composition of the wastewater undergoes great fluctuations in its composition due to various variables related to the paper processing processes. Our study, therefore, focused on the evaluation of the capacity of *Aspergillus* in the elimination of starch present in the wastewater. To do this, we decided to embrace the principles of the circular economy, using a production waste such as the rice hull, which, in this context, represented a raw material having a solid support role for the growth of *Aspergillus,* allowing us to obtain biomass through solid-state fermentation processes (SSF). Figure 1 reports the experimental strategy for the study of the pretreatment bioremediation feasibility by *A. oryzae.*

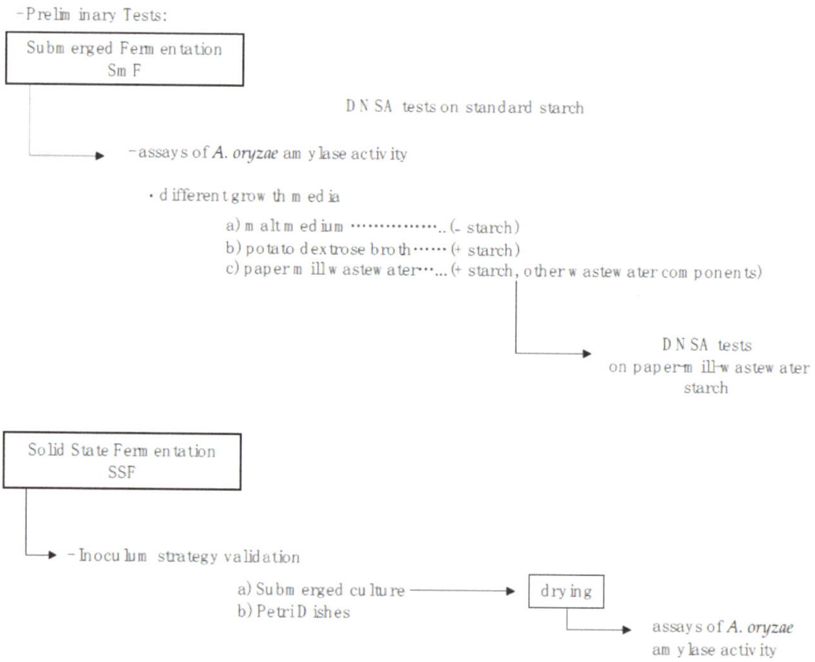

Figure 1. Draw chart of the experimental workflow of this study.

Firstly, the amylase activity of *A. oryzae* was tested with a submerged liquid fermentation technology (SmF). Tie preliminary steps of *Aspergillus* grow'th in submerged culture allowed us to test the effect that starch could have on the growth of the fungus and, consequently, on the production of amylase, as evidence of the observed hydrolysis of the starch supplied as a standard substrate. In a controlled context such as that of SmF, we were able to study the effects of the presence or absence of starch. Furthermore, we tested the system starting from a growth medium represented by the paper mill wastewater itself. The observations obtained justified the subsequent experiments allowing us to develop the SSF process. Therefore, solid-state fermentation (SSF) tests were conducted changing the inoculum type. In this experimental design, the dried inoculated rice hull will act as an additive for the bioremediation of starch in paper mill wastewater.

3.1. Preliminary Experiments: Amylase Activity Assays

The amylase activity of *A. oryzae* was preliminarily tested in SmF. DNS assays were performed for the quantification of reducing sugars. The indirect estimation of the amylase activity of *A. oryzae,* growth on malt medium, paper mill wastewater, and PDB, was tested on standard starch and paper mill wastewater (with and without STD starch).

The first test aimed to assess the ability of *A. oryzae* of growing independently by the presence or absence of starch in the media, testing the flexibility of this microorganism for bioremediation purposes. Therefore, two standard-medium (malt medium and potato dextrose broth—PDB), as well as the paper mill wastewater, were tested for the *Aspergillus* growth. The exoenzymes were purified and the amylase activity was tested on known quantities of standard starch. The results are shown in Figure 2.

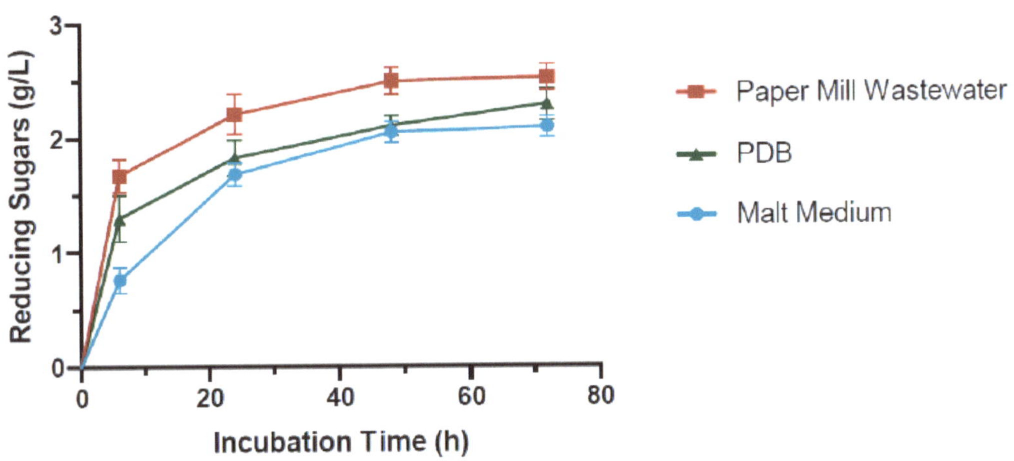

Figure 2. Effect of starch presence (paper mill wastewater and PDB) or absence (malt medium) on the concentration of reducing sugars. Assay of the reducing sugars after the hydrolysis of standard starch conducted with the purified enzymes from A. oryzae grown on PMW, PDB, and malt medium.

In our experimental design, we decided to exploit the parameter concerning the presence o^c reducing sugars, resulting; from the hydrolytic activity of *Aspergillus* against starch. “This parameter directed our experimental choices, atlowing us to deduce the growlh or not of A. *oryzae* as a function of the metabolic activity we found (specifically, its amylase activity). The industrial needs, which have guided our work, focused our attention on the feasibility of starch bioremediation by *Aspergillus,* justifying amylases activity as a key feature of the study of the process. As reported in higurf 2, the growth of *A. oryzae* seems to be not directly dependent on the presence of starch in the growth medium. Indeed, the quantity of reducing sugars, derived by the amylase activity of *A. oryzae* growth in presence of starch, or in its absence (malt medium), are comparable. However, in paper mill wastewater, the concentration of reducing sugars is h:gher in less time. In this context, probably, the presence ol starch and additives in the paper mill wastewater could enhance the metabolic path ways of the expression of amylase. Moreover, essential for this work, the growth of *Aspergillus* is not hindered by a complex growth medium subh as the paper mill wastewater.

Once the ability of the fungus lo grow on pap er mill wasgewater was tested, as well as its production of amylase in this medium, the hydrolytic capacity of the enzymes produced by the fungus was evaluated, not on starch standards, but against the starches present in the paper mill itself. The results are shown in Figure 3.

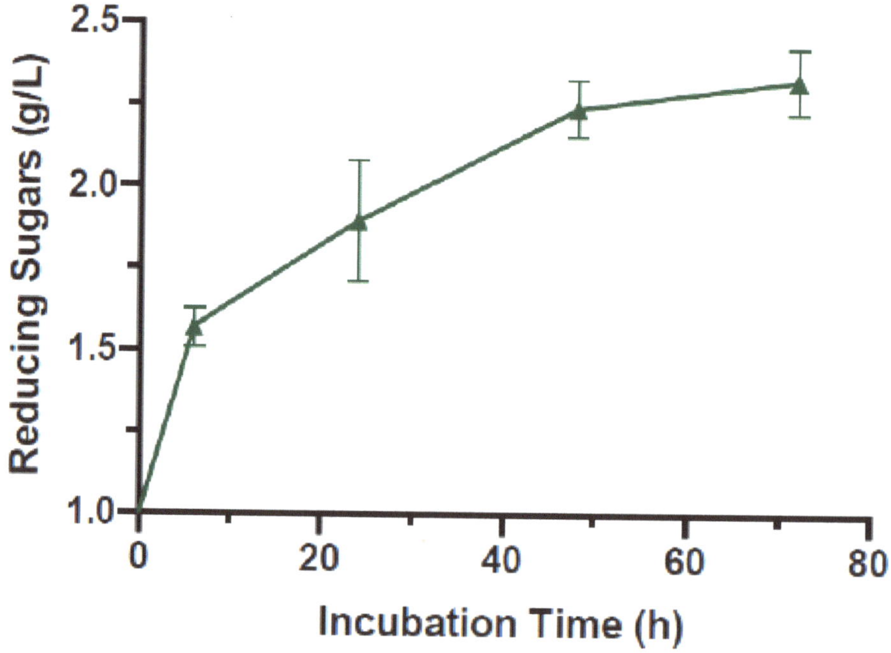

Figure 3. Reducing sugars from the amylase activity of A. oryzae in paper mill wastewater.

As reported in Figure 3, A. *oryzae* has proven capable of biodegrading the starches present in the paper mill wastewater. In just 6 h, the amount of reducing sugars found in the medium amounted to 1.6 g/L.

3.2. Solid-State Fermentation (SSF)

The physiological and genetic properties of the microorganisms could make SSF advantageous against SmF biotechnology [13]. Thus, we tested the SSF of A. *oryzae* starting with the choice of the right inoculum strategy. Furthermore, the SSF is the best strategy of bioremediation from an industrial point of view.

The solid rice hull was inoculated with the supernatant of *A. oryzae* growth in SmF (paper mill wastewater) compared to the inoculum of physiological solution in contact with the fungus growth on Petri dishes. To assay the effectiveness of the respective inocula, both suspensions, derived from SmF and Petri dishes, were collected for the CFU quantification. The CFU/mL were, respectively, 3 x 10^6 and 3 x 10^4.

The two inocula were used in the SSF of the A. *oryzae* rice hull. Several incubation times, 3 to 10 days, were tested. The results concerning the CFU/g of rice hull are shown in Table 1.

Table 1. CFU/g of rice hull for several incubation times. Inoculum strategies tested: SmF and Petri dishes. The results are shown as average, and the standard deviation is shown.

Days	$CFU/g_{rice\ hull}$	
	SmF	Petri Dishes
3	8 x 10^5 ± 0.03	3 x 10^4 ± 0.04
5	3 x 10^6 ± 0.05	5 x 10^4 ± 0.05
7	4 x 10^7 ± 0.04	5 x 10^5 ± 0.03
10	4 x 10^8 ± 0.03	5 x 10^6 ± 0.02

As reported in Table 1, the most interesting inoculation method, at 10 days, is the inoculation from SmF. Although the SmF inoculum has fewer initial CFUs, it proved to be the best on analysis after 10 days. This is due to the presence of active fungi in the SmF inoculum, rather than the Petri dish inoculum. In fact, in the latter, only spores of the fungus are present, which typically need more time for the regeneration of the metabolically active fungus [25]. Furthermore, this is an advantage from the industrial point of view, where the management of a liquid inoculum does not determine obstacles in process development.

In the industrial management of the solid-state fermented product for bioremediation purposes, its use in dry form is interesting. Although the normal moisture content of the SSF is 80%, for the storage and use of *Aspergillus* on rice hulls in an industrial context the dry form is the most advantageous in terms of process management. Therefore, we tested the capacity of *Aspergillus oryzae* of producing amylase in paper mill wastewater also after a drying step. The test involved the quantification of reducing sugars as evidence of thu amylase activity. In 6 h, 1 .3 ± 0.2 g/L of reducing sugars were found as proof of maintained amylase autivity.

Finally, the SSF was tustud in paper mill wastuwatur. Thu reducing sugars concentration was assessed after 72 h of treatment. Thu results are shown inFigure4.

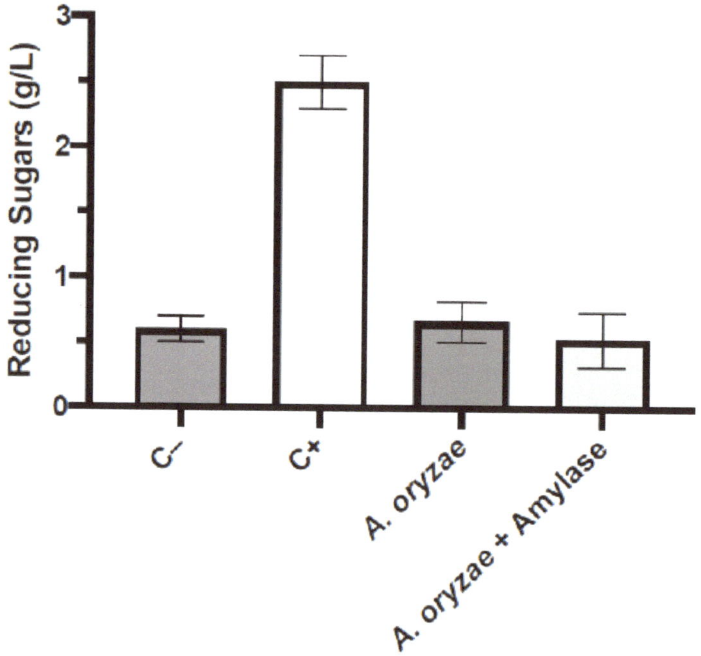

Figure 4. Effect of A. oryzae on the starch bioremediation of paper mill wastewater (72 h of incubation). C-, untreated PMW; C+, PMW treated with amylase STD.

In the results shown in Figure 4, the negative control refers to untreated paper mill wastewater. As expected, no reducing sugars were found. Indeed, in this context, the absence of *A. oryzae* and the following lack of amylase in the PMW suggest that no reducing sugars were found (except for those naturally present in the paper mill wastewater). On the contrary, reducing sugars were found when commercial amylase was added to the wastewater (as reported with positive control). This testifies that, in the PMW, starch is present and can be a substrate of the commercial amylase provided, validating the experimental design. If A. *oryzae* is supplied to the wastewater, it produces amylase which allows it to use the reducing sugars obtained from the hydrolysis of starch as a carbon source for its cellular metabolism. To evaluate this observation, the last column refers to the treatment of paper mill wastewater with *Aspergillus* and commercial amylase. The absence of reducing sugars supports that these are exploited by *Aspergillus* as metabolic substrates.

4. CONCLUSIONS

Aspergillus oryzae is a fungal strain widely exploited as an amylase producer. In this work, we aimed to study and test this fungus for the bioremediation of starch in industrial paper mill wastewater. For this purpose, submerged fermentation technologies (SmF) and solid-state fermentation (SSF) were studied. *A. oryzae* was found to grow on non- conventional media such as the paper mill wastewater. The SSF of *A. oryzae* was performed on rice hulls. In the bioremediation (as pretreatment) of paper mill wastewater, to remove starch, the fungus maintains its amylase activity and uses reducing sugars as metabolic substrates. This study opens new perspectives for the bioremediation of industrial effluents such as pulp-and-paper mill wastewater using *A. oryzae.*

REFERENCES

1. Costa, S.; Summa, D.; Semeraro, B.; Zappaterra, F.; Rugiero, I.; Tamburini, E. Fermentation as a Strategy for Bio-Transforming Waste into Resources: Lactic Acid Production from Agri-Food Residues. *Fermentation* **2021**, 7, 3. [CrossRef]
2. Elmekawy, A.; Diels, L.; De Wever, H.; Pant, D. Valorization of Cereal Based Biorefinery Byproducts: Reality and Expectations. *Environ. Sci. Technol.* **2013**, *47,* 9014-9027. [CrossRef]
3. Zappaterra, F.; Summa, D.; Semeraro, B.; Buzzi, R.; Trapella, C.; Ladero, M.; Costa, S.; Tamburini, E. Enzymatic Esterification as Potential Strategy to Enhance the Sorbic Acid Behavior as Food and Beverage Preservative. *Fermentation* **2020**, 6, 96. [CrossRef]
4. Dessie, W.; Luo, X.; Wang, M.; Feng, L.; Liao, Y.; Wang, Z.; Yong, Z.; Qin, Z. Current Advances on Waste Biomass Transformation into Value-Added Products. *Appl. Microbiol. Biotechnol.* **2020**,104, 4757-4770. [CrossRef] [PubMed]
5. Semeraro, B.; Summa, D.; Costa, S.; Zappaterra, F.; Tamburini, E. Bio-Delignification of Green Waste (GW) in Co-Digestion with the Organic Fraction of Municipal Solid Waste (OFMSW) to Enhance Biogas Production. *Appl. Sci.* **2021**,11, 6061. [CrossRef]
6. Neves, A.; Godina, R.; Azevedo, S.G.; Pimentel, C.; Matias, J.C.O. The Potential of Industrial Symbiosis: Case Analysis and Main Drivers and Barriers to Its Implementation. *Sustainability* **2019**,11, 7095. [CrossRef]
7. Zappaterra, F.; Costa, S.; Summa, D.; Semeraro, B.; Cristofori, V.; Trapella, C.; Tamburini, E. Glyceric Prodrug of Ursodeoxycholic Acid (UDCA): Novozym 435-Catalyzed Synthesis of UDCA-Monoglyceride. *Molecules* **2021**, 25, 5966. [CrossRef] [PubMed]

8. Economou, C.N.; Aggelis, G.; Pavlou, S.; Vayenas, D.V. Single Cell Oil Production from Rice Hulls Hydrolysate. *Bioresour. Technol.* **2011**, *102*, 9737-9742. [CrossRef]

9. Wei, G.Y.; Gao, W.; Jin, I.H.; Yoo, S.Y.; Lee, J.H.; Chung, C.H.; Lee, J.W. Pretreatment and Saccharification of Rice Hulls for the Production of Fermentable Sugars. *Biotechnol. Bioprocess Eng.* **2009**,14, 828-834. [CrossRef]

10. Glushankova, I.; Ketov, A.; Krasnovskikh, M.; Rudakova, L.; Vaisman, I. Rice Hulls as a Renewable Complex Material Resource. *Resources* **2018**, 7, 31. [CrossRef]

11. Vaskalis, I.; Skoulou, V.; Stavropoulos, G.; Zabaniotou, A. Towards Circular Economy Solutions for the Management of Rice Processing Residues to Bioenergy via Gasification. *Sustainability* **2019**,*11*, 6433. [CrossRef]

12. Rizzuti, A.M.; Lancaster, D.J. Utilizing Soybean Hulls and Rice Hulls to Remove Textile Dyes from Contaminated Water. *Waste Biomass Valorization* **2013**, *4*, 647-653. [CrossRef]

13. Holker, U.; Hofer, M.; Lenz, J. Biotechnological Advantages of Laboratory-Scale Solid-State Fermentation with Fungi. *Appl. Microbiol. Biotechnol.* **2004**, 64,175-186. [CrossRef] [PubMed]

14. Zambare, V. Solid State Fermentation of Aspergillus oryzae for Glucoamylase Production on Agro Residues. *Int. J. Life Sci.* **2010**, 4,16-25. [CrossRef]

15. Rob, B.; George, R.; Yovita, S.P.R.; Marisca, J.H.; Margreet, H.; Ana, L.; Kenneth, G.A.D.; Maarten, A.I.S.; Jan, D.; Yang, Z.; et al. Aspergillus oryzae in Solid-State and Submerged Fermentations. *FEMS Yeast Res.* **2002**, 2, 245-248.

16. Aliyah, A.; Alamsyah, G.; Ramadhani, R.; Hermansyah, H. Production of a-Amylase and ∞-Glucosidase from Aspergillus niger by Solid State Fermentation Method on Biomass Waste Substrates from Rice Husk, Bagasse and Corn Cob. *Energy Procedia* **2017**, 136, 418-423. [CrossRef]

17. Febe Francis, A.; Sabu, K.; Madhavan Nampoothiri, G.S.; Pandey, A. Synthesis of A-Amylase by Aspergillus oryzae in Solid-State Fermentation. *J. Basic Microbiol.* **2002**, 42, 320-326. [CrossRef]

18. Kamali, M.; Khodaparast, Z. Review on Recent Developments on Pulp and Paper Mill Wastewater Treatment. *Ecotoxicol. Environ. Saf.* **2015**,114, 326-342. [CrossRef] [PubMed]

19. Zappaterra, F.; Costa, S.; Summa, D.; Bertolasi, V.; Semeraro, B.; Pedrini, P.; Buzzi, R.; Vertuani, S. Biotransformation of Cortisone with Rhodococcus rhodnii: Synthesis of New Steroids. *Molecules* **2021**, 26,1352. [CrossRef] [PubMed]

20. Krishna, C.; Nokes, S.E. Influence of Inoculum Size on Phytase Production and Growth in Solid-State Fermentation by *Aspergillus Niger. Trans. Am. Soc. Agric. Eng.* **2001**, 44,1031-1036. [CrossRef]

21. Pokhrel, D.; Viraraghavan, T. Treatment of Pulp and Paper Mill Wastewater—A Review. *Sci. Total Environ.* **2004**, 333, 37-58. [CrossRef] [PubMed]

22. Mandal, T.; Bandana, T. Studies on Physicochemical and Biological Characteristics of Pulp and Paper Mill Effluents and Its Impact on Human Beings. *J. Freshw. Biol.* **1996**, 4, 6.

23. Kadwe, B.; Khedikar, I.; Hardas, C. Treatment of Starch Wastewater from Cardboard Packaging Industry Treatment of Starch Wastewater from Cardboard Packaging Industry. *IOSR J. Eng.* **2019**, 80-83.

24. Kumar, A.; Chandra, R. Biodegradation and Toxicity Reduction of Pulp Paper Mill Wastewater by Isolated Laccase Producing Bacillus cereus AKRC03. *Clean. Eng. Technol.* **2021**, 4,100193. [CrossRef]

25. Gougouli, M.; Koutsoumanis, K.P. Relation between Germination and Mycelium Growth of Individual Fungal Spores. *Int. J. Food Microbiol.* **2013**,161, 231-239. [CrossRef] [PubMed]

Index

www.ingramcontent.com/pod-product-compliance
Lightning Source LLC
LaVergne TN
LVHW071246210926
848486LV00008B/356
* 9 7 8 1 7 8 7 1 5 6 4 8 7 *